Springer Theses

Recognizing Outstanding Ph.D. Research

Aims and Scope

The series "Springer Theses" brings together a selection of the very best Ph.D. theses from around the world and across the physical sciences. Nominated and endorsed by two recognized specialists, each published volume has been selected for its scientific excellence and the high impact of its contents for the pertinent field of research. For greater accessibility to non-specialists, the published versions include an extended introduction, as well as a foreword by the student's supervisor explaining the special relevance of the work for the field. As a whole, the series will provide a valuable resource both for newcomers to the research fields described, and for other scientists seeking detailed background information on special questions. Finally, it provides an accredited documentation of the valuable contributions made by today's younger generation of scientists.

Theses may be nominated for publication in this series by heads of department at internationally leading universities or institutes and should fulfill all of the following criteria

- They must be written in good English.
- The topic should fall within the confines of Chemistry, Physics, Earth Sciences, Engineering and related interdisciplinary fields such as Materials, Nanoscience, Chemical Engineering, Complex Systems and Biophysics.
- The work reported in the thesis must represent a significant scientific advance.
- If the thesis includes previously published material, permission to reproduce this must be gained from the respective copyright holder (a maximum 30% of the thesis should be a verbatim reproduction from the author's previous publications).
- They must have been examined and passed during the 12 months prior to nomination.
- Each thesis should include a foreword by the supervisor outlining the significance of its content.
- The theses should have a clearly defined structure including an introduction accessible to new PhD students and scientists not expert in the relevant field.

Indexed by zbMATH.

Yu Ozawa

Copper(I)-Catalyzed Stereoselective Borylation Reactions

Multisubstituted Alkenyl and Allylic Boronates

Doctoral Thesis accepted by
Hokkaido University, Sapporo, Japan

Springer

Author
Dr. Yu Ozawa
Synthetic Organic Chemistry Laboratories
FUJIFILM Corporation
Shizuoka, Japan

Graduate School of Chemical Sciences and Engineering
Hokkaido University
Sapporo, Japan

Supervisor
Prof. Hajime Ito
Division of Applied Chemistry, Graduate School of Engineering
Hokkaido University
Sapporo, Japan

Institute for Chemical Reaction Design and Discovery (WPI-ICReDD)
Hokkaido University
Sapporo, Japan

ISSN 2190-5053 ISSN 2190-5061 (electronic)
Springer Theses
ISBN 978-981-99-1097-7 ISBN 978-981-99-1098-4 (eBook)
https://doi.org/10.1007/978-981-99-1098-4

This Springer imprint is published by the registered company Springer Nature Singapore Pte Ltd.
The registered company address is: 152 Beach Road, #21-01/04 Gateway East, Singapore 189721, Singapore

Supervisor's Foreword

Since their first emergence in 1860, organoboron compounds, especially organoboronic acids and those esters, have occupied an important position in organic synthesis because of the outstanding reliability and versatility of their transformation reactions. Furthermore, their high chemical stability and low toxicity make them highly practical. The derivatization reactions of organoboronates into alcohols and amines show good reactivity, stereospecificity, and functional group tolerance. Another typical application is the palladium-catalyzed Suzuki–Miyaura cross-coupling reaction between organoboron compounds and organohalides developed by Prof. Akira Suzuki, the 2010 Novel Prize winner in Chemistry. The coupling reaction demonstrates that the organoboronates can be used for the construction of carbon skeletons under mild reaction conditions instead of other organometallic reagents such as organolithium, -zinc, and -magnesium. Thus, the organoboron compounds gain attention from both academics and industry in organic synthesis and pharmaceuticals.

To date, many stoichiometric and catalytic reactions for the preparation of organoboron compounds have been developed. Although the accessible structures are increasingly spread, the synthesis of multi-substituted organoboronates is still challenging due to the requirement of sufficient reactivity and the difficulty of enantio- and stereo-control. Since 2000, our group has developed nucleophilic borylation reactions by means of copper(I)/diboron catalyst systems and demonstrated the catalyst is suitable for the synthesis of multi-substituted aliphatic boronates.

The Ph.D. thesis of Dr. Yu Ozawa focuses on copper(I) catalysis for the synthesis of functionalized multi-substituted organoboron compounds, alkenyl, and allylic boronates. In Chap. 2, he developed the new chiral ligands for the enantio-convergent borylation reactions of racemic allylic electrophiles. A through-space interaction in the ligand improves the reactivity of the catalyst to afford the product, α-chiral allylic boronates, in high yield with high stereo- and enantioselectivity. In Chap. 3, he achieved the synthesis of alkenyl boronates bearing a unique cyclobutane ring via regioselective borylative cyclization of allene substrates. In Chap. 4, he synthesized allylic boronates which have sterically congested tetrasubstituted alkene structures constructed via regio- and stereoselective alkylboration of allene substrates.

Those new organoboron compounds are rarely accessible with other borylation reactions and have great potential utility for the synthesis of complex molecules such as pharmaceutical drugs. Through Chaps. 2–4, he analyzed the reaction mechanisms by means of computational methods such as density functional theory calculation. The detailed analysis of the transition states revealed the selectivity-determining mechanisms and key role of catalysts and ligands.

In the thesis, he achieved the development of novel borylation reactions suited to synthesizing functionalized multi-substituted organoboronates. These results afford the guide to identify the effective catalyst and control selectivities in copper(I)/diboron catalyst systems.

Sapporo, Japan
Sapporo, Japan
December 2022

Prof. Hajime Ito

List of Publications

Parts of this thesis have been published in the following journal articles:

Chapter 2:
Modification of QuinoxP-Type Bisphosphine Ligands for High-Performance Asymmetric Boryl Substitution of Racemic Allyl Electrophiles*
Iwamoto, H.;[†] Ozawa, Y.;[†] Takenouchi, Y.; Imamoto, T.; Ito, H. *J. Am. Chem. Soc.* **2021**, *143*, 6413–6422.
[†]Co-first author.

Chapter 3:
Allylcopper(I) Isomerization-Enabled Copper(I)-Catalyzed Intramolecular Alkylboration of Terminal Allenes
Ozawa, Y.; Iwamoto, H.; Ito, H. *Chem. Commun.* **2018**, *54*, 4991–4994.

Chapter 4:
Intermolecular Alkylboration of gem-Disubstituted Allenes for Stereoselective Synthesis of Multi-alkylated Allylic Boronates
Ozawa, Y.; Endo, K.; Ito, H. *J. Am. Chem. Soc.* **2021**, *143*, 13865–13877.

Chapter 5:
Computational Investigation on Copper(I)-Catalyzed Enantioselective Radical Borylation of Benzyl Halides
Iwamoto, H.; Endo, K.; Ozawa, Y.; Watanabe, Y.; Kubota, K.; Imamoto, T.; Ito, H. *Angew. Chem. Int. Ed.* **2019**, *58*, 11112–11117.

Acknowledgements

The studies presented in this thesis have been carried out under the direction of Professor Hajime Ito, Associate Professor Tatsuo Ishiyama, Associate Professor Koji Kubota, Lecturer Tomohiro Seki (currently at Shizuoka University), Assistant Professor Mingoo Jin, and Guest Professor Tsuneo Imamoto (Emeritus Professor at Chiba University) at the Division of Applied Chemistry, Graduate School of Engineering, Hokkaido University, during 2016–2021. I was financially supported by scholarships from the Japan Student Services Organization (JASSO), the Inoue Scholarship Foundation, the Program for Leading Graduate Schools (Hokkaido University "Ambitious Leaders Program"), and the Research Fellowship (DC1) of the Japan Society for the Promotion of Science (JSPS). This work was supported by the JSPS KAKENHI (DC1, Grant number: 19J20823) and the Ambitious Leaders Program at Hokkaido University. Parts of the computational calculations were carried out on the supercomputer of the Research Center for Computational Science, Okazaki, Japan.

I would like to express my gratitude to Professor Hajime Ito for his continuous guidance and encouragement with kindness throughout these works. His guidance extended to organic chemistry, organic synthesis techniques, technical writing techniques, and computational chemistry. I also wish to express my thanks to Associate Professor Tatsuo Ishiyama and Associate Professor Koji Kubota for their great suggestions and fruitful discussions on the copper(I) catalysis and borylation reactions. I want to thank Lecturer Tomohiro Seki and Assistant Professor Mingoo Jin for their instruction and support of single-crystal X-ray analysis. I would like to acknowledge Guest Professor Tsuneo Imamoto for his kind advice and discussions. I would like to express my special thanks to Dr. Iwamoto Hiroaki for his instruction and encouragement with kindness and enthusiasm. I acknowledge Mr. Yuta Hayashi, Mr. Kohei Endo, Mr. Hisao Koriyama, Mr. Yuma Shiratori, and Mr. Rikuro Takahashi for our good collaborations on research during my Ph.D. course. I also want to express my earnest thanks to Dr. Rina Takahashi and all the members of the Ito group for putting up with me. I really enjoyed our good collaborations and discussions. I also want to thank Professor Satoshi Maeda for giving me an opportunity for a short-term visit to his laboratory. I thank Associate

Professor Kimichi Suzuki, Assistant Professor Yu Harabuchi, Assistant Professor Kenichiro Saita, Dr. Tomoya Ichino, and all the members of the Maeda group for their kind instructions on computational chemistry for me. I would like to thank Mr. Satoru Kimura and Mr. Tomohiro Hirose for their technical support on NMR and HRMS measurements. I wish to thank Professor Peter R. Schreiner, Mr. Finn Wilming, and the group members in the Institute of Organic Chemistry, Justus Liebig University Giessen, for giving me an opportunity to join their group for two months. I could expand my viewpoint on organic chemistry. I would like to thank Professor Atsushi Yokota and the mates in the Inoue Scholarship Foundation for their kind encouragement.

Finally, I would like to express my deepest gratitude to my family for their constant understanding, support, and encouragement with kindness.

Contents

Chapter 1
General Introduction

1.1 Organoboron Compounds

An organoboron represents a multi-functional organic compound that has C–B bond(s) on the carbon skeleton [1, 2]. Since the Frankland group achieved the first preparation of an organoboron compound in 1860 [3, 4] their reactivities in the transformation reactions, optical properties, electronic properties, and bioactivities have been gathering attention from researchers in a variety of research fields.

The organoboron compounds play essential roles in organic synthesis [1, 2, 5–10]. A slightly polarized and electron-rich nature of the C–B bond and a Lewis acidity at the boron center can show nucleophilic and/or electrophilic reactivities according to the reaction conditions and substituents on the boron atom (Fig. 1.1). The electron-rich C–B bond is readily oxidized to a boronic acid ester (C–O–B) by treatment with peroxides or oxygen, which was originally reported by the Johnson group in 1938 (Fig. 1.2A) [11–14]. The C–B bond can formally be transformed into the C–O bond. In 1956, the Brown group developed a sequential procedure of hydroboration/oxidation of double bonds (Fig. 1.2B) [15–17]. This reaction is a reliable *anti*-Markovnikov hydroxylation of alkenes via the corresponding organoboron compounds. Recently, the C–B bonds were found to be transformed into the other C–heteroatom bonds in one step, i.e., C–N and C–halogen bonds (Fig. 1.2C) [9]. Another essential role of the organoboron compounds is a source of alkyl or aryl groups through a transmetallation process in transition metal catalysis (Fig. 1.3) [5]. The pioneering work on palladium catalysis is known as the Suzuki–Miyaura reaction reported in 1979 [18, 19]. The substituent on the boron atom can be transferred onto the palladium or other transition metal species via activation of the boron center with a Lewis base, which has been commonly used for the generation of the organometallic species (Fig. 1.3) [1, 2, 20–22]. Thus, the organoboron compounds are nowadays recognized as versatile intermediates in organic synthesis.

The boron-containing organic molecules have also been well explored as functional organic materials for optical, electronic, and pharmaceutical applications as well as catalysts [23–29]. Boron moieties having triaryl groups have a high electron

Y. Ozawa, *Copper(I)-Catalyzed Stereoselective Borylation Reactions*, Springer Theses,
https://doi.org/10.1007/978-981-99-1098-4_1

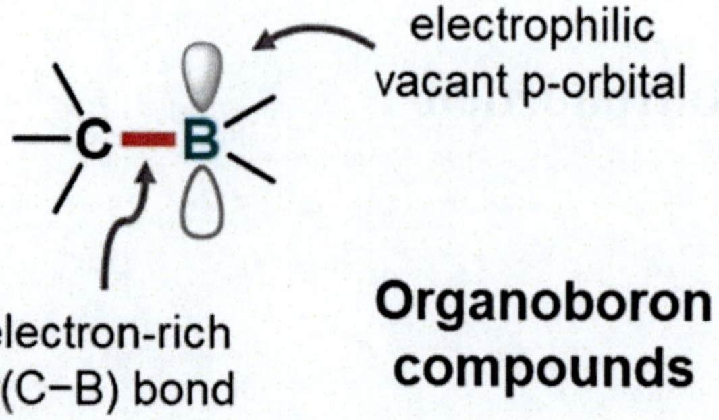

Fig. 1.1 Nucleophilicity and electrophilicity derived from the boryl group in organoboron compounds

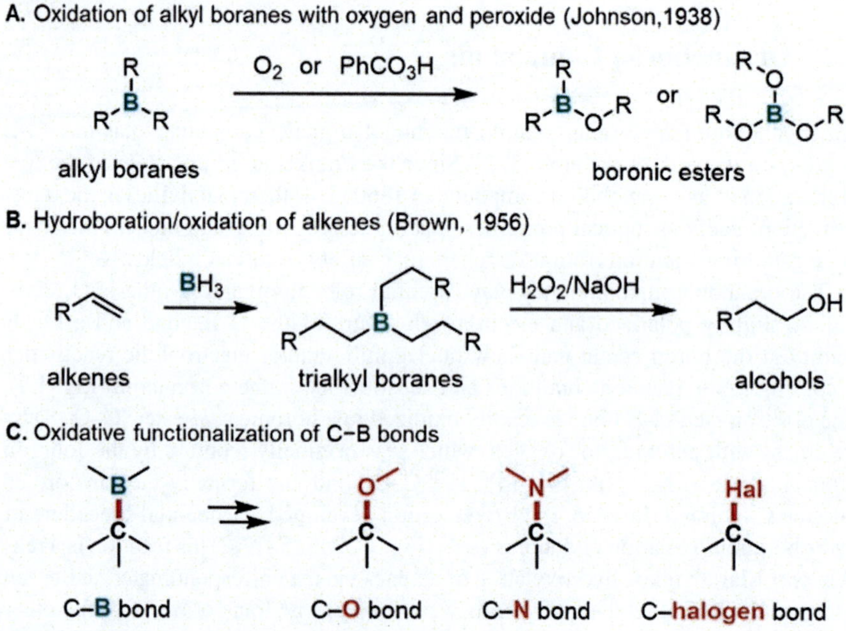

Fig. 1.2 Oxidation reactions of organoboron compounds

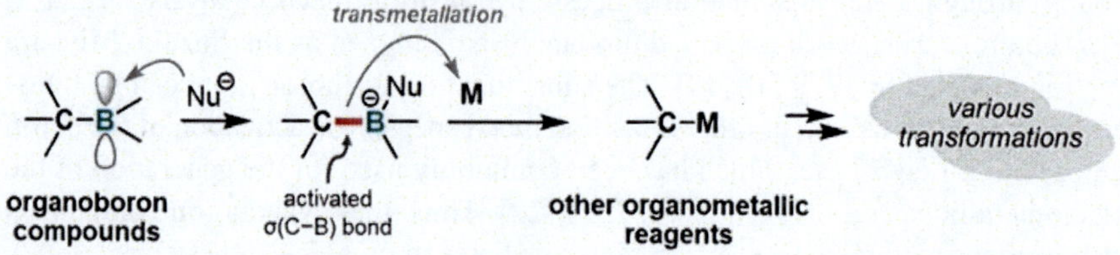

Fig. 1.3 Organoboron compounds as transmetallation reagents. M=Pd, Ni, Rh, Cu, Fe, etc.

affinity. They are thus used as electron-acceptor units in optical and electronic materials, i.e., fluorescent materials, electron transport materials, and molecular sensors (Fig. 1.4) [23–27, 30]. The innovative and practical example of boron-containing drugs is the bortezomib, an anti-cancer agent that acts as a proteasome inhibitor (Fig. 1.5) [28, 31]. The introduction of boron groups in drug molecules can sometimes give positive modifications to the drug. Furthermore, many types of boron catalysts are developed (Fig. 1.6) [29]. For instance, chiral oxazaborolidines are used as the catalyst for asymmetric hydrogenation of ketones, which is known as Corey–Bakshi–Shibata reduction [32–36].

The organoboron compounds are attractive multi-functional organic materials. The development of novel organoboron compounds should contribute to numerous

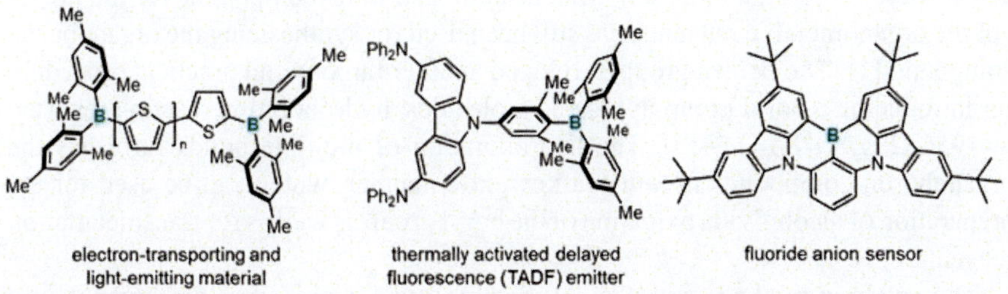

Fig. 1.4 Applications of organoboron compounds in functional materials

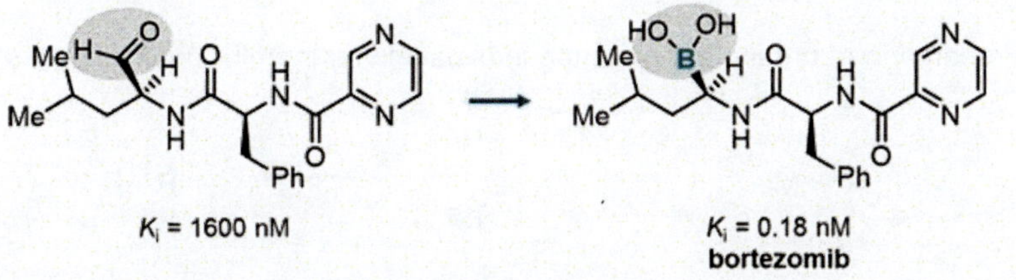

Fig. 1.5 Application of organoboron compounds in pharmaceuticals

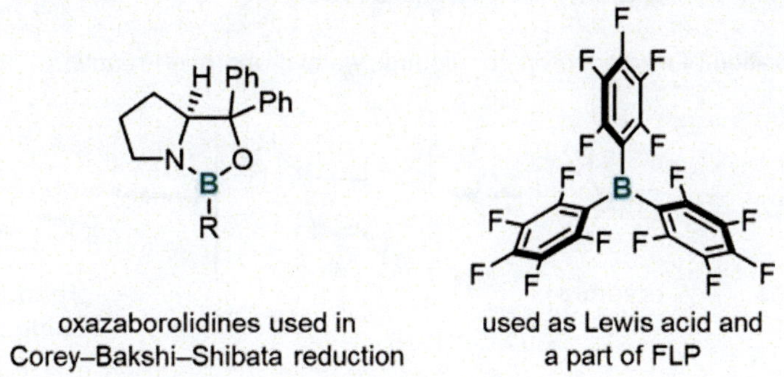

Fig. 1.6 Applications of organoboron compounds in catalysts

research fields. In this thesis, I focus on the synthesis of aliphatic organoboron compounds functionalized as alkenyl and allylic boronates, or chiral boronates.

1.2 Borylation Reactions of Organic Molecules

Traditionally, non-catalytic procedures of organoboron compounds have been developed [1, 2]. In 1860, the Frankland group first synthesized an organoboron compound via nucleophilic substitution on electrophilic boronic ester with an organozinc reagent (Fig. 1.7A) [3, 4]. This classical procedure and similar combination of reagents, including boronic esters and halo boron compounds as the electrophilic boron source, and organolithium, -magnesium, and -zinc compounds for the nucleophilic organometallic reagents, has still been used for synthesizing the organoboron compound [1]. The Brown group developed another famous and practical procedure for introducing a boryl group in organic molecules, hydroboration reaction reported in 1956 (Fig. 1.7B) [15–17]. The hydroboration of multiple bonds furnishes the organoboron compounds in *anti*-Markovnikov manner, which can be used for the preparation of alcohols via oxidation of the boryl group as well as for transmetallation reagents.

The catalytic methods are also practically used for preparing the organoboron compounds. The first catalytic reaction is a rhodium(I)-catalyzed hydroboration of unactivated alkenes using catechol boronate as the boron source, which was reported

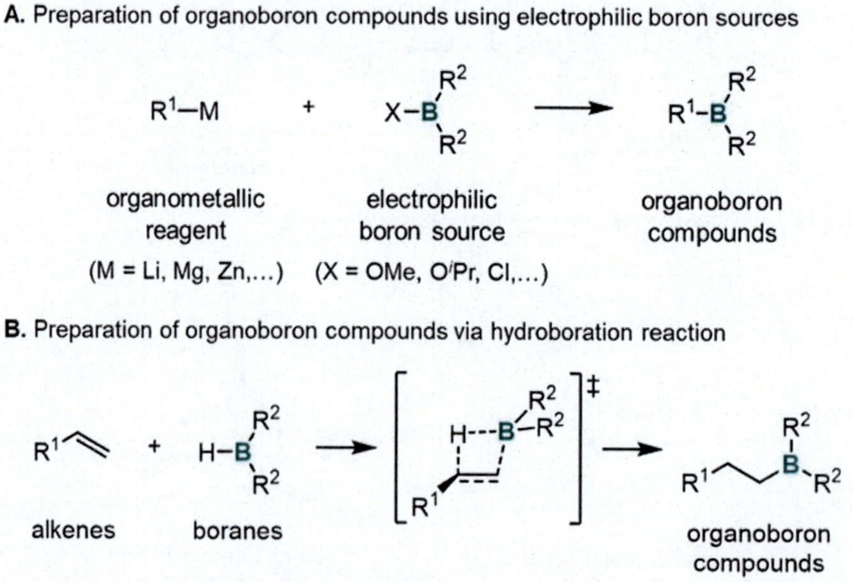

Fig. 1.7 Non-catalytic preparation method of organoboron compounds

by the Nöth group in 1985 (Fig. 1.8A) [37]. This reaction proceeds toward terminal alkenes with *anti*-Markovnikov selectivity to produce the corresponding terminal alkyl boronates [38, 39]. In 1995, the group of Miyaura and Ishiyama developed a palladium(II)-catalyzed borylative substitution of aryl halides using a diboron reagent as the boron source, which is known as Miyaura–Ishiyama borylation reaction (Fig. 1.8B) [40, 41]. This reaction is useful and commonly used for the preparation of aryl boron compounds [42, 43]. Also, the Hartwig group reported the first transition metal-mediated borylation of C(sp^3)–H bonds using boryl-manganese, -rhenium, and -iron complexes in 1995 [44, 45] and then developed the catalytic version of the C(sp^3)–H bond borylation reaction using iridium and rhodium catalyst in 2000 (Fig. 1.8C) [46]. Furthermore, in 2002, the Ishiyama and Hartwig group found the iridium can catalyze the C(aryl)–H bond borylation reaction (Fig. 1.8D) [47, 48]. After these pioneering works, many types of catalytic borylation reactions have been developed to date [1, 2, 49–54].

1.3 Copper(I)-Catalyst/Diboron System for Borylation Reactions

In 2000, the Hosomi and Ito group and the Miyaura and Ishiyama group independently developed the copper(I)-catalyzed borylation reaction of multiple bonds using a diboron reagent, bis(pinacolate) diboron which possesses B–B bond, as the boron source (Fig. 1.9A) [55–58]. The active catalyst species, boryl copper(I) intermediate, is generated in situ via σ-bond metathesis between the diboron reagent and a copper(I) alkoxide (Fig. 1.9B) [59–61]. The 1,2-insertion of the boryl copper(I) species to enone substrates furnishes the alkyl copper(I) intermediate, and the subsequent protonation gives the corresponding protoboration product with concomitant regeneration of the copper(I) alkoxide. This formal conjugate addition of the boryl group indicates the nucleophilicity of the boryl group, even though boryl groups have electrophilicity in general mainly due to the vacant p-orbital of the boron atom [59]. To date, a variety of electrophiles has been found to be suitable to the borylation reactions, e.g., conjugated esters, allylic electrophile, carbonyl compounds, and imines (Fig. 1.10) [59–61]. In 2005 and 2007, the Ito and Sawamura group developed the copper(I)-catalyzed enantiospecific and enantioselective γ-borylation of allylic electrophiles for synthesizing allylic boronates (Fig. 1.11) [62, 63]. Recently, the Ito group also reported that electron-deficient aromatic rings were also able to be borylated using the boryl copper(I) species (Fig. 1.12) [64–66].

In 2000 and 2001, the Miyaura and Ishiyama group reported that the boryl copper(I) species has reactivity to unactivated multiple bonds such as terminal alkynes to produce alkenyl boronates (Fig. 1.13) [56, 57]. This reaction proceeds via the borylcupration (1,2-di-metallation) of the multiple bonds and the subsequent generation of alkenyl copper(I) intermediate, which represents the boryl copper(I) species has the reactivity of reduction of multiple bonds. Afterward, these types

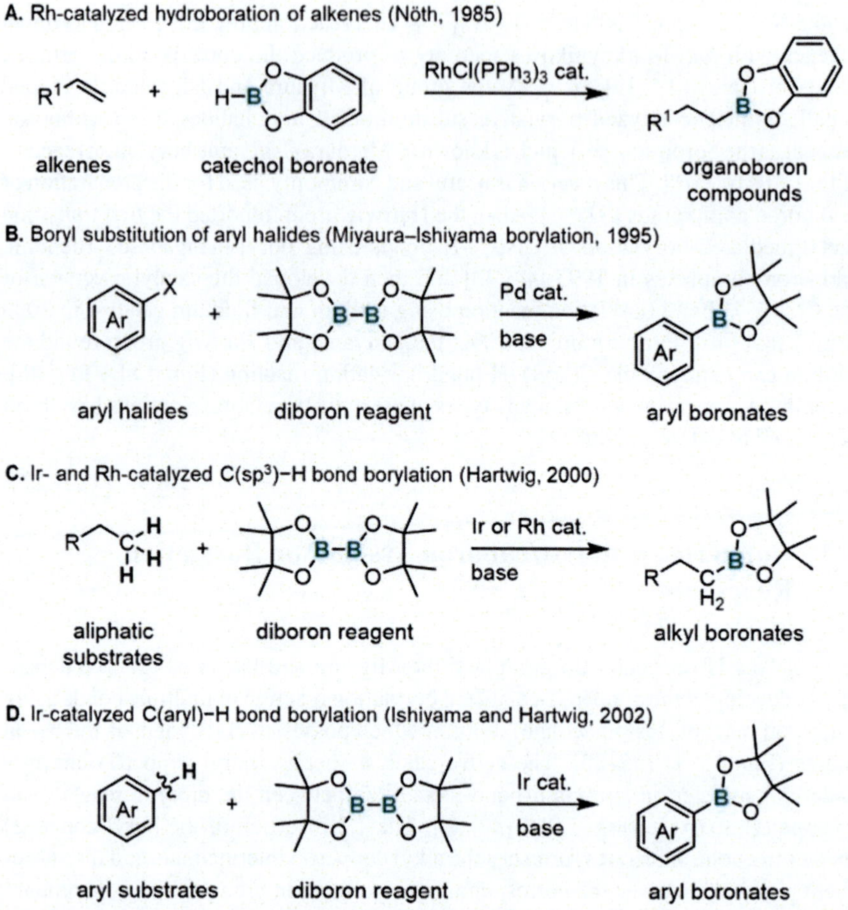

Fig. 1.8 Pioneering works on catalytic borylation reactions

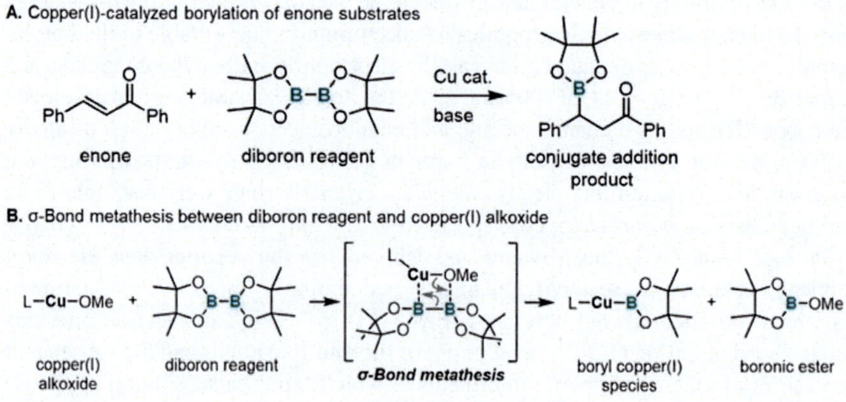

Fig. 1.9 Pioneering works on copper(I)-catalyzed borylation reactions using a diboron reagent

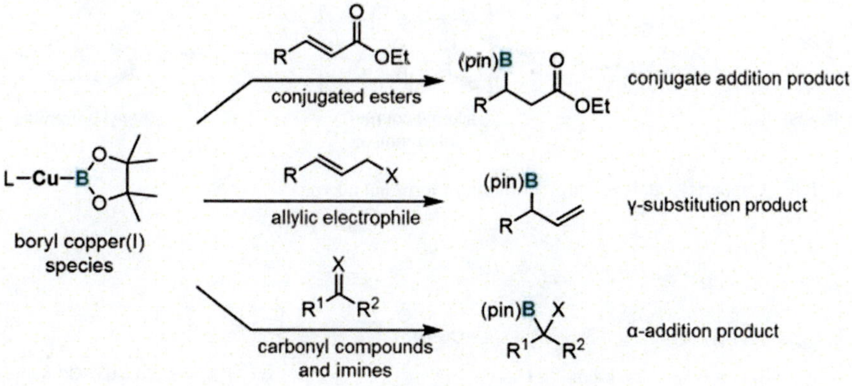

Fig. 1.10 Reactions of boryl copper(I) species and various electrophiles

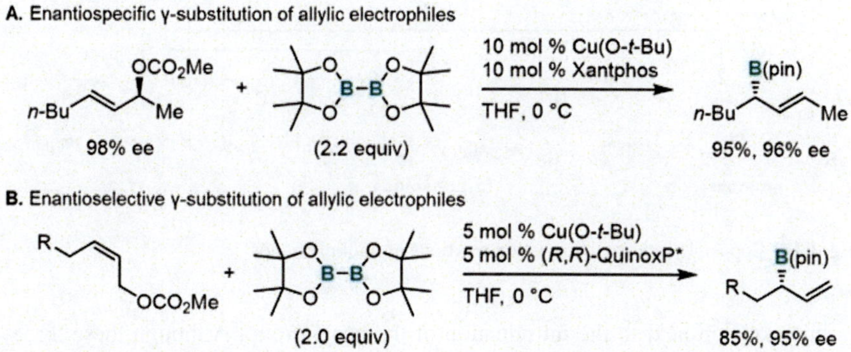

Fig. 1.11 γ-Borylation of allylic electrophiles

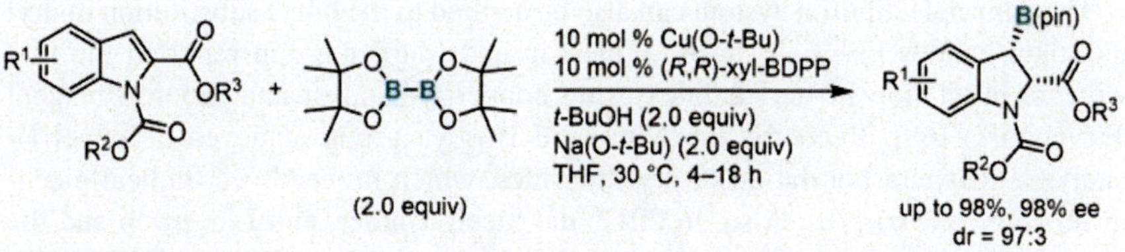

Fig. 1.12 Asymmetric dearomative borylation of indoles

of borylation reactions of the other multiple bonds have been developed, including alkenes, allenes, dienes, and enynes (Fig. 1.14A) [59–61]. Furthermore, the use of organocopper(I) intermediates generated via the borylcupration of the multiple bonds has been studied (Fig. 1.14B, C) [67, 68]. The trapping of the intermediate with electrophiles can introduce an additional functional group, including a

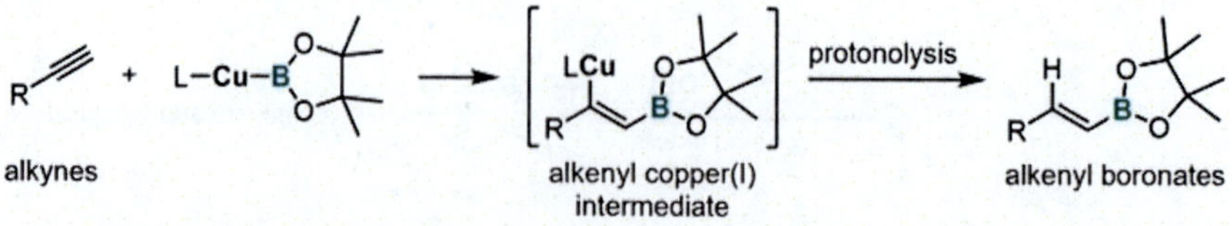

Fig. 1.13 Copper(I)-catalyzed protoboration of terminal alkynes

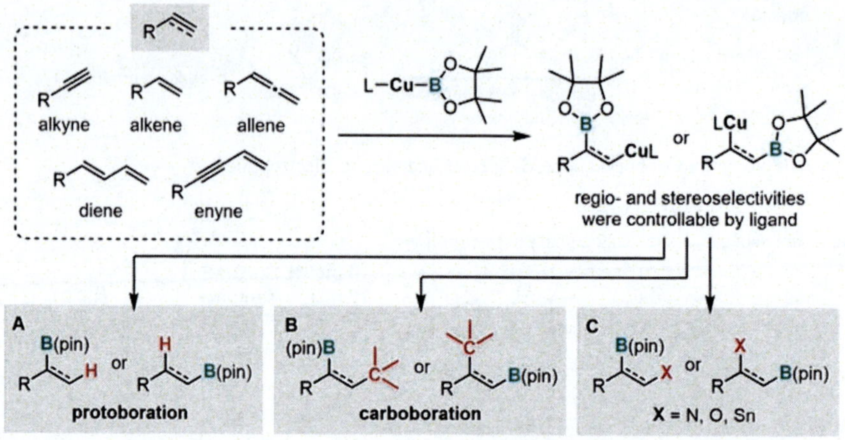

Fig. 1.14 Copper(I)-catalyzed borylation reactions of multiple bonds

carbon skeleton next to the introduction of the boryl group. Although these three-component couplings are challenging owing to the chemoselectivity issue, the reactions are useful for the preparation of highly functionalized and densely substituted organoboron compounds.

The copper(I)/diboron system can also be applied to the boryl substitution of aryl and alkyl halides [59–61]. In 2009, the Lin and Marder group reported the first Miyaura borylation of aryl halides using copper(I) catalyst and diboron reagent (Fig. 1.15A) [69]. Recently, the Niwa and Hosoya group achieved a copper(I)-catalyzed Miyaura borylation of aryl fluorides, which proceeds via radical mechanism (Fig. 1.15B) [70]. Also, in 2012, the Steel, Marder, and Liu group and the Ito group independently reported the boryl substitution reaction of alkyl halides (Fig. 1.15C) [71, 72]. In spite of these pioneering works, the enantioselective borylative substitution reaction of alkyl halides has not been developed.

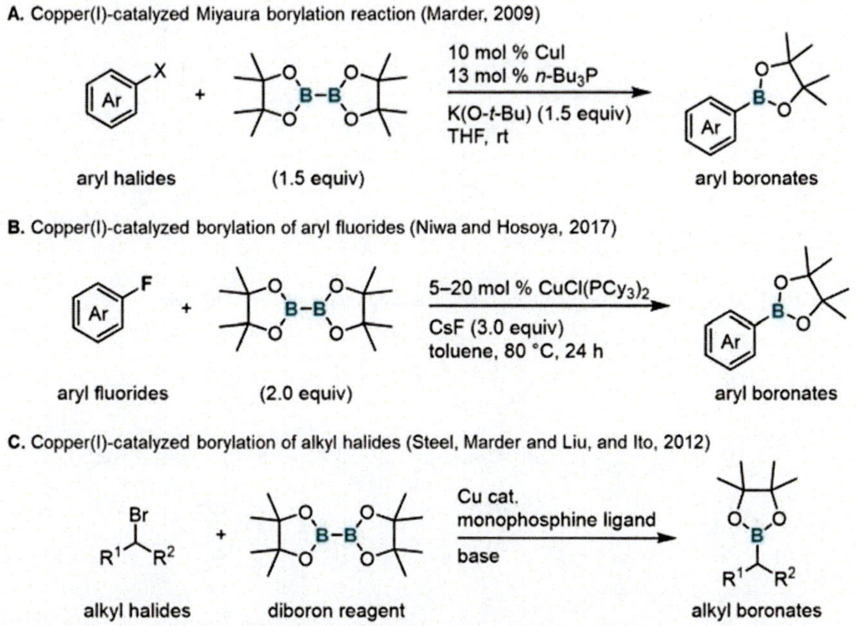

Fig. 1.15 Copper(I)-catalyzed borylation reactions of aryl and alkyl halides

1.4 Characterization of Boryl Copper(I) Species

In 2005, the Sadighi group first isolated a boryl copper(I) complex having a bulky *N*-heterocyclic carbene (NHC) ligand, IPr (Fig. 1.16) [73]. The complex was fully characterized via NMR and single-crystal X-ray analysis. The di-coordinated copper(I) center was found to be a linear geometry. The structurally related complexes were isolated and characterized as the similar linear geometry of the copper(I) center (Fig. 1.17) [74–77]. Di-coordinated linear boryl copper(I) complexes could also be prepared via boryl anions and copper(I) salts reported by the Yamashita and Nozaki group (Fig. 1.18) [78–80]. Although boryl copper(I) complexes that have other ligands, including phosphine ligands, are rarely isolated due to their instability under ambient conditions, similar structures have been computationally predicted by many researchers (Fig. 1.19). Such linear monomeric boryl copper(I) complexes are commonly proposed as catalysis active species.

In 2017 and 2018, the Kleeberg group revealed that boryl copper(I) complexes having dppf or less-bulky NHC ligand form the dimer, characterized by single-crystal X-ray analysis (Fig. 1.20) [81, 82]. Also, in 2017, Tobisch reported that the dimerization of dppf–boryl copper(I) complex is thermodynamically favored by 11.0 kcal/mol, which was estimated by density functional theory (DFT) calculation [83]. However, the in situ dimerization of boryl copper species and its influence in catalysis have not been studied well.

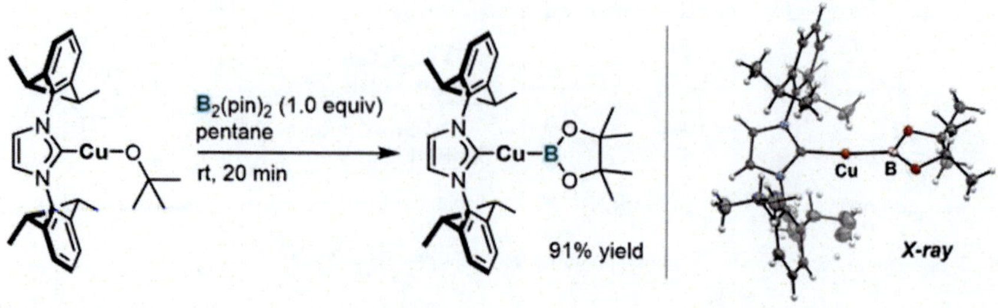

Fig. 1.16 First isolation of boryl copper(I) species supported with NHC ligand

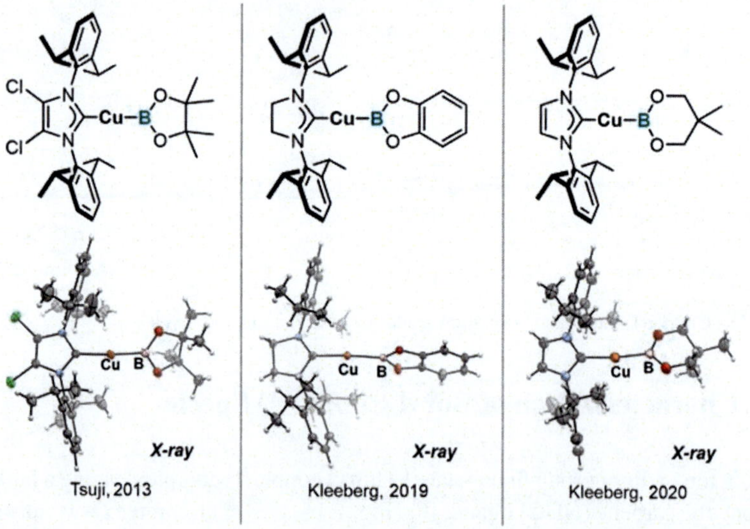

Fig. 1.17 X-ray structures of boryl copper(I) species supported with NHC ligand

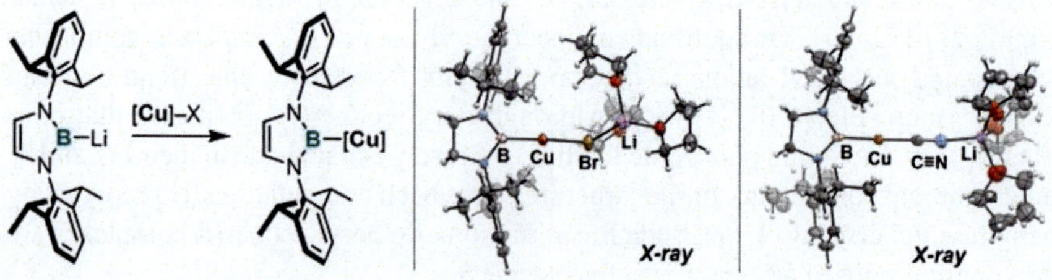

Fig. 1.18 Preparation of boryl copper(I) complex from boryl anion and copper(I) salt

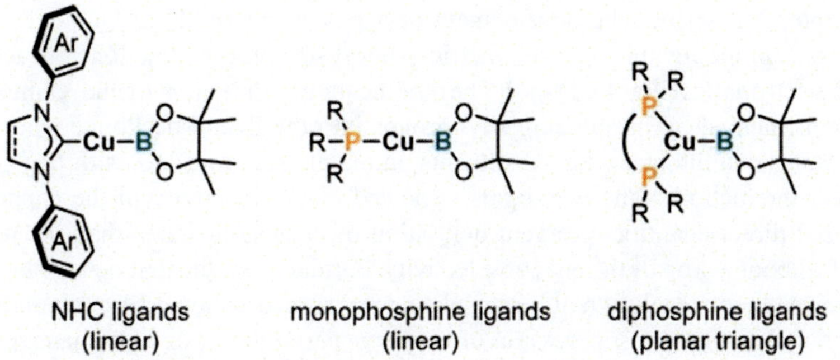

Fig. 1.19 Geometries of boryl copper(I) complex related to the ligands

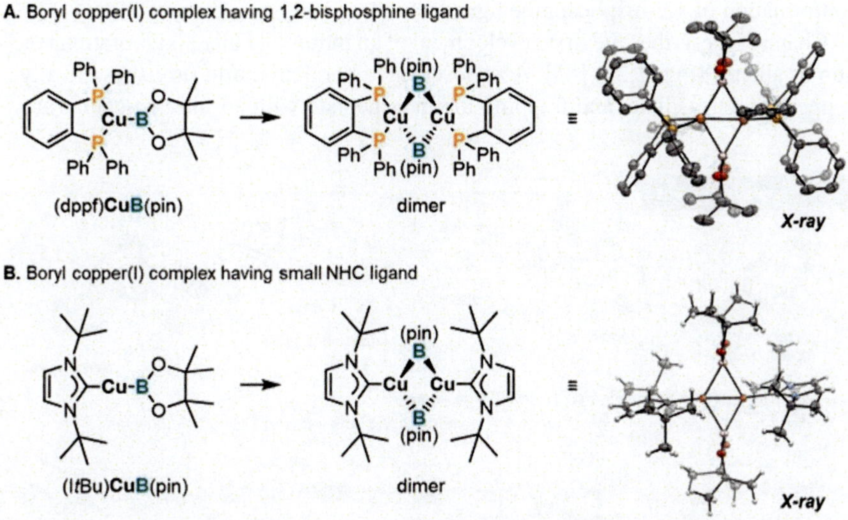

Fig. 1.20 Isolated boryl copper(I) dimers

1.5 Overview

Copper(I)-catalyzed borylation reactions using diboron reagents are recognized as a powerful and reliable way for the preparation of organoboron compounds. After careful designing and optimizing catalysts, the system can afford the functionalized organoboronates in a stereo- and enantiocontrolled manner. In this thesis, I focus on the development of novel stereo- and enantioselective synthetic methods of multi-substituted allylic and alkenyl boronates. Furthermore, the stereo- and enantiodetermining mechanisms are exhaustively explored by DFT calculation analysis.

Chapter 2 describes the development of a new series of chiral QuinoxP*-type 1,2-bisphosphine ligands for asymmetric γ-boryl substitution reactions of racemic allylic electrophiles (Fig. 1.21) [84]. The modification of the parent ligand, QuinoxP*, was implemented by introducing silyl groups into the ligand backbone where the silyl groups can interact with substituents on the phosphorous atoms through non-covalent interactions. The new ligands showed higher reactivity in the copper(I)-catalyzed direct enantioconvergent borylation of cyclic substrates than the parent ligand without a drop of the enantioselectivity. Furthermore, the first development of borylative kinetic resolution of linear substrates was also achieved. A computational study revealed that the suppression of boryl copper(I) dimer as a dormant species in catalysis improves the reactivity of the catalyst using the new ligand. Moreover, the detailed analysis on the enantiodetermining mechanism of the direct enantioconvergent borylation reactions was conducted using newly defined descriptors for conformation of 1,2-bisphosphine ligands.

Chapter 3 describes the first development of an intramolecular alkylboration reaction of allenes (Fig. 1.22) [85]. A regioselective borylcupration furnishes the allylcopper(I) intermediate, and trapping the intermediate with an intramolecular alkyl

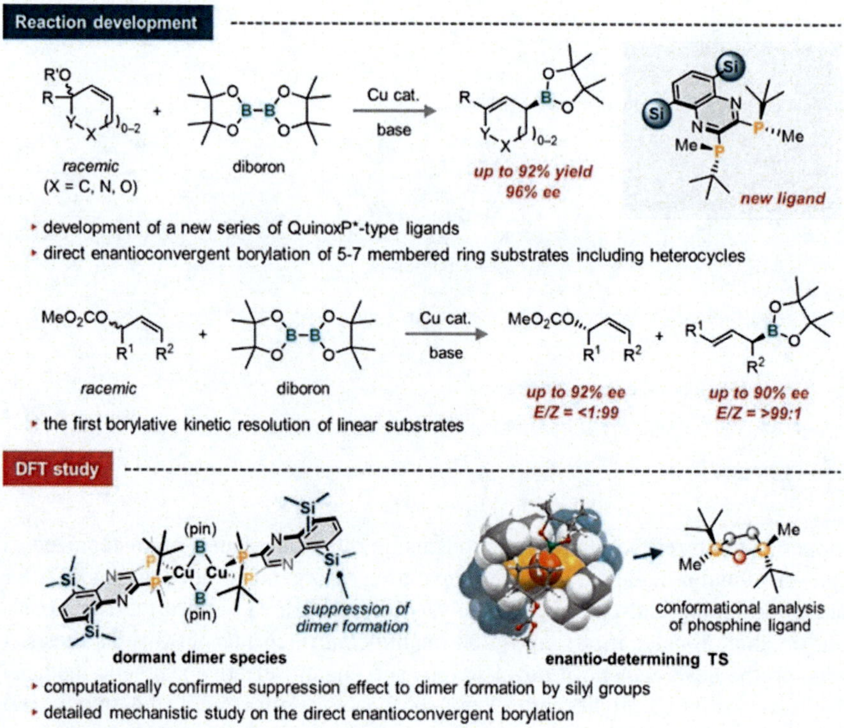

Fig. 1.21 Graphical abstract of Chap. 2

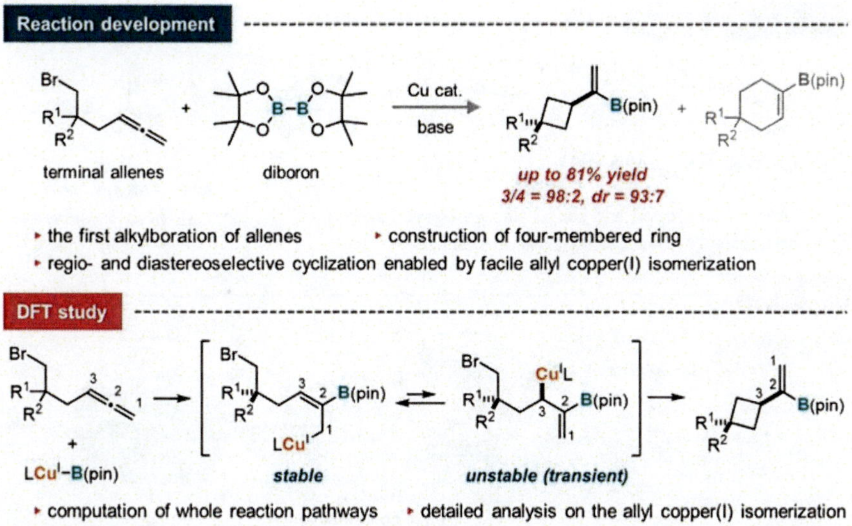

Fig. 1.22 Graphical abstract of Chap. 3

halides moiety produces the cyclization product containing alkenyl boron and four-membered ring structures. Furthermore, products were obtained in high diastereoselectivity, enabled by facile allylic isomerization of allylcopper(I) species, which was confirmed by DFT calculations of the cyclization step.

Chapter 4 describes the synthesis of multi-substituted allylic boronates, which contains a differentially tetrasubstituted alkene moiety, via a copper(I)-catalyzed stereoselective alkylboration of *gem*-disubstituted allenes (Fig. 1.23) [86]. Ligands having two pocket-like structures realized the regioselective 1,2-borylcupration of allenes and their first use in carboboration reactions. The reaction could be applied to a variety of allene substrates, including exo-cyclic allenes. Furthermore, allylboration of aldehydes with the borylation products afforded the densely substituted homoallylic alcohols with high diastereoselectivity. A computational investigation revealed an unprecedented regio- and stereoselectivity-determining mechanism, which involves the coordination step of the boryl copper(I) species to allenes in addition to the borylcupration step.

Chapter 5 describes the first computational investigation on an enantioface recognition of benzyl radical species with the boryl copper(II) intermediate (Fig. 1.24) [87]. The transition state analysis indicated that the reaction proceeds via the concerted mechanism of boryl transfer to the radical species. Furthermore, an analysis focused on non-covalent interactions revealed that the enantioselectivity is determined by a combination of attractive C–H/π interaction and repulsive steric repulsion.

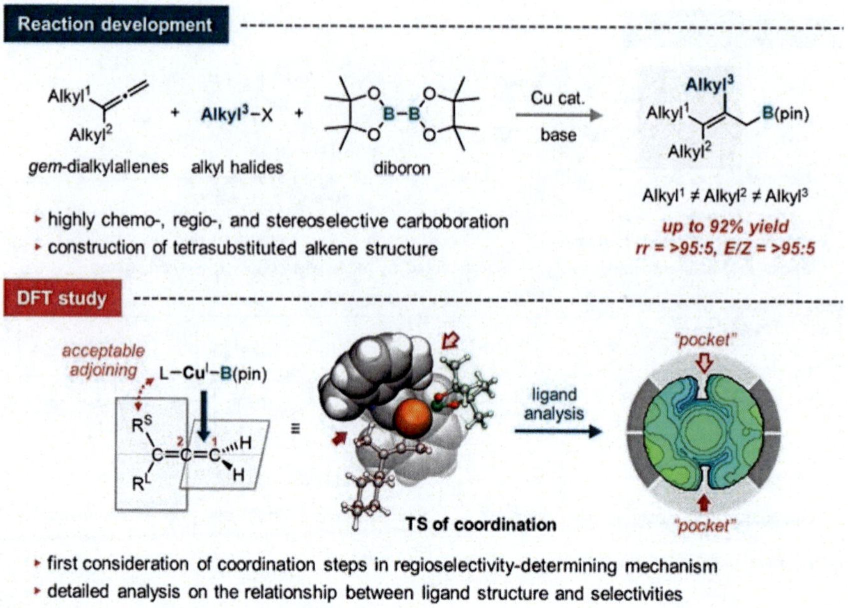

Fig. 1.23 Graphical abstract of Chap. 4

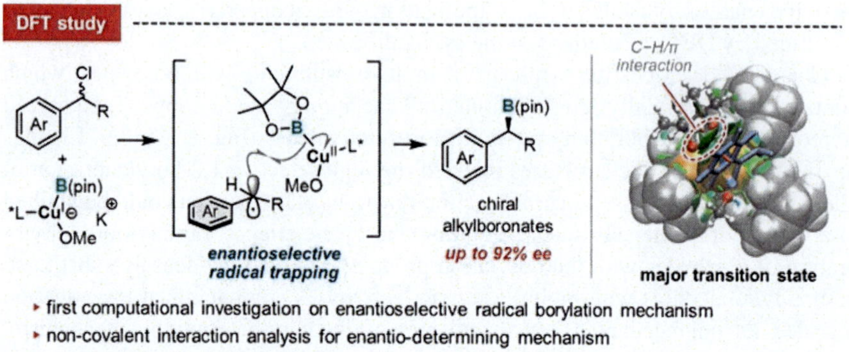

Fig. 1.24 Graphical abstract of Chap. 5

References

1. Hall DG (ed) (2011) Boronic acids: preparation, applications in organic synthesis and medicine, vols 1–2. Wiley-VCH, Weinheim
2. Fyfe JWB, Watson AJB (2017) Recent developments in organoboron chemistry: old dogs, new tricks. Chem 3:31–55
3. Frankland E, Duppa B (1860) On boric ethide. Proc R Soc London 10:568–570
4. Frankland E, Duppa B (1860) Vorläufige Notiz Über Boräthyl. Justus Liebigs Ann Chem 115:319–322

5. Miyaura N, Suzuki A (1995) Palladium-catalyzed cross-coupling reactions of organoboron compounds. Chem Rev 95:2457–2483
6. Crudden CM, Glasspoole BW, Lata CJ (2009) Expanding the scope of transformations of organoboron species: carbon-carbon bond formation with retention of configuration. Chem Commun 6704–6716
7. . Leonori D, Aggarwal VK (2015) Stereospecific couplings of secondary and tertiary boronic esters. Angew Chem Int Ed 54:1082–1096
8. Diner C, Szabó KJ (2017) Recent advances in the preparation and application of allylboron species in organic synthesis. J Am Chem Soc 139:2–14
9. Sandford C, Aggarwal VK (2017) Stereospecific functionalizations and transformations of secondary and tertiary boronic esters. Chem Commun 53:5481–5494
10. Carreras J, Caballero A, Pérez PJ (2019) Alkenyl boronates: synthesis and applications. Chem Asian J 14:329–343
11. Snyder HR, Kuck JA, Johnson RJ (1938) Organoboron compounds, and the study of reaction mechanisms. Primary aliphatic boronic acids. J Am Chem Soc 60:105–111
12. Johnson JR, Van Campen MG, Grummitt O (1938) Organoboron compounds. II. The reducing action of some organoboronic acids. J Am Chem Soc 60:111–115
13. Johnson JR, Snyder HR, Van Campen MG (1938) Organoboron compounds. III. Reactions of tri-n-butylborine. J Am Chem Soc 60:115–121
14. Johnson JR, Van Campen MG (1938) Organoboron compounds. IV. Reaction of tri-n-butylborine with peroxides and with oxygen. Mechanism of autoöxidation. J Am Chem Soc 60:121–124
15. Brown HC, Subba Rao BC (1956) A New technique for the conversion of olefins into organoboranes and related alcohols. J Am Chem Soc 78:5694–5695
16. Brown HC, Subba Rao BC (1957) Hydroboration of olefins. A remarkably fast room-temperature addition of diborane to olefins. J Org Chem 22:1136–1137
17. Brown HC, Zweifel G (1961) Hydroboration. IX. The hydroboration of cyclic and bicyclic olefins-stereochemistry of the hydroboration reaction. J Am Chem Soc 83:2544–2551
18. Miyaura N, Suzuki A (1979) Stereoselective synthesis of arylated (*E*)-alkenes by the reaction of alk-1-enylboranes with aryl halides in the presence of palladium catalyst. J Chem Soc Chem Commun 866–867
19. Miyaura N, Yamada K, Suzuki A (1979) A New stereospecific cross-coupling by the palladium-catalyzed reaction of 1-alkenylboranes with 1-alkenyl or 1-alkynyl halides. Tetrahedron Lett 20:3437–3440
20. Miyaura N (2002) Cross-coupling reaction of organoboron compounds via base-assisted transmetalation to palladium(II) complexes. J Organomet Chem 653:54–57
21. Lennox AJJ, Lloyd-Jones GC (2014) Selection of boron reagents for Suzuki-Miyaura coupling. Chem Soc Rev 43:412–443
22. Enantioselective, rhodium-catalyzed 1,4-addition of organoboron reagents to electron-deficient alkenes. In: Organic reactions, vol 93. Wiley, Hoboken, NJ, USA, pp 1–686
23. Aydin M (ed) (2019) Recent advances in boron-containing materials. IntechOpen
24. Ren Y, Jäkle F (2016) Merging thiophene with boron: new building blocks for conjugated materials. Dalton Trans 45:13996–14007
25. Haque A, Al-balushi RA, Raithby PR, Khan MS (2020) Recent advances in π-conjugated N^C-chelate. Molecules 25:2645
26. Lee H, Karthik D, Lampande R, Ryu JH, Kwon JH (2020) Recent advancement in boron-based efficient and pure blue thermally activated delayed fluorescence materials for organic light-emitting diodes. Front Chem 8:1–16
27. Dhiman A, Giribabu L, Trivedi R (2021) π-Conjugated materials derived from boron-chalcogenophene combination. A brief description of synthetic routes and optoelectronic applications. Chem Rec 21:1738–1770
28. Song S, Gao P, Sun L, Kang D, Kongsted J, Poongavanam V, Zhan P, Liu X (2021) Recent developments in the medicinal chemistry of single boron atom-containing compounds. Acta Pharm Sin B 11:3035–3059

29. Pineschi M (2021) Boron reagents and catalysts for the functionalization of strained heterocycles. Adv Synth Catal 2325–2339
30. Qi Y, Cao X, Zou Y, Yang C (2021) Multi-resonance organoboron-based fluorescent probe for ultra-sensitive, selective and reversible detection of fluoride ions. J Mater Chem C 9:1567–1571
31. Adams J, Behnke M, Chen S, Cruickshank AA, Dick LR, Grenier L, Klunder JM, Ma YT, Plamondon L, Stein RL (1998) Potent and selective inhibitors of the proteasome: dipeptidyl boronic acids. Bioorganic Med Chem Lett 8:333–338
32. Corey EJ, Bakshi RK, Shibata S (1987) Highly enantioselective borane reduction of ketones catalyzed by chiral oxazaborolidines. Mechanism and synthetic implications. J Am Chem Soc 109:5551–5553
33. Corey EJ, Helal CJ (1998) Remarkable advances in catalytic methods for enantio- selective synthesis of chiral organic molecules have chemistry. Great progress has been made not only in reduction of carbonyl compounds with chiral oxazaborolidine catalysts: a new paradigm for enan. Angew Chem Int Ed 37:1986–2012
34. Cho BT (2006) Recent advances in the synthetic applications of the oxazaborolidine-mediated asymmetric reduction. Tetrahedron 62:7621–7643; Cho BT (2009) Recent development and improvement for boron hydride-based catalytic asymmetric reduction of unsymmetrical ketones. Chem Soc Rev 38:443–452
35. Shende VS, Singh P, Bhanage BM (2018) Recent trends in organocatalyzed asymmetric reduction of prochiral ketones. Catal Sci Technol 8:955–969
36. Schwinger DP, Bach T (2020) Chiral 1,3,2-oxazaborolidine catalysts for enantioselective photochemical reactions. Acc Chem Res 53:1933–1943
37. Männig D, Nöth H (1985) Catalytic hydroboration with rhodium complexes. Angew Chem Int Ed 24:878–879
38. Burgess K, Ohlmeyer MJ (1991) Transition-metal-promoted hydroborations of alkenes, emerging methodology for organic transformations. Chem Rev 91:1179–1191
39. Carroll AM, O'Sullivan TP, Guiry PJ (2005) The development of enantioselective rhodium-catalysed hydroboration of olefins. Adv Synth Catal 347:609–631
40. Ishiyama T, Murata M, Miyaura N (1995) Palladium(0)-catalyzed cross-coupling reaction of alkoxydiboron with haloarenes: a direct procedure for arylboronic esters. J Org Chem 60:7508–7510
41. Takagi J, Takahashi K, Ishiyama T, Miyaura N (2002) Palladium-catalyzed cross-coupling reaction of bis(pinacolato)diboron with 1-alkenyl halides or triflates: convenient synthesis of unsymmetrical 1,3-dienes via the borylation-coupling sequence. J Am Chem Soc 124:8001–8006
42. Ishiyama T, Miyaura N (2004) Metal-catalyzed reactions of diborons for synthesis of organoboron compounds. Chem Rec 3:271–280
43. Chow WK, Yuen OY, Choy PY, So CM, Lau CP, Wong WT, Kwong FY (2013) A decade advancement of transition metal-catalyzed borylation of aryl halides and sulfonates. RSC Adv 3:12518–12539
44. Waltz KM, He X, Muhoro C, Hartwig JF (1995) Hydrocarbon functionalization by transition metal boryls. J Am Chem Soc 117:11357–11358
45. Waltz KM, Hartwig JF (2000) Functionalization of alkanes by isolated transition metal boryl complexes. J Am Chem Soc 122:11358–11369
46. Chen H, Schlecht S, Semple TC, Hartwig JF (2000) Thermal, catalytic, regiospecific functionalization of alkanes. Science 287:1995–1997
47. Takagi J, Sato K, Hartwig JF, Ishiyama T, Miyaura N (2002) Iridium-catalyzed c-h coupling reaction of heteroaromatic compounds with bis(pinacolato)diboron: regioselective synthesis of heteroarylboronates. Tetrahedron Lett 43:5649–5651
48. Ishiyama T, Takagi J, Ishida K, Miyaura N, Anastasi NR, Hartwig JF (2002) Mild iridium-catalyzed borylation of arenes. high turnover numbers, room temperature reactions, and isolation of a potential intermediate. J Am Chem Soc 124:390–391
49. Mkhalid IAI, Barnard JH, Marder TB, Murphy JM, Hartwig JF (2010) C-H activation for the construction of C–B bonds. Chem Rev 110:890–931

50. Hartwig JF (2011) Regioselectivity of the borylation of alkanes and arenes. Chem Soc Rev 40:1992–2002
51. Xu L, Wang G, Zhang S, Wang H, Wang L, Liu L, Jiao J, Li P (2017) Recent advances in catalytic C−H borylation reactions. Tetrahedron 73:7123–7157
52. Kuroda Y, Nakao Y (2019) Catalyst-enabled site-selectivity in the iridium-catalyzed CH borylation of arenes. Chem Lett 48:1092–1100
53. Wang M, Shi Z (2020) Methodologies and strategies for selective borylation of C-Het and C–C bonds. Chem Rev 120:7348–7398
54. Yadagiri B, Daipule K, Singh SP (2021) Photoinduced borylation reactions: an overview. Asian J Org Chem 10:7–37
55. Ito H, Yamanaka H, Tateiwa J, Hosomi A (2000) Boration of an α, β-enone using a diboron promoted by a copper(I)–phosphine mixture catalyst. Tetrahedron Lett 41:6821–6825
56. Takahashi K, Ishiyama T, Miyaura N (2000) Addition and coupling reactions of bis(pinacolato)diboron mediated by CuCl in the presence of potassium acetate. Chem Lett 57:982–983
57. Takahashi K, Ishiyama T, Miyaura N (2001) A borylcopper species generated from bis(pinacolato)diboron and its additions to α, β-unsaturated carbonyl compounds and terminal alkynes. J Organomet Chem 625:47–53
58. Neeve EC, Geier SJ, Mkhalid IAI, Westcott SA, Marder TB (2016) Diboron(4) compounds: from structural curiosity to synthetic workhorse. Chem Rev 116:9091–9161
59. Kubota K, Iwamoto H, Ito H (2017) Formal nucleophilic borylation and borylative cyclization of organic halides. Org Biomol Chem 15:285–300
60. Hemming D, Fritzemeier R, Westcott SA, Santos WL, Steel PG (2018) Copper-boryl mediated organic synthesis. Chem Soc Rev 47:7477–7494
61. Takale BS, Thakore RR, Etemadi-Davan E, Lipshutz BH (2020) Recent advances in Cu-catalyzed C(sp3)–Si and C(sp3)–B bond formation. Beilstein J Org Chem 16:691–737
62. Ito H, Kawakami C, Sawamura M (2005) Copper-catalyzed γ-selective and stereospecific substitution reaction of allylic carbonates with diboron: efficient route to chiral allylboron compounds. J Am Chem Soc 127:16034–16035
63. Ito H, Ito S, Sasaki Y, Matsuura K, Sawamura M (2007) Copper-catalyzed enantioselective substitution of allylic carbonates with diboron: an efficient route to optically active α-chiral allylboronates. J Am Chem Soc 129:14856–14857
64. Kubota K, Hayama K, Iwamoto H, Ito H (2015) Enantioselective borylative dearomatization of indoles through copper(I) catalysis. Angew Chem Int Ed 54:8809–8813
65. Hayama K, Kubota K, Iwamoto H, Ito H (2017) Copper(I)-catalyzed diastereoselective dearomative carboborylation of indoles. Chem Lett 46:1800–1802
66. Hayama K, Kojima R, Kubota K, Ito H (2020) Synthesis of chiral *N*-heterocyclic allylboronates via the enantioselective borylative dearomatization of pyrroles. Org Lett 22:739–744
67. Shimizu Y, Kanai M (2014) Recent progress in copper-catalyzed difunctionalization of unactivated carboncarbon multiple bonds. Tetrahedron Lett 55:3727–3737
68. Liu Z, Gao Y, Zeng T, Engle KM (2020) Transition-metal-catalyzed 1,2-carboboration of alkenes: strategies, mechanisms, and stereocontrol. Isr J Chem 60:219–229
69. Kleeberg C, Dang L, Lin Z, Marder TB (2009) A facile route to aryl boronates: room-temperature, copper-catalyzed borylation of aryl halides with alkoxy diboron reagents. Angew Chem 121:5454–5458
70. Niwa T, Ochiai H, Hosoya T (2017) Copper-catalyzed *ipso*-borylation of fluoroarenes. ACS Catal 7:4535–4541
71. Yang CT, Zhang ZQ, Tajuddin H, Wu CC, Liang J, Liu JH, Fu Y, Czyzewska M, Steel PG, Marder TB, Liu L (2012) Alkylboronic esters from copper-catalyzed borylation of primary and secondary alkyl halides and pseudohalides. Angew Chem Int Ed 51:528–532
72. Ito H, Kubota K (2012) Copper(I)-catalyzed boryl substitution of unactivated alkyl halides. Org Lett 14:890–893
73. Laitar DS, Müller P, Sadighi JP (2005) Efficient homogeneous catalysis in the reduction of CO_2 to CO. J Am Chem Soc 127:17196–17197

74. Semba K, Shinomiya M, Fujihara T, Terao J, Tsuji Y (2013) Highly selective copper-catalyzed hydroboration of allenes and 1,3-dienes. Chem Eur J 19:7125–7132
75. Semba K, Fujihara T, Terao J, Tsuji Y (2015) Copper-catalyzed borylative transformations of non-polar carbon-carbon unsaturated compounds employing borylcopper as an active catalyst species. Tetrahedron 71:2183–2197
76. Drescher W, Kleeberg C (2019) Terminal versus bridging boryl coordination in *N*-heterocyclic carbene copper(I) boryl complexes: syntheses, structures, and dynamic behavior. Inorg Chem 58:8215–8229
77. Drescher W, Borner C, Kleeberg C (2021) Stability and decomposition of copper(I) boryl complexes: [(IDipp)Cu-Bneop], [(IDipp∗)Cu-Bneop] and copper clusters. New J Chem 45:14957–14964
78. Segawa Y, Yamashita M, Nozaki K (2007) Boryl anion attacks transition-metal chlorides to form boryl complexes: syntheses, spectroscopic, and structural studies on group 11 borylmetal complexes. Angew Chem Int Ed 46:6710–6713
79. Kajiwara T, Terabayashi T, Yamashita M, Nozaki K (2008) Syntheses, structures, and reactivities of borylcopper and -zinc compounds: 1,4-silaboration of an α, β-unsaturated ketone to form a γ-siloxyallylborane. Angew Chem Int Ed 47:6606–6610
80. Okuno Y, Yamashita M, Nozaki K (2011) One-Pot carboboration of alkynes using lithium borylcyanocuprate and the subsequent Suzuki-Miyaura cross-coupling of the resulting tetra-substituted alkenylborane. Eur J Org Chem 3951–3958
81. Borner C, Anders L, Brandhorst K, Kleeberg C (2017) Elusive phosphine copper(I) boryl complexes: synthesis, structures, and reactivity. Organometallics 36:4687–4690
82. Kleeberg C, Borner C (2018) Syntheses, structures, and reactivity of NHC copper(I) boryl complexes: a systematic study. Organometallics 37:4136–4146
83. Tobisch S (2017) Copper-catalysed aminoboration of vinylarenes with hydroxylamine esters-A computational mechanistic study. Chem Eur J 23:17800–17809
84. Iwamoto H, Ozawa Y, Takenouchi Y, Imamoto T, Ito H (2021) Backbone-modified C_2-symmetrical chiral bisphosphine TMS-QuinoxP*: asymmetric borylation of racemic allyl electrophiles. J Am Chem Soc 143:6413–6422
85. Ozawa Y, Iwamoto H, Ito H (2018) Copper(I)-catalysed regio- and diastereoselective intramolecular alkylboration of terminal allenes via allylcopper(I) isomerization. Chem Commun 54:4991–4994
86. Ozawa Y, Endo K, Ito H (2021) Regio- and stereoselective synthesis of multi-alkylated allylic boronates through three-component coupling reactions between allenes, alkyl halides, and a diboron reagent. J Am Chem Soc 143:13865–13877
87. Iwamoto H, Endo K, Ozawa Y, Watanabe Y, Kubota K, Imamoto T, Ito H (2019) Copper(I)-catalyzed enantioconvergent borylation of racemic benzyl chlorides enabled by quadrant-by-quadrant structure modulation of chiral bisphosphine ligands. Angew Chem Int Ed 58:11112–11117

Z

Chapter 2
Modification of QuinoxP*-Type Bisphosphine Ligands for High-Performance Asymmetric Boryl Substitution of Racemic Allyl Electrophiles

2.1 Introduction

α-Chiral allylic boronates play essential roles in the field of organic synthesis and pharmaceuticals [1, 2]. Various methods for stereospecific transformation of the boryl group have been developed and are now available, including allylboration, oxidation, amination, halogenation, and homologation reactions, as well as cross-coupling reactions, which proceed at α- or γ-position (Fig. 2.1) [3–18]. Therefore, the chiral boronates are recognized as versatile intermediates. Thus, the development of novel α-chiral allylic boronates and their synthetic approach is important research target.

Chiral bisphosphine ligands have made a significant contribution to the evolution of enantioselective organic transformations [19–21]. Focused on structural modification of the ligand for improvement of the catalyst performance (Fig. 2.2) [22–24], an exchange of the backbone structure is one of the successful strategies, e.g., BINAP to SEGPHOS in catalytic hydrogenation reaction [25–27]. Also, an exchange or modification of substituents on the phosphorous atom(s) is effective because the substituents are positioned around the metal center and directly interact with the contacting substrate, e.g., a series of Duphos ligand [28–32]. However, this drastic structural change sometimes requires laborious reconsideration of synthetic route and may cause unexpected influence on the asymmetric reaction space of the catalyst.

In 2005, the Imamoto group developed a C_2-symmetrical P-chirogenic bisphosphine ligand (*R*,*R*)-QuinoxP* (Fig. 2.3) [33–35]. This ligand can be prepared in one-pot from commercially available 2,3-dichloroquinoxaline and a chiral phosphine anion. In 2018, the Ito group reported that the modification of alkyl substituents on the phosphorous atoms in this ligand was able to improve the regio- and enantioselectivities in Markovnikov asymmetric protoboration of terminal alkenes (Fig. 2.4) [35–38]. For further development of QuinoxP*-type ligand for high performance in copper(I)-catalyzed borylation reactions, I planned a new strategy for the modification of the ligand in such a way that the supporting groups are introduced into the ligand backbone and interact with the substituents on the phosphorous atoms from the back side

Y. Ozawa, *Copper(I)-Catalyzed Stereoselective Borylation Reactions*, Springer Theses,
https://doi.org/10.1007/978-981-99-1098-4_2

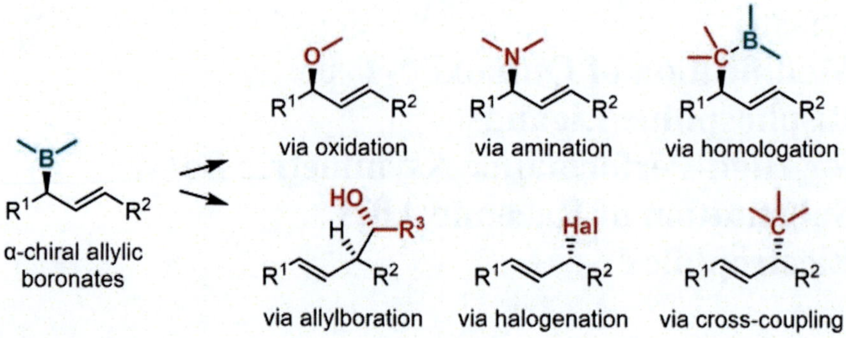

Fig. 2.1 Derivatization reactions of α-chiral allylic boronates

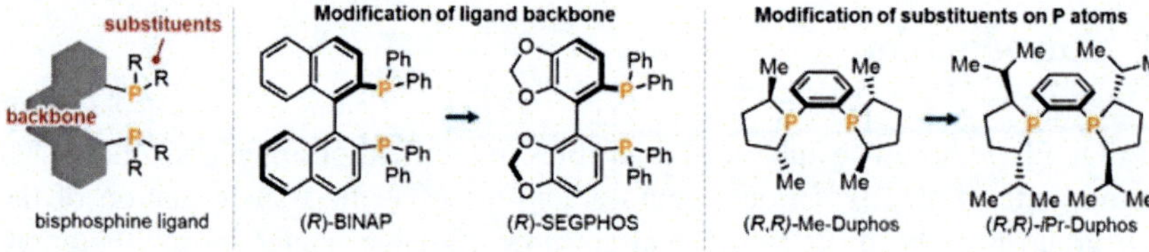

Fig. 2.2 C_2-symmetrical chiral bisphosphine ligands

through non-covalent (through-space) interaction (Fig. 2.5A) [39–42]. As the asymmetric reaction space would not be drastically changed with this strategy, reactions that QuinoxP* ligand originally works well might be improved. In this study, I demonstrated that this strategy was successfully implemented in the QuinoxP* ligand and improved the reactivity and enantioselectivity of the asymmetric borylative substitution reaction of racemic allylic electrophiles. The reactivity of boration reaction in a copper(I)-catalyzed direct enantioconvergent borylation of six-membered substrates including heterocycles was increased without drop of enantioselectivity (Fig. 2.5B) [43]. Furthermore, the new ligands enabled the first borylative kinetic resolution of the linear substrates (Fig. 2.5C).

Fig. 2.3 Synthesis of QuinoxP* ligand

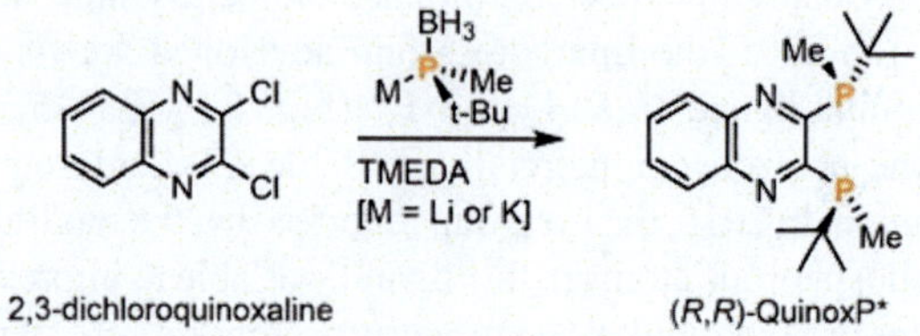

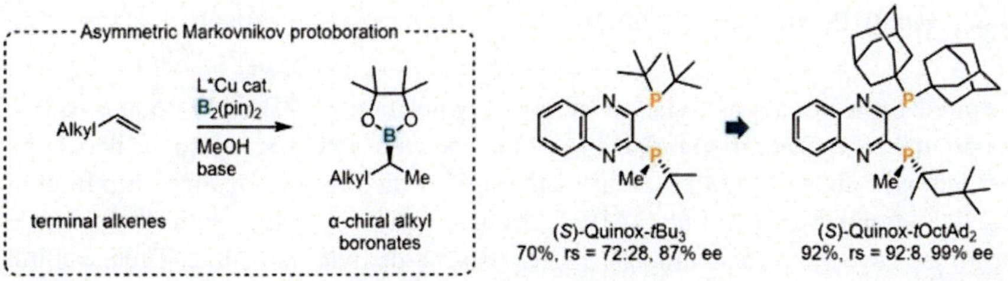

Fig. 2.4 Modification of QuinoxP*-type ligand and its application to asymmetric Markovnikov protoboration reaction

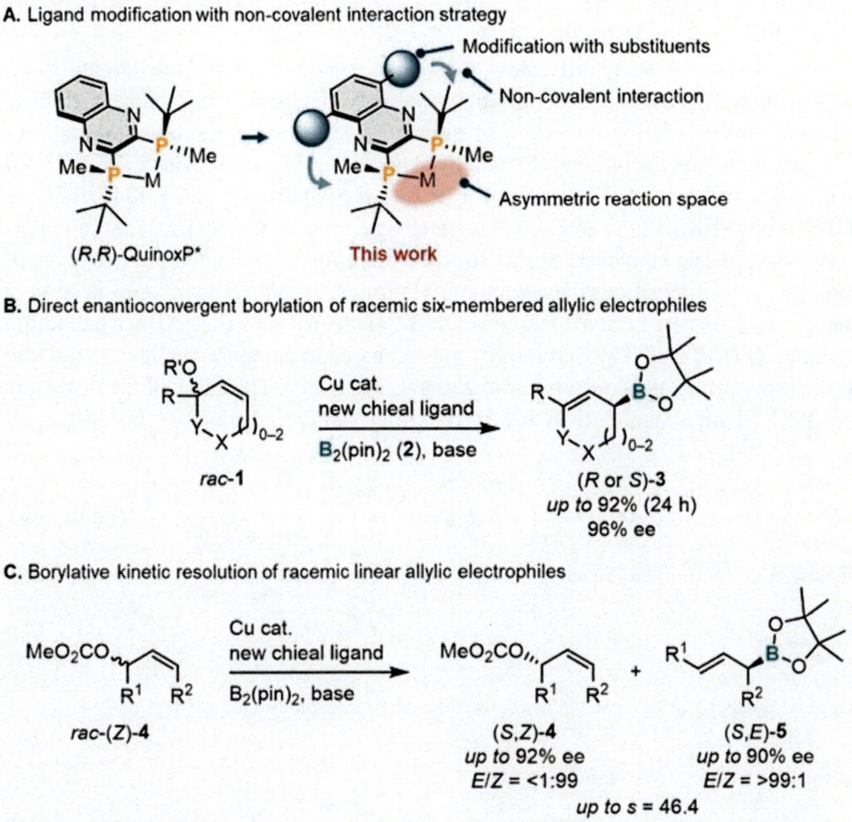

Fig. 2.5 Modification of QuinoxP*-type ligand with non-covalent interaction strategy and its application to asymmetric borylation of racemic allylic electrophiles

2.2 Results and Discussion

Inspired by a lithiation/silylation at 5 and 8-position of 2,3-dichloroquinoxaline reported by Knochel group in 2014 [44], I introduced silyl groups into the backbone of QuinoxP* ligand (Fig. 2.6). Dilithiation of 2,3-dichloroquinoxaline with lithium 2,2,6,6-tetramethylpiperidine (TMPLi) and subsequent silylation with silyl chloride or triflate gave the corresponding 2,3-dichloro-5,8-disilylquinoxaline. Then, a chiral phosphine blocks were introduced into the quinoxaline derivative using phosphine anion reagent, which was prepared from (*S*)-*tert*-butylmethylphosphine–borane and *n*-butyl lithium or potassium *tert*-butoxide [35, 45]. Accordingly, I synthesized a series of novel QuinoxP*-type chiral bisphosphine ligands, (*R*,*R*)-5,8-*Si*-QuinoxP* (*Si* = TMS, TES, or TIPS) (Fig. 2.7).

A single-crystal X-ray diffraction analysis of those ligands and the parent ligand, (*R*,*R*)-QuinoxP*, was carried out to investigate the ligand structures (Fig. 2.8). Focused on the twisting distortion of quinoxaline backbone, the dihedral angle of 7.4° was found in (*R*,*R*)-QuinoxP* when the angle was measured with C1-C2-C3-C4 (Fig. 2.8A, B). Contrary, the angles for (*R*,*R*)-5,8-*Si*-QuinoxP* (*Si* = TMS, TES, or TIPS) were 11.8°, 12.0°, and 14.6°, respectively (Fig. 2.8C–E). Those large distortions of backbone compared to that for the parent ligand indicated the existence of through-space interactions between the silyl groups and phosphine groups in spite of that those groups are located at three bonds distant from each other. Also, a palladium complex of (*R*,*R*)-5,8-TMS-QuinoxP* was prepared to compare the ligand structure in the form of free and coordinated to a metal (Fig. 2.8F). The angle of the distortion was 8.8°, which is smaller than that of free (*R*,*R*)-5,8-TMS-QuinoxP*, but still larger

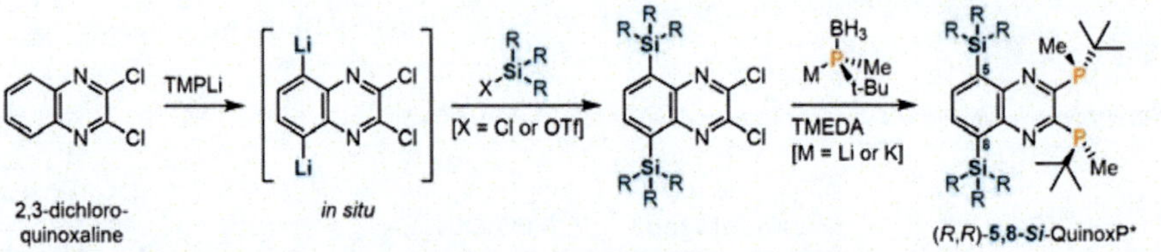

Fig. 2.6 Synthesis of silylated QuinoxP*-type ligands: (*R*,*R*)-5,8-*Si*-QuinoxP*

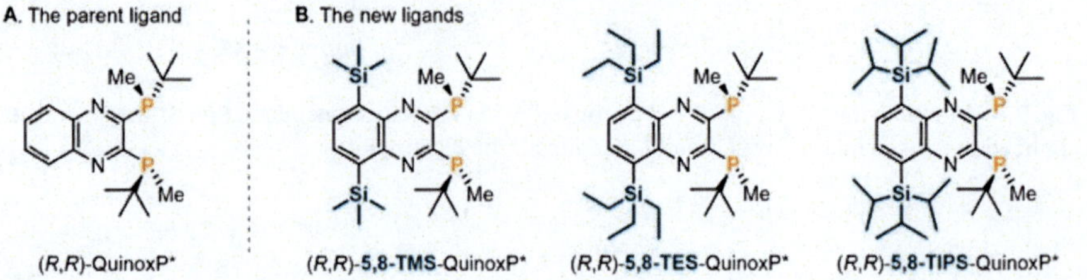

Fig. 2.7 A series of QuinoxP*-type ligands: (*R*,*R*)-5,8-*Si*-QuinoxP*

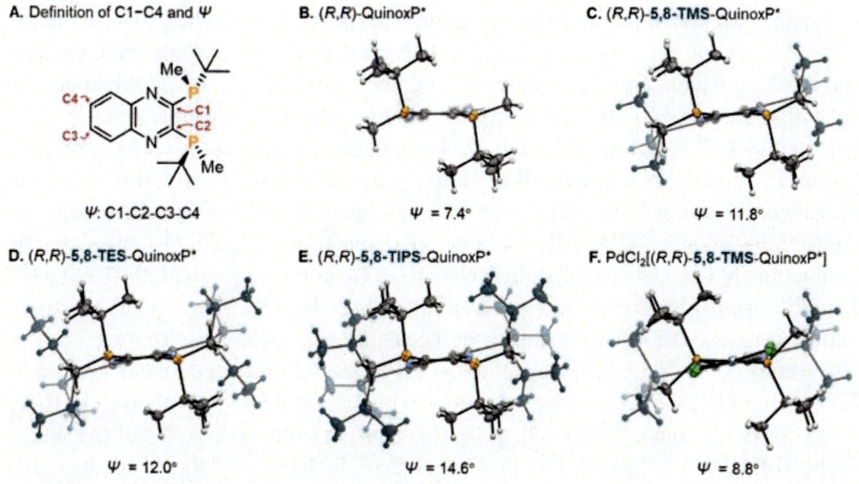

Fig. 2.8 X-ray structure of QuinoxP*-type ligands

than that of (*R*,*R*)-QuinoxP*. Thus, the through-space interactions between the silyl and phosphine groups in (*R*,*R*)-5,8-*Si*-QuinoxP* are valid in metal–ligand complex.

For a benchmark of the catalytic performance of those ligands, I chose the copper(I)-catalyzed direct enantioconvergent borylation of racemic six-membered cyclic allyl electrophiles. This type of enantioconvergent transformation of racemic substrate is rare and proceeds via different mechanism from other enantioconvergent reactions that the substrate is racemized or converted to an achiral intermediate in situ, which was first reported by the Ito and Sawamura group in 2010 (Fig. 2.9) [43, 46–49]. The enantioenriched borylation product is afforded via the *anti*-S_N2' pathway from one enantiomer of the substrate, and *syn*-S_N2' pathway from another enantiomer, which has been realized by using a copper(I)-catalyst and (*R*,*R*)-QuinoxP* ligand. However, the reactivity of a six-membered ring substrate was low, although the reaction of five-membered ring substrates proceeded smoothly at room temperature and was completed within 24 h [43]. Thus, I decided to explore the ligand performance using the six-membered substrate.

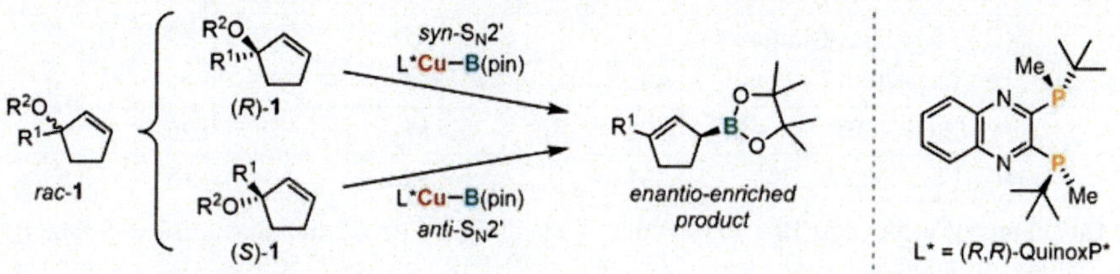

Fig. 2.9 Direct enantioconvergent borylation of racemic allyl electrophile using a copper(I)/(*R*,*R*)-QuinoxP* catalyst reported by the Ito and Sawamura group

For the baseline of the ligand investigation, the borylation reaction was conducted with (*R*,*R*)-QuinoxP* using [Cu(MeCN)$_4$]BF$_4$ as the precursor of copper(I) catalyst and semi-equivalent amount of base at 50 °C for 24 h (Table 2.1). The substrate was not fully consumed to afford the product in low yield with moderate enantioselectivity (entry 1: 39% conv., 12% yield, 79% ee). In contrast, the use of (*R*,*R*)-5,8-TMS-QuinoxP* drastically improved the reactivity and enantioselectivity of the reaction to produce the chiral cyclic allylic boronate in high yield with high enantioselectivity (entry 2: 94% conv., 92% yield, 93% ee). The reaction using Cu(O-*t*-Bu) gave the product in slightly decreased yield (entry 3: >95% conv., 77% yield, 94% ee) [43]. The TES-group introduced ligand, (*R*,*R*)-5,8-TES-QuinoxP*, showed high reactivity to fully consume the substrate; however, the enantioselectivity was dropped (entry 4: >95% conv., 94% yield, 85% ee). With the TIPS-group introduced ligand, (*R*,*R*)-5,8-TIPS-QuinoxP*, the product was obtained in high yield with high enantioselectivity (entry 5: >95% conv., 95% yield, 93% ee). Thus, it could be concluded that introduction of bulky silyl groups into the backbone of the QuinoxP* ligand enhances the reactivity and enantioselectivity of the direct enantioconvergent borylation reaction of the six-membered allyl electrophile. Then, I also investigated other commercially available chiral bisphosphine ligands, i.e., (*R*,*R*)-BenzP*, (*R*)-SEGPHOS and (*R*,*R*)-Me-Duphos, for comparing the performance against the QuinoxP*-type ligands. The (*R*,*R*)-BenzP* ligand was found to be suitable for the direct enantioconvergent borylation of *rac*-**1a** to give the product in high yield with high enantioselectivity (entry

Table 2.1 Chiral ligand screening[a]

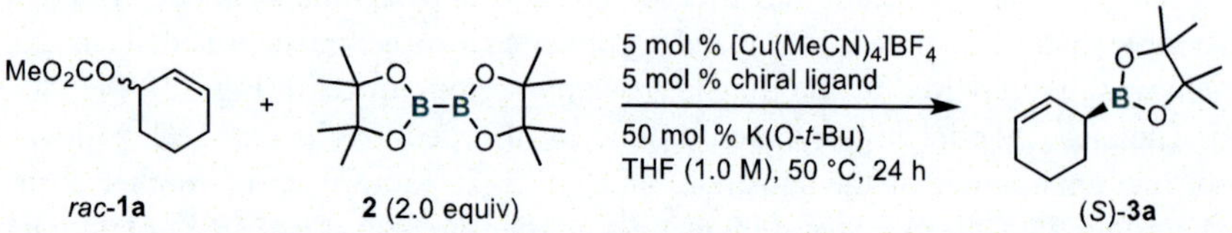

Entry	Chiral ligand	Conv. of **1a** [%][b]	Yield of **3a** [%][b]	ee of **3a** [%][c]
1	(*R*,*R*)-QuinoxP*	39	12	79
2	(*R*,*R*)-5,8-TMS-QuinoxP*	94	92	93
3[d]	(*R*,*R*)-5,8-TMS-QuinoxP*	>95	77	94
4	(*R*,*R*)-5,8-TES-QuinoxP*	>95	94	85
5	(*R*,*R*)-5,8-TIPS-QuinoxP*	>95	95	93
6	(*R*,*R*)-BenzP*	>95	92	93
7	(*R*)-SEGPHOS	92	57	25
8	(*R*,*R*)-Me-Duphos	94	89	45

[a] Conditions: [Cu(MeCN)$_4$]BF$_4$ (0.025 mmol), chiral ligand (0.025 mmol), *rac*-**1a** (0.5 mmol), B$_2$(pin)$_2$ (1.0 mmol), and K(O-*t*-Bu) (0.25 mmol) in THF (0.5 mL) at 50 °C for 24 h. [b] Determined by quantitative ^{1}H NMR analysis of the crude material using an internal standard. [c] Determined by HPLC analysis after allylboration of benzaldehyde with the borylation product. [d] Conditions: Cu(O-*t*-Bu) (0.05 mmol), chiral ligand (0.05 mmol), *rac*-**1a** (0.5 mmol), B$_2$(pin)$_2$ (0.75 mmol) in Et$_2$O (0.5 mL) at 30 °C for 24 h

6: >95% conv., 92% yield, 93% ee). The other ligands, which are chiral ligands possessing C_2-symmetric structure as well as QuinoxP*, showed high reactivity in the reaction. However, the enantioselectivities were lower than (*R*,*R*)-QuinoxP* and the series of (*R*,*R*)-5,8-*Si*-QuinoxP* (entry 7: 92% conv., 57% yield, 25% ee; entry 8: 94% conv., 89% yield, 45% ee).

Then, I investigated the substrate scope of this reaction by employing commercially available $[Cu(MeCN)_4]BF_4$ as the catalyst precursor, and (*R*,*R*)-QuinoxP* and (*R*,*R*)-5,8-TMS-QuinoxP* as the ligand for comparing the ligand performance (Table 2.2). The model substrate (*S*)-**3a**, which is an aliphatic six-membered ring, was obtained in high yield with high enantioselectivity when (*R*,*R*)-5,8-TMS-QuinoxP* was employed (92%, 93% ee). However, (*R*,*R*)-QuinoxP* showed low reactivity and enantioselectivity (12%, 79% ee). Then, hetero-cyclic substrates were investigated in this reaction, which have not been examined in the direct enantioconvergent borylation reaction in the previous report from the Ito and Sawamura group. With the parent ligand, (*R*,*R*)-QuinoxP*, the tetrahydropyridine substrates bearing tosyl- (Ts), *tert*-butoxycarbonyl- (Boc), and benzyloxycarbonyl- (Cbz) protected amine group were converted to the corresponding cyclic allylic boronates in moderate to good yield with good enantioselectivity [(*R*)-**3b**: 50%, 76% ee; (*R*)-**3c**: 67%, 83% ee; (*R*)-**3d**: 38%, 82% ee]. In contrast, (*R*,*R*)-5,8-TMS-QuinoxP* showed higher reactivity and enantioselectivity for those tetrahydropyridine substrates to give the product in high yield with high enantioselectivity [(*R*)-**3b**: 85%, 95% ee; (*R*)-**3c**: 84%, 90% ee; (*R*)-**3d**: 84%, 90% ee]. Also, for the dihydropyran substrate *rac*-**1e**, the new ligand afforded the product (*R*)-**3e** in higher yield with higher enantioselectivity than the parent ligand [(*R*,*R*)-QuinoxP*: 84%, 80% ee; (*R*,*R*)-5,8-TMS-QuinoxP*: 92%, 96% ee]. Then, I examined the substrate which has substituents at the "X"-position as shown in Table 2.2. Although the borylation product (*S*)-**3f** was not obtained using the parent ligand, the new ligand gave the product in high yield even with slightly lowered enantioselectivity [(*R*,*R*)-QuinoxP*: <5%; (*R*,*R*)-5,8-TMS-QuinoxP*: 73%, 84% ee]. The substrates possessing a heteroatom at the "Y"-position are then investigated. For these substrates, methoxy and benzyloxy group was adopted as the leaving group because of the high stability of the substrate. However, the *N*,*O*-acetal substrate **1g** showed high reactivity than the other substrates, and thus, the reaction was performed at the lowered temperature (0 °C). At this temperature, the product was obtained in low yield with the parent ligand, while the new ligand furnished the product in high yield [(*R*,*R*)-QuinoxP*: 10%; (*R*,*R*)-5,8-TMS-QuinoxP*: 90%, 79% ee]. Also, similar difference in the reactivity was observed for the *O*,*O*-acetal substrate **1h**, although the product was obtained merely in moderate yield even with the new ligand [(*R*,*R*)-QuinoxP*: <5%; (*R*,*R*)-5,8-TMS-QuinoxP*: 40%, 56% ee]. Next, the effect of the ring size was explored. The five-membered ring substrates have already been studied in the report from the Ito and Sawamura group and found to show high reactivity under conditions with (*R*,*R*)-QuinoxP*. A simple substrate **1i** was successfully converted to the corresponding allylic boronate (*S*)-**3i** in high yield with high enantioselectivity using the both ligands [(*R*,*R*)-QuinoxP*: 81%, 91% ee; (*R*,*R*)-5,8-TMS-QuinoxP*: 80%, 92% ee]. In the case of the substrate bearing

Table 2.2 Substrate scope[a]

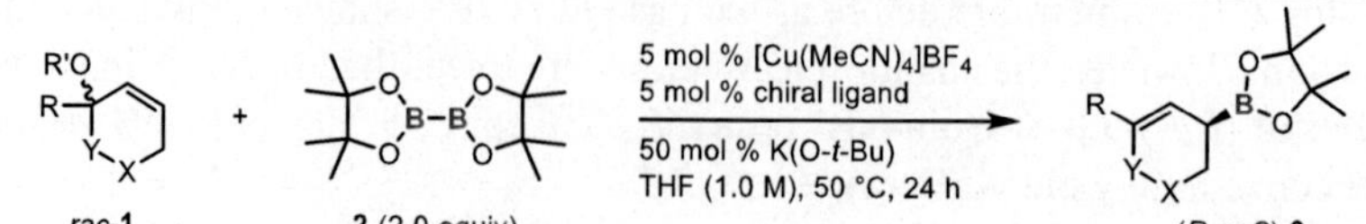

rac-**1** + **2** (2.0 equiv) → (*R* or *S*)-**3**

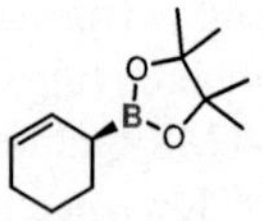

(*S*)-**3a**
[OR' = OCO$_2$Me]
(*R*,*R*)-QuinoxP*
12%, 79% ee
(*R*,*R*)-**5,8-TMS**-QuinoxP*
92% (86%), 79% ee

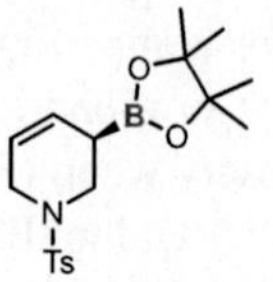

(*R*)-**3b**
[OR' = OCO$_2$Me]
(*R*,*R*)-QuinoxP*
50%, 76% ee
(*R*,*R*)-**5,8-TMS**-QuinoxP*
85% (77%), 95% ee

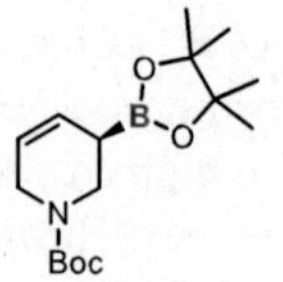

(*R*)-**3c**
[OR' = OCO$_2$Me]
(*R*,*R*)-QuinoxP*
67%, 83% ee
(*R*,*R*)-**5,8-TMS**-QuinoxP*
84% (52%), 90% ee

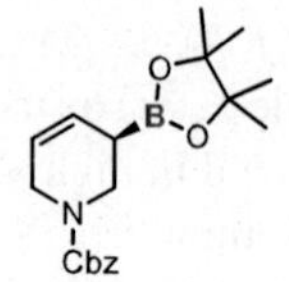

(*R*)-**3d**
[OR' = OCO$_2$Me]
(*R*,*R*)-QuinoxP*
38%, 82% ee
(*R*,*R*)-**5,8-TMS**-QuinoxP*
84% (79%), 90% ee

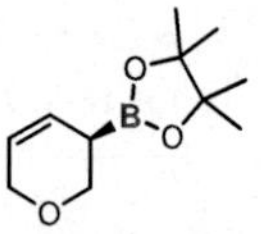

(*R*)-**3e**
[OR' = OCO$_2$Me]
(*R*,*R*)-QuinoxP*
84%, 80% ee
(*R*,*R*)-**5,8-TMS**-QuinoxP*
92% (57%), 96% ee

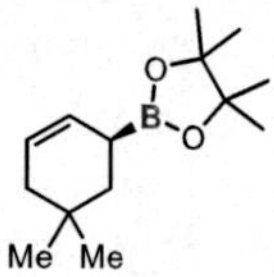

(*S*)-**3f**
[OR' = OCO$_2$Me]
(*R*,*R*)-QuinoxP*
<5%
(*R*,*R*)-**5,8-TMS**-QuinoxP*
73% (58%), 84% ee

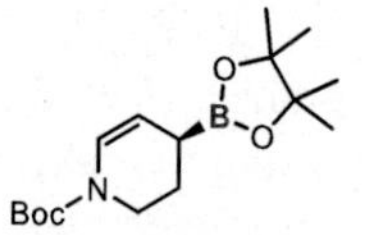

(*S*)-**3g**[b,c]
[OR' = OMe]
(*R*,*R*)-QuinoxP*
10%
(*R*,*R*)-**5,8-TMS**-QuinoxP*
90%, 79% ee

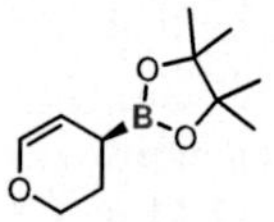

(*S*)-**3h**[b]
[OR' = OBn]
(*R*,*R*)-QuinoxP*
<5%
(*R*,*R*)-**5,8-TMS**-QuinoxP*
40%, 56% ee

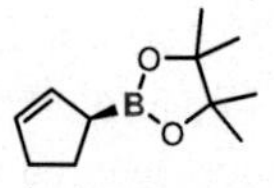

(*R*)-**3e**
[OR' = OBn]
(*R*,*R*)-QuinoxP*
81% (57%), 91% ee
(*R*,*R*)-**5,8-TMS**-QuinoxP*
80% (51%), 92% ee

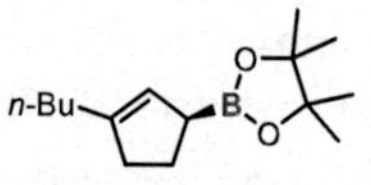

(*S*)-**3f**[d]
[OR' = OMe]
(*R*,*R*)-QuinoxP*
48%, 97% ee
(*R*,*R*)-**5,8-TMS**-QuinoxP*
92% (79%), 94% ee

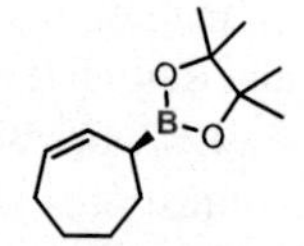

(*S*)-**3g**
[OR' = OCO$_2$Me]
(*R*,*R*)-QuinoxP*
27%, 63% ee
(*R*,*R*)-**5,8-TMS**-QuinoxP*
79% (72%), 68% ee

[a] Conditions: [Cu(MeCN)$_4$]BF$_4$ (0.025 mmol), chiral ligand (0.025 mmol), *rac*-**1** (0.5 mmol), B$_2$(pin)$_2$ (1.0 mmol), and K(O-*t*-Bu) (0.25 mmol) in THF (0.5 mL) at 50 °C for 24 h. The yield was determined by quantitative ^{1}H NMR analysis of the crude material using an internal standard. The isolated yield is given in parenthesis. The enantioselectivity was determined by HPLC analysis after allylboration of benzaldehyde with the borylation product. [b] The reaction was conducted at 0 °C. [c] The reaction was conducted with 5 mol % of K(O-*t*-Bu) (0.025 mmol). [d] The reaction was conducted at 30 °C

substituent R, (*R*,*R*)-QuinoxP* gave the product in moderate yield with high enantioselectivity (48%, 97% ee), while (*R*,*R*)-5,8-TMS-QuinoxP* afforded the product in higher yield with slightly decreased enantioselectivity (92%, 94% ee). Then, the seven-membered substrate was investigated. The reaction with the parent ligand showed insufficient reactivity to produce the allylic boronate in low yield. With the new ligand, the product was obtained in good yield, although the enantioselectivity was not improved compared to the parent ligand. In summary, the use of the new ligand instead of the parent ligand improves the reactivity for all types of the substrate investigated here, while the obvious increase of the enantioselectivity was found in the six-membered ring substrates.

Thus, I anticipated that this catalyst system with the new ligand, (*R*,*R*)-5,8-TMS-QuinoxP*, can be applied to the direct enantioconvergent borylation or the borylative kinetic resolution of the internal linear allyl electrophiles [50], which has no reactivity in the copper(I)-catalyzed borylation reactions with (*R*,*R*)-QuinoxP* ligand (Table 2.3). At first, the direct enantioconvergent borylation of the linear substrate (*Z*)-**4** was attempted under the standard conditions applied in the investigation of the substrate scope. However, the reaction resulted in insufficient conversion of the substrate and low enantioselectivity (74%, 48% ee). Then, the reaction was examined with the standard conditions but at low temperature. The reaction of the racemic mixture of the internal linear allyl carbonates *rac*-(*Z*)-**4a**–**4d** successfully furnished the enantioenriched substrates (*S*,*Z*)-**4a**–**4d** and products (*S*,*E*)-**5a**–**5d** with high enantioselectivity and perfect stereoselectivity [(*S*,*Z*)-**4a**: 58%, 63% ee; (*S*,*E*)-**5a**: 42%, 86% ee, *s* = 21.2; (*S*,*Z*)-**4b**: 51%, 74% ee; (*S*,*E*)-**5b**: 45%, 89% ee, *s* = 16.9; (*S*,*Z*)-**4c**: 44%, 92% ee; (*S*,*E*)-**5c**: 51%, 83% ee, *s* = 21.3; (*S*,*Z*)-**4d**: 45%, 81% ee; (*S*,*E*)-**5d**: 45%, 88% ee, *s* = 12.0] [51]. For the substrate bearing *iso*-propyl group [*rac*-(*Z*)-**4e**], the product was obtained with high enantioselectivity, while the enantioselectivity of the recovered substrate was moderate [(*S*,*Z*)-**4e**: 36%, 88% ee; (*S*,*E*)-**5e**: 51%, 55% ee, *s* = 6.3]. The reaction was able to apply the substrate possessing MOM-protected hydroxyl group and proceeded with high selectivity factor of 46.4 [(*S*,*Z*)-**4f**: 42%, 90% ee; (*S*,*E*)-**5f**: 57%, 69% ee, *s* = 46.4]. Contrary, (*R*,*R*)-QuinoxP* and similar ligand, (*R*,*R*)-BenzP* showed low conversion of *rac*-(*Z*)-**4** [52].

As the cyclic allylic boronates synthesized in this method were sometimes unstable to silica gel column chromatography, a one-pot sequence of the borylation and allylboration was employed (Fig. 2.10) [2, 53]. The direct enantioconvergent borylation of *rac*-**1c** was performed under the standard reaction conditions. After stirring for 24 h, an aryl aldehyde was added to the reaction mixture to produce the corresponding homoallylic alcohol (*S*,*S*)-**6** in improved yield with slightly decreased enantioselectivity compared to the two-step procedure described in Table 2.2 (87%, dr = >95:5, 87% ee) [54].

To gain insight into the reaction mechanism of the direct enantioconvergent borylation reaction, the deuterium label study was conducted with the model six-membered substrate having deuterium atom at the α-position of the leaving group (*rac*-**1a-D**) using (*R*,*R*)-5,8-TMS-QuinoxP* ligand (Fig. 2.11). The reaction under the standard conditions gave the product with the same enantioselectivity as the non-labeled substrate (84%, 93% ee). In the product, the deuterium atom was found

Table 2.3 Substrate scope for the Borylative Kinetic resolution[a]

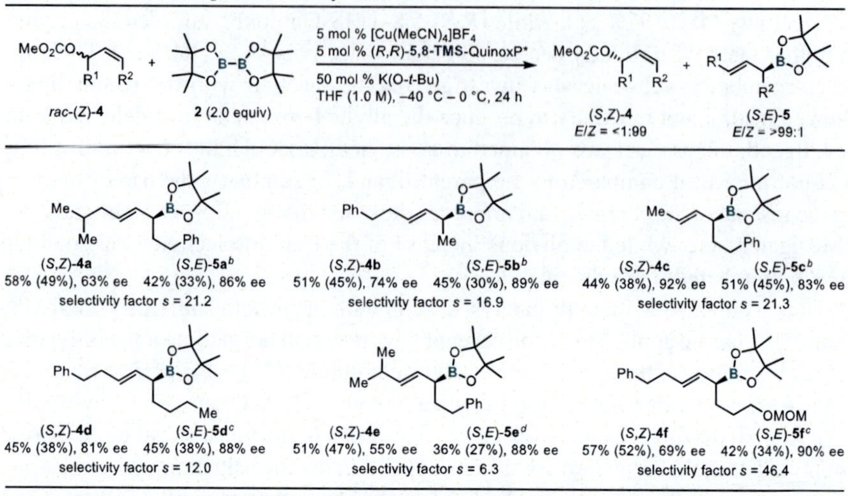

[a] Conditions: $[Cu(MeCN)_4]BF_4$ (0.025 mmol), chiral ligand (0.025 mmol), *rac*-(*Z*)-**4** (0.5 mmol), $B_2(pin)_2$ (1.0 mmol), and K(O-*t*-Bu) (0.25 mmol) in THF (0.5 mL) for 24 h. The yield was determined by quantitative ^{1}H NMR analysis of the crude material using an internal standard. The isolated yield is given in parenthesis. The enantioselectivity was determined by HPLC analysis. The selectivity factor was calculated from $s = \ln\{(1–C)/(1–ee)\}/\ln\{(1–C)/(1+ee)\}$, where C is the conversion of the substrate, and *ee* is the enantioselectivity of the recovered substrate. [b] The reaction was conducted at –40 °C. [c] The reaction was conducted at –20 °C. [d] The reaction was conducted at 0 °C

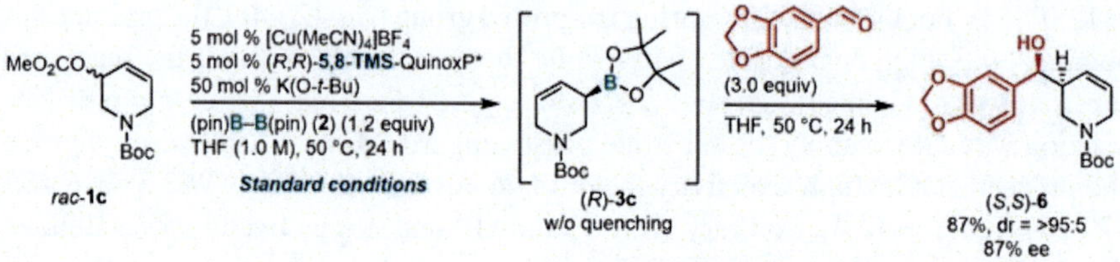

Fig. 2.10 One-pot sequence of the direct enantioconvergent borylation and allylboration

at γ-position of the boryl group. Thus, the reaction was proceeded through *syn*- and *anti*-S_N2' mechanism rather than S_N2 and S_N2' mechanism or summarization via π-allylcopper(III) intermediate, [49, 55–58] which is in good agreement with the mechanistic study using a five-membered substrate reported by the Ito and Sawamura group [43].

The proposed reaction mechanism was shown in Fig. 2.12 [43, 59–61]. At first, the copper(I) alkoxide **I** is generated from the off-cycle copper(I) salt and alkoxide base. The σ-bond metathesis between **I** and the diboron reagent **2** furnishes the boryl copper(I) species **II**. After the subsequent coordination of substrate to **II**, the selectivity-determining borylcupration via the corresponding TS (**TS1**) occurs to generate alkyl copper(I) intermediate **IV**. The following reductive elimination

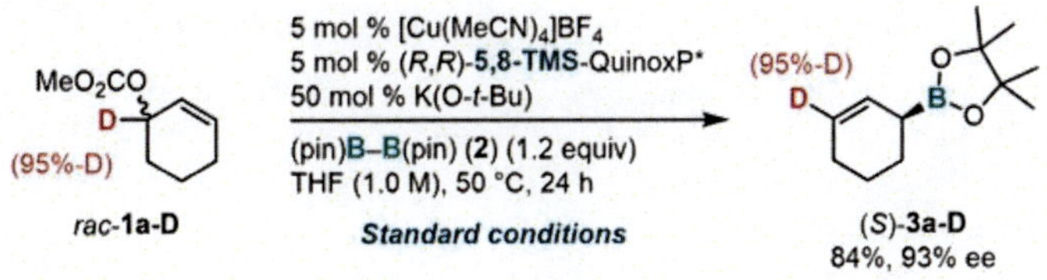

Fig. 2.11 Deuterium labeling study

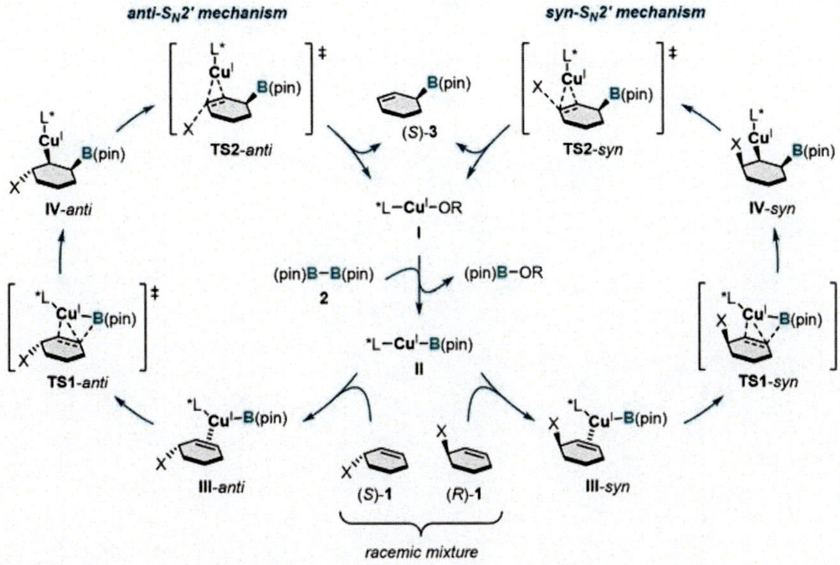

Fig. 2.12 Proposed reaction mechanism of the direct enantioconvergent borylation

produces the borylation product **3a** and copper(I) alkoxide **I** via the corresponding TS (**TS2**).

To investigate the origin of reactivity enhancement by the introduction of silyl groups into the ligand backbone, I focused on the dormant species in boryl copper(I) catalysis. Recently, several research groups were found that the dimeric boryl copper(I) species is thermodynamically more stable than the monomer, although the monomeric form is known to be the active species in copper(I)/diboron system-catalyzed borylation reactions. In 2017, the Kleeberg group reported the dppf–boryl copper(I) complex could be isolated as the dimeric form, which was confirmed by a single-crystal X-ray diffraction analysis [62]. Also, in 2017, Tobisch reported the dimeric dppf–boryl copper(I) complex is more stable than the monomeric form and can be an off-cycle dormant species in the catalytic reaction, which was estimated by DFT calculations [63]. The dppf, which is a 1,2-bisphosphine bearing aryl backbone, is structurally related to the QuinoxP* and 5,8-TMS-QuinoxP*. Therefore, I speculated that the stability of the dimeric boryl copper(I) intermediate as the

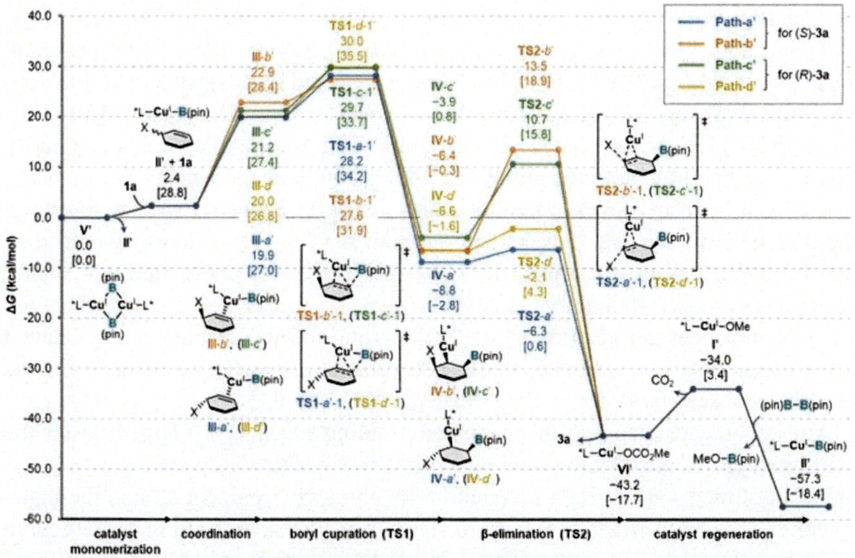

Fig. 2.14 Free energy diagram of the direct enantioconvergent borylation of **1a**

For enantioconvergent transformations of racemic substrate, four paths should be considered as the individual reaction paths: the reactions of (*R*)- and (*S*)-**1a** give the major product (*S*)-**3a** [path-*a*' and path-*b*', respectively] and the reactions of (*R*)- and (*S*)-**1a** give the minor product (*R*)-**3a** [path-*c*' and path-*d*', respectively].

The boryl copper(I) dimer **V′** was used for the energy reference of the relative Gibbs free energies ($\Delta G = 0.0$ kcal/mol). The dissociation of one molecule of the boryl copper(I) species from the dimer forms the slightly unstable monomeric species **II′** ($\Delta G = 2.4$ kcal/mol). Then, substrate coordination forms the transient species **III** ($\Delta G = 19.9$–22.9 kcal/mol), followed by borylcupration via **TS1** ($\Delta G^{\ddagger} = 27.6$–$30.0$ kcal/mol) to generate alkyl copper(I) intermediates **IV** ($\Delta G = -8.8$ to -3.9 kcal/mol). This activation barrier of approximately 30 kcal/mol is assumed to be gone over at the reaction temperature of 50 °C. For the next β-elimination step, *anti*-eliminations on path-*a*′ and -*d*′ ($\Delta G^{\ddagger} = -6.3$ to -2.1 kcal/mol) are more favored than the *syn*-eliminations on path-*b*′ and -*c*′ ($\Delta G^{\ddagger} = 10.7$–$13.5$ kcal/mol). Regardless of those differences, the activation energies are smaller than that of **TS1**, and the next restoration of copper(I) carbonate **VI′** is highly exergonic ($\Delta G = -43.2$ kcal/mol). Through the following dissociation of carbon dioxide to form copper(I) alkoxide **I′** ($\Delta G = -34.0$ kcal/mol). Finally, the boryl copper(I) species **II′** is regenerated via σ-bond metathesis with the diboron reagent ($\Delta G = -57.3$ kcal/mol). According to those energy profiles, the enantioselectivity-determining step should be **TS1** because the subsequent steps are irreversible.

Focused on the TSs of enantioselectivity-determining step (**TS1**), I conducted the comprehensive conformational search of the TS and picked up the most stable conformers for the respective paths *a*′ to *d*′ (Fig. 2.15) (for details, see Sect. 2.5). The *anti*- and *syn*-borylcupration of (*S*)-**1a** produce (*S*)- and (*R*)-**3a** via **TS1**-*S*-*anti* (**TS1**-*a*′-1) and **TS1**-*S*-*syn* (**TS1**-*c*′-1), respectively [**TS1**-*S*-*anti*: $\Delta\Delta G = 0.6$ kcal/mol, **TS1**-*S*-*syn*: $\Delta\Delta G = 2.1$ kcal/mol]. Also, the *syn*- and *anti*-borylcupration of (*R*)-**1a** produce (*S*)- and (*R*)-**3a** via **TS1**-*R*-*syn* (**TS1**-*b*′-1) and **TS1**-*R*-*anti* (**TS1**-*d*′-1), respectively [**TS1**-*S*-*syn*: $\Delta\Delta G = 0.0$ kcal/mol, **TS1**-*S*-*anti*: $\Delta\Delta G = 2.4$ kcal/mol]. The barrier height is not affected by the configuration of the leaving group, while that is controlled by directions from which the boryl copper(I) species inserts to the double bond; e.g., the borylcupration from the top side is more favored than that from the bottom side shown in Fig. 2.15. Thus, it is considered that such a strong enantioface recognition ability of the boryl copper(I) species realizes the enantioconvergent introduction of the boryl group to the substrate. The predicted enantioselectivity considering all the TS conformers was in good agreement with the experimental selectivity (ee_{pre}: 95% ee; ee_{exp}: 93% ee) [64].

For better understanding of enantiodetermining mechanism, I focused on the conformations of the bisphosphine ligand in **TS1** because the orientations of phosphine substituents are directly affected to the chiral environment around the metal center created by the ligand (Fig. 2.16). The comprehensive conformational search of **TS1** produced 38 different conformers (for details, see Sect. 2.5). For the categorization of those structures, I defined a new descriptor based on the dihedral angles in core five-membered ring found in 1,2-bisphosphine/metal complex (Fig. 2.16A, B: $\Psi_1 = C^1–C^2–P^2–[M]$; Ψ_2: $C^2–C^1–P^1–[M]$) [65, 66]. Using a combination of Ψ_1 and Ψ_2, the conformation of the ligand can be described and categorized into four conformer groups (Fig. 2.16C).

Then, distribution of the conformer structures is analyzed using scatter plot on the dihedral angles Ψ_1 and Ψ_2, which was colored based on the activation barrier energy (Fig. 2.17). The conformers were discretely distributed in this plot and could

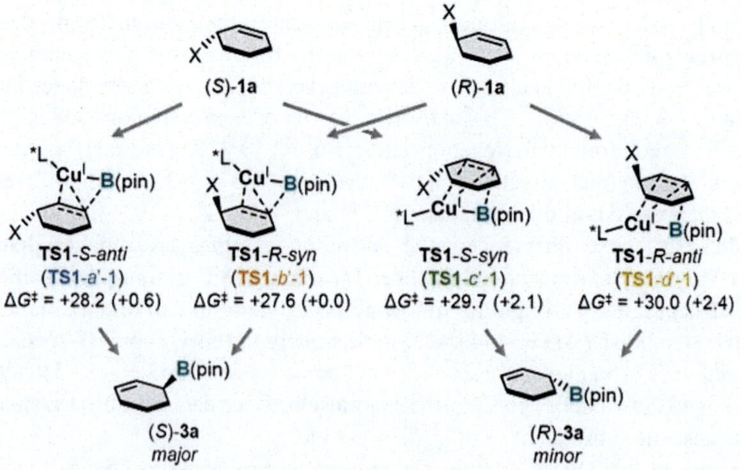

Fig. 2.15 Free energies of most stable conformers of **TS1**

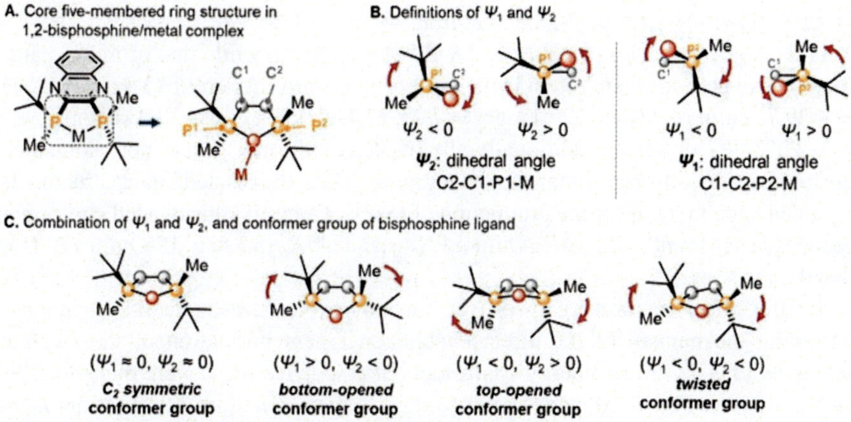

Fig. 2.16 Descriptors for conformation of 1,2-bisphosphine ligands

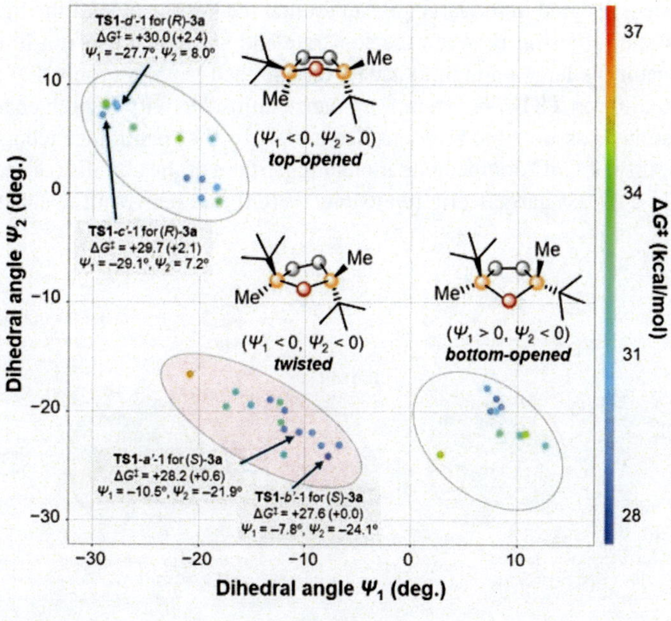

Fig. 2.17 Scatter plot of conformations of **TS1** based on dihedral angles Ψ_1 and Ψ_2

be categorized into three groups: (1) bottom-opened conformer group with positive Ψ_1 and negative Ψ_2 ($\Psi_1 \geq 0$, $\Psi_2 < 0$) values; (2) top-opened conformer group with negative Ψ_1 and positive Ψ_2 ($\Psi_1 < 0$, $\Psi_2 \geq 0$) values; (3) twisted conformer group, in which both dihedral angles are negative ($\Psi_1 < 0$, $\Psi_2 < 0$). The lowest-energy TS conformers of major paths (path-a' and path-b') were found in the twisted conformer group (**TS1**-*a*'-1: $\Delta G^{\ddagger}$ = +28.2 kcal/mol, $\Psi_1 = -10.5°$, $\Psi_2 = -21.9°$; **TS1**-*b*'-1: $\Delta G^{\ddagger}$ = +27.6 kcal/mol, $\Psi_1 = -7.8°$, $\Psi_2 = -24.1°$), while that of minor paths (path-*c*′ and path-*d*′) were found in the top-opened conformer group (**TS1**-*c*'-1: $\Delta G^{\ddagger}$ = +29.7 kcal/mol, $\Psi_1 = -29.1°$, $\Psi_2 = 7.2°$; **TS1**-*d*′-1: $\Delta G^{\ddagger}$ = +30.0 kcal/mol, $\Psi_1 = -27.7°$, $\Psi_2 = 8.0°$). Additionally, the twisted conformer group more contained thc low-barrier conformers than the other groups. Thus, the twisted ligand structure is assumed to be favored for the production of (*S*)-**3a**. Contrary, this twisted conformer group was not found with (*R*,*R*)-QuinoxP* instead of (*R*,*R*)-5,8-TMS-QuinoxP* (for details, see Sect. 2.5).

Finally, the structural analysis was carried out to investigate the enantiodetermining mechanism of the direct enantioconvergent borylation of the racemic substrate **1a**. The lowest-energy conformers of respective paths were picked up (as well as those in Fig. 2.15), and the steric maps of ligands in those TSs were prepared (Fig. 2.18). For the twisted conformers of **TS1**-*a*′-1 and **TS1**-*b*′-1 (Fig. 2.18A, B), the B(pin) group is located at top side without significant steric repulsion because there is no sterically hindered substituent around there shown in their steric maps. Also, the substrate is located in the large pocket around the bottom side of the ligand, and thus, the steric repulsion between the substrate and the ligand was not found. This pocket structure is large enough to accept the inclined leaving group **X** even in the *cis*-configuration in **TS1**-*b*′-1. Therefore, the enantioselectivity of both enantiomers of the substrate is assumed to be controlled merely by the enantioface recognition of the prochirality of C=C double bond moiety regardless of the chirality of the leaving group **X**. In contrast, large steric repulsions were found between the substrate and

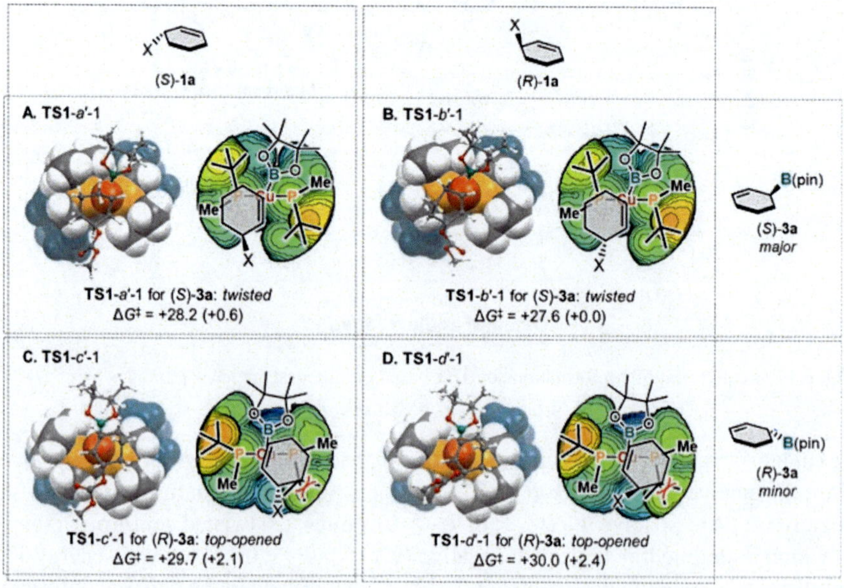

Fig. 2.18 Structural analysis of the lowest-energy conformers of **TS1**

one of *tert*-butyl group of the ligand in the top-opened conformers of **TS1**-*c*′-1 and **TS1**-*d*′-1, although the B(pin) group is inclined into the pocket around the top side (Fig. 2.18C, D).

2.3 Conclusions

A novel series of chiral 1,2-bisphosphine ligands, (*R*,*R*)-5,8-*Si*-QuinoxP*, was developed based on the non-covalent interaction strategy demonstrated by the introduction of silyl groups on the ligand backbone. The ligand could be synthesized in two steps from commercially available 2,3-dichloroquinoxaline. The ligand showed high reactivity and enantioselectivity in the direct enantioconvergent borylation of racemic and cyclic allylic electrophiles including five- to seven-membered ring and heterocycles (up to 92% yield and 96% ee). Furthermore, the ligand enabled the first borylative kinetic resolution of racemic linear allylic electrophiles with high selectivity factor of *s* (up to 90% ee, $s = 46.4$). A computational study revealed the introduction of the silyl group contributed to suppress the dimer formation of boryl copper(I) species and thus promote the generation of active monomeric boryl copper(I) species. Also, the mechanism of the direct enantioconvergent borylation using (*R*,*R*)-5,8-TMS-QuinoxP* was also studied in detail. Conformers of the enantioselectivity-determining TS were found to be categorized into three groups: top-opened, bottom-opened, and twisted conformer groups. The lowest-energy TS conformers producing the major enantiomer (*S*)-**3** were categorized in the twisted conformer group, and the absence of significant steric repulsions in the TS makes these structures more favored than the minor TSs categorized in the top-opened conformer group and having the steric congestion between the substrate and ligand.

2.4 Experimental Details

2.4.1 Instrumentation and Chemicals

Materials were obtained from commercial suppliers and purified by standard procedures unless otherwise noted. Solvents were also purchased from commercial suppliers, degassed via three freeze–pump–thaw cycles, and further dried over molecular sieves (MS 4A). NMR spectra were recorded on JEOL JNM-ECX400P and JNM-ECS400 spectrometers (^{1}H: 400 MHz, ^{13}C: 100 MHz). Tetramethylsilane (^{1}H, 0.00 ppm) and $CDCl_3$ (^{13}C, 77.0 ppm) were employed as the external standards, respectively. $[Cu(MeCN)_4]BF_4$ (T2666-1G, ≥98%) was purchased from Tokyo Chemical Industry Co. and stored in an argon-filled glove box. The clear crystal of this copper(I) salt should be used until it turns a dull green color. K(O-*t*-Bu)/THF (1.0 M, 328,650-50ML) was purchased from Sigma-Aldrich Co. and used as received. GLC analyses were conducted with a Shimadzu GC-2014 or GC-2025 equipped with a ULBON HR-1 glass capillary column (Shinwa Chemical Industries) and a FID detector. HPLC analyses with chiral stationary phase were carried out using a Hitachi LaChrom Elite HPLC system with a L-2400 UV detector. Specific optical rotations were measured with HORIBA SEPA-300 and a Rudolph Research Analytical Autopol IV Polarimeter. High-resolution mass spectra were recorded at the Global Facility Center, Hokkaido University or GC–MS & NMR Lab., Research Faculty of Agriculture, Hokkaido University. Single-crystal X-ray structural analyses were carried out on a Rigaku XtaLAB PRO MM007 diffractometer using graphite monochromated Mo-K_α or Cu-K_α radiation. The structure was solved by direct methods and expanded using Frontier techniques. Non-hydrogen atoms were refined anisotropically. Hydrogen atoms were refined using the riding model. All calculations were performed using Olex crystallographic software package except for refinement, which was performed using SHELXL [67].

2.4.2 *General Experimental Procedure*

Procedure A: Procedure for the Copper(I)-Catalyzed Direct Enantioconvergent Borylation of 1a (Fig. 2.19)

$[Cu(MeCN)_4]BF_4$ (7.9 mg, 0.025 mmol) and bis(pinacolato)diboron (**2**) (253.9 mg, 1.00 mmol), (*R*,*R*)-5,8-TMS-QuinoxP* (12.0 mg, 0.025 mmol) were placed in an oven-dried reaction vial in an argon-filled glove box. After the vial was sealed with a screw cap containing a Teflon®-coated rubber septum, and the vial was taken from the glove box. Dry THF (0.25 mL) and K(O-*t*-Bu)/THF (1.0 M, 0.25 mL, 0.25 mmol) were added in the vial through the rubber septum using a syringe. After stirring for 15 min, *rac*-**1a** (77.5 mg, 0.50 mmol) was added to the mixture at 50 °C. After the reaction was completed, the reaction mixture was passed through a short silica gel column (*Φ*: 10 mm, the height of the silica gel column: ca. 30 mm) eluting with Et_2O/hexane 5:95. The crude material was purified by flash column chromatography (SiO_2, Et_2O/hexane, typically 0:100–3:97) to give the corresponding borylation product (*S*)-**3a** as a colorless oil.

In the reaction vial, allyl boronate (*S*)-**3a** (41.5 mg, 0.20 mmol) and benzaldehyde (40.8 μL, 0.40 mmol) were dissolved in dry toluene (0.4 mL). After stirred for 16 h at 60 °C, the reaction mixture was quenched by 10% triethanolamine solution in DCM (1.0 mL) and H_2O (1.0 mL) and extracted with Et_2O three times (2.0 mL ×

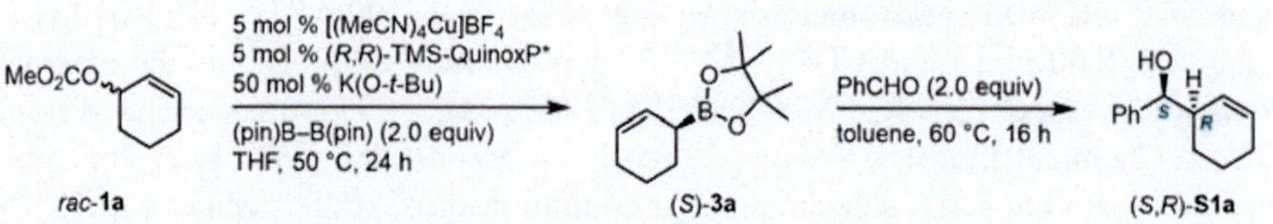

Fig. 2.19 Scheme of procedure A

3). The combined organic layer was then dried over $MgSO_4$. After filtration, the solvents were removed by evaporation. The crude product was purified by flash column chromatography (SiO_2, EtOAc/hexane, typically 0:100–6:94) to obtain the corresponding alcohol (*S*,*R*)-**S1a** as a colorless oil.

Procedure B: Procedure for the Copper(I)-Catalyzed Direct Enantioconvergent Borylation of 1g (Fig. 2.20)

[Cu(MeCN)$_4$]BF_4 (7.9 mg, 0.025 mmol) and bis(pinacolato)diboron (**2**) (253.9 mg, 1.00 mmol), (*R*,*R*)-5,8-TMS-QuinoxP* (12.0 mg, 0.025 mmol) were placed in an oven-dried reaction vial in an argon-filled glove box. After the vial was sealed with a screw cap containing a Teflon®-coated rubber septum, the vial was taken from the glove box. Dry THF (475 μL) and K(O-*t*-Bu)/THF (1.0 M, 25 μL, 0.025 mmol) were added in the vial through the rubber septum using a syringe. After stirring for 15 min, *rac*-**1g** (106.5 mg, 0.50 mmol) was added to the mixture at 0 °C. After the reaction was completed, the reaction mixture was passed through a short silica gel column (*Φ*: 10 mm, the height of the silica gel column: ca. 30 mm) eluting with Et_2O/hexane 50:50. The crude material was purified by flash column chromatography (SiO_2, Et_2O/hexane, typically 0:100–10:90) to give the corresponding borylation product (*S*)-**3g** with a small amount of bis(pinacolato)diboron. In the reaction vial, the allyl boronate (*S*)-**3g** and benzaldehyde (102.0 μL, 1.00 mmol) were dissolved in dry toluene (1.0 mL). After stirred for 17 h at 60 °C, the reaction mixture was quenched by 10% triethanolamine solution in DCM (1.0 mL) and H_2O (1.0 mL) and extracted with Et_2O three times (2.0 mL × 3). The combined organic layer was then dried over $MgSO_4$. After filtration, the solvents were removed by evaporation. The crude product was purified by flash column chromatography (SiO_2, EtOAc/hexane, typically 5:95–20:80) to obtain the corresponding alcohol (*R*,*R*)-**S1g** as a colorless oil with a small amount of pinacol.

Procedure C: Procedure for the Copper(I)-Catalyzed Direct Enantioconvergent Borylation of 1i (Fig. 2.21)

[Cu(MeCN)$_4$]BF_4 (7.9 mg, 0.025 mmol) and bis(pinacolato)diboron (**2**) (253.9 mg, 1.00 mmol), (*R*,*R*)-5,8-TMS-QuinoxP* (12.0 mg, 0.025 mmol) were placed in an oven-dried reaction vial in an argon-filled glove box. After the vial was sealed with a screw cap containing a Teflon®-coated rubber septum, the vial was taken from the glove box. Dry THF (0.25 mL) and K(O-*t*-Bu)/THF (1.0 M, 0.25 mL, 0.25 mmol) were added in the vial through the rubber septum using a syringe. After stirring for 15 min, *rac*-**1i** (87.2 mg, 0.50 mmol) was added to the mixture at 0 °C. After

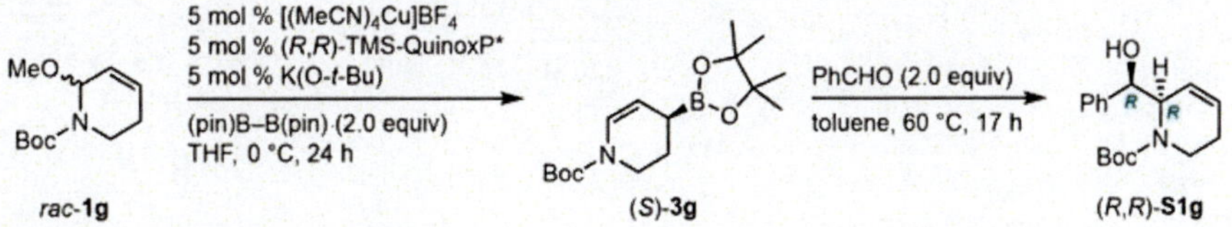

Fig. 2.20 Scheme of procedure B

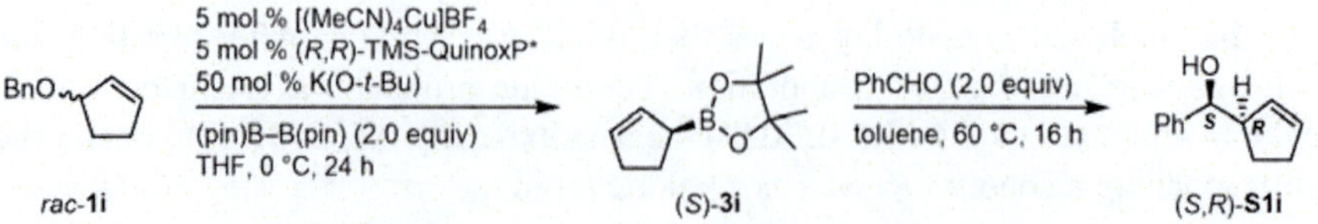

Fig. 2.21 Scheme of procedure C

the reaction was completed, the reaction mixture was passed through a short silica gel column (*Φ*: 10 mm, the height of the silica gel column: ca. 30 mm) eluting with Et_2O/hexane 5:95. The crude material was purified by flash column chromatography (SiO_2, Et_2O/hexane, typically 0:100–3:97) to give the corresponding borylation product (*S*)-**3i** as a colorless oil.

In the reaction vial, allyl boronate (*S*)-**3i** (21.9 mg, 0.11 mmol) and benzaldehyde (23.1 μL, 0.23 mmol) were dissolved in dry toluene (0.22 mL). After stirred for 16 h at 60 °C, the reaction mixture was quenched by 10% triethanolamine solution in DCM (1.0 mL) and H_2O (1.0 mL) and extracted with Et_2O three times (2.0 mL × 3). The combined organic layer was then dried over $MgSO_4$. After filtration, the solvents were removed by evaporation. The crude product was purified by flash column chromatography (SiO_2, EtOAc/hexane, typically 0:100–6:94) to obtain the corresponding alcohol (*S*,*R*)-**S1i** as a colorless oil.

Procedure D: Procedure for the Copper(I)-Catalyzed Direct Enantioconvergent Borylation of 1j (Fig. 2.22)

$[Cu(MeCN)_4]BF_4$ (7.9 mg, 0.025 mmol) and bis(pinacolato)diboron (**2**) (253.9 mg, 1.00 mmol), (*R*,*R*)-5,8-TMS-QuinoxP* (12.0 mg, 0.025 mmol) were placed in an oven-dried reaction vial in an argon-filled glove box. After the vial was sealed with a screw cap containing a Teflon®-coated rubber septum, the vial was taken from the glove box. Dry THF (0.25 mL) and K(O-*t*-Bu)/THF (1.0 M, 0.25 mL, 0.25 mmol) were added in the vial through the rubber septum using a syringe. After stirring

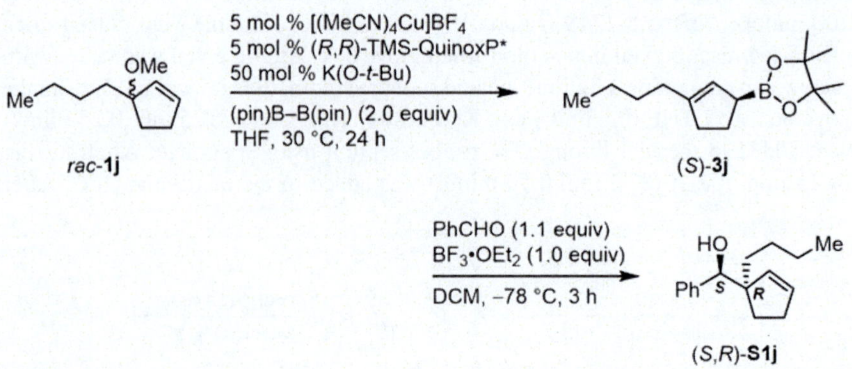

Fig. 2.22 Scheme of Procedure D

for 15 min, *rac*-**1j** (76.8 mg, 0.50 mmol) was added to the mixture at 30 °C. After the reaction was completed, the reaction mixture was passed through a short silica gel column (*Φ*: 10 mm, the height of the silica gel column: ca. 30 mm) eluting with Et_2O/hexane 5:95. The crude material was purified by flash column chromatography (SiO_2, Et_2O/hexane, typically 0:100–3:97) to give the corresponding borylation product (*S*)-**3j** as a colorless oil.

In the reaction vial, allyl boronate (*S*)-**3j** (50.9 mg, 0.20 mmol) was dissolved in dry DCM (0.4 mL). After the mixture was cooled to −78 °C, benzaldehyde (23.0 μL, 0.22 mmol) and $BF_3 \cdot OEt_2$ (26.0 μL, 0.20 mmol) were added to the mixture. After stirred for 3 h at −78 °C, the reaction mixture was quenched by 10% triethanolamine solution in DCM (1.0 mL) and H_2O (1.0 mL) and extracted with Et_2O three times (2.0 mL × 3). The combined organic layer was then dried over $MgSO_4$. After filtration, the solvents were removed by evaporation. The crude product was purified by flash column chromatography (SiO_2, EtOAc/hexane, typically 0:100–6:94) to obtain the corresponding alcohol (*S*,*R*)-**S1j** as a colorless oil.

Procedure E: Procedure for the Copper(I)-Catalyzed Enantioselective Borylation of (Z)-4a through Kinetic Resolution (Fig. 2.23)

$[Cu(MeCN)_4]BF_4$ (7.9 mg, 0.025 mmol) and bis(pinacolato)diboron (**2**) (253.9 mg, 1.00 mmol), (*R*,*R*)-5,8-TMS-QuinoxP* (12.0 mg, 0.025 mmol) were placed in an oven-dried reaction vial in an argon-filled glove box. After the vial was sealed with a screw cap containing a Teflon®-coated rubber septum, the vial was taken from the glove box. Dry THF (0.25 mL) and K(O-*t*-Bu)/THF (1.0 M, 0.25 mL, 0.25 mmol) were added in the vial through the rubber septum using a syringe. After stirring for 15 min, *rac*-(Z)-**4a** (138.1 mg, 0.50 mmol) was added to the mixture at −40 °C. After the reaction was completed, the reaction mixture was passed through a short silica gel column (*Φ*: 10 mm, the height of the silica gel column: ca. 30 mm) eluting with Et_2O/hexane 5:95. The crude material was purified by flash column chromatography (SiO_2, Et_2O/hexane, typically 0:100–3:97) to give the mixture of the corresponding borylation product (*S*,*E*)-**5a** and the substrate (*S*,*Z*)-**4a**. In the reaction vial, the mixture of the products was dissolved in THF (0.75 mL) and H_2O (0.5 mL). To the mixture, $NaBO_3 \cdot 4H_2O$ (153.9 mg, 1.0 mmol) was added and the mixture

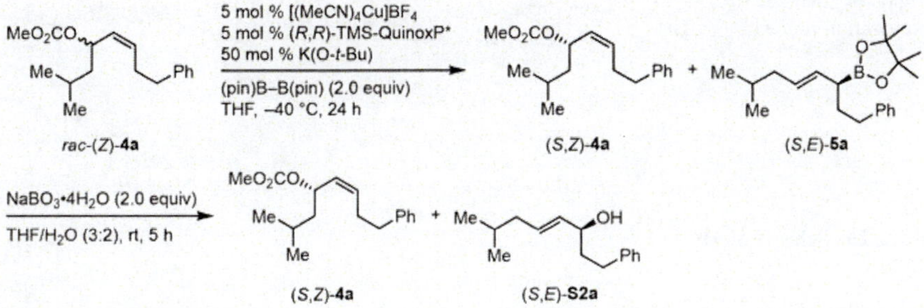

Fig. 2.23 Scheme of Procedure E

was stirred for 5 h at room temperature. Then, the reaction mixture was extracted with Et_2O three times (2.0 mL × 3). The combined organic layer was then dried over $MgSO_4$. After filtration, the solvents were removed by evaporation. The crude product was purified by flash column chromatography (SiO_2, EtOAc/hexane, typically 0:100–10:90) to obtain the corresponding oxidation product (*S*,*E*)-**S2a** as a colorless oil and the recovered substrate (*S*,*Z*)-**4a**.

2.4.3 Substrate Preparation Procedure

Preparation of cyclohex-2-en-1-yl methyl carbonate (1a, Fig. 2.24) [68]

In a vacuum-dried 30 mL two-neck round-bottomed flask, cyclohex-2-en-1-ol (0.99 mL, 10 mmol) and pyridine (3.63 mL, 45 mmol) were dissolved in dry DCM (10 mL), and the mixture was cooled to 0 °C under a nitrogen atmosphere. Methyl chloroformate (1.69 mL, 22 mmol) was added to the mixture dropwise, and the mixture was warmed to room temperature. After the mixture was stirred for 7 h, the reaction mixture was quenched by H_2O (10 mL) and extracted with hexane three times (10 mL × 3). The combined organic layer was then dried over $MgSO_4$. After filtration, the solvents were removed by evaporation. The crude product was purified by flash column chromatography to obtain the corresponding allyl carbonate (1.33 g, 8.50 mmol, 85%) as a colorless oil.

^{1}H NMR (400 MHz, $CDCl_3$, δ): 1.58–1.68 (m, 1H), 1.70–2.04 (m, 4H), 2.05–2.15 (m, 1H), 3.78 (s, 3H), 5.09–5.15 (m, 1H), 5.78 (ddt, J = 10.0, 3.9, 2.0 Hz, 1H), 5.99 (dtd, J = 10.1, 3.6, 0.8 Hz, 1H). ^{13}C NMR (100 MHz, $CDCl_3$, δ): 18.4 (CH_2), 24.7 (CH_2), 28.1 (CH_2), 54.4 (CH_3), 71.7 (CH), 124.8 (CH), 133.2 (CH), 155.4 (C). HRMS-EI (*m*/*z*): $[M]^+$ calcd for $C_8H_{12}O_3$, 156.0786; found, 156.0788.

Methyl (1-tosyl-1,2,3,6-tetrahydropyridin-3-yl) carbonate (1b, Fig. 2.25) [69]

1b was prepared from the corresponding alcohol according to the procedure described above. The product **1b** was obtained in 87% yield (800.0 mg, 2.6 mmol, white solid).

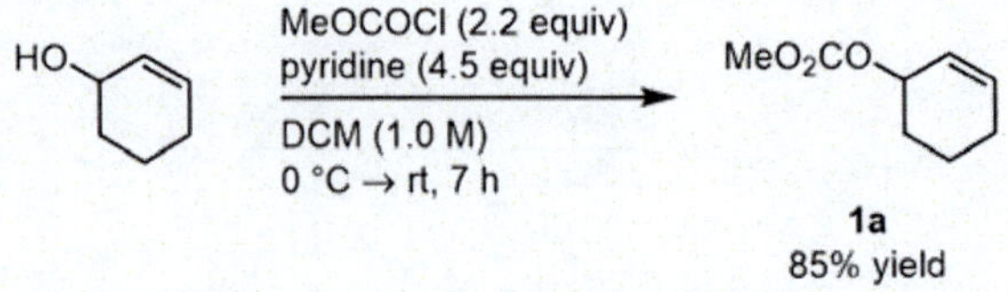

Fig. 2.24 Scheme of preparation procedure of **1a**

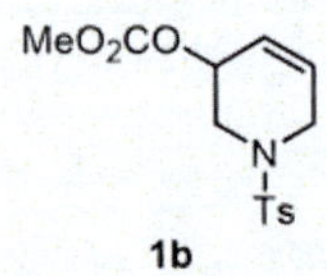

Fig. 2.25 Scheme of **1b**

^{1}H NMR (400 MHz, $CDCl_3$, δ): 2.44 (s, 3H), 3.31 (dd, J = 12.3, 5.2 Hz, 1H), 3.42 (dd, J = 12.3, 4.3 Hz, 1H), 3.58 (dq, J = 17.2, 2.3 Hz, 1H), 3.70 (ddt, J = 17.1, 3.5, 1.8 Hz, 1H), 3.80 (s, 3H), 5.11–5.18 (m, 1H), 5.84 (ddt, J = 10.3, 4.2, 1.8 Hz, 1H), 5.94 (dtd, J = 10.1, 3.2, 1.1 Hz, 1H), 7.33 (d, J = 8.1 Hz, 2H), 7.69 (d, J = 8.1 Hz, 2H). ^{13}C NMR (100 MHz, $CDCl_3$, δ): 21.3 (CH_3), 44.2 (CH_2), 46.1 (CH_2), 54.7 (CH_3), 68.6 (*C*H), 123.2 (*C*H), 127.4 (*C*H), 128. 2 (*C*H), 129.6 (*C*H), 132.9 (*C*), 143.7 (*C*), 154.8 (*C*). HRMS-ESI (*m*/*z*): $[M + Na]^+$ calcd for $C_{14}H_{17}NO_5SNa$, 334.0720; found, 334.0717.

tert-Butyl 3-[(methoxycarbonyl)oxy]-3,6-dihydropyridine-1(2*H*)-carboxylate (1c, Fig. 2.26) [69]

1c was prepared from the corresponding alcohol according to the procedure described above. The product **1c** was obtained in 84% yield (248.6 mg, 0.97 mmol, colorless oil).

^{1}H NMR (400 MHz, $CDCl_3$, δ): 1.46 (s, 9H), 3.53–3.67 (m, 1H), 3.71–3.88 (m, 2H), 3.79 (s, 3H), 3.95–4.19 (m, 1H), 4.98–5.16 (m, 1H), 5.90 (ddt, J = 10.2, 4.1, 2.1 Hz, 1H), 5.93–6.03 (m, 1H). ^{13}C NMR (100 MHz, $CDCl_3$, δ): 28.3 (CH_3), 43.0 (br, CH_2), 44.5 (br, CH_2), 54.6 (CH_3), 69.2 (*C*H), 80.0 (*C*), 123.4 (*C*H), 130.7 (br, *C*H), 154.6 (*C*), 155.2 (*C*). HRMS-ESI (*m*/*z*): $[M + Na]^+$ calcd for $C_{12}H_{19}NO_5Na$, 280.1155; found, 280.1153.

Benzyl 3-[(methoxycarbonyl)oxy]-3,6-dihydropyridine-1(2*H*)-carboxylate (1d, Fig. 2.27) [69]

1d was prepared from the corresponding alcohol according to the procedure described above. The product **1d** was obtained in 84% yield (983.3 mg, 3.4 mmol, colorless oil).

^{1}H NMR (400 MHz, $CDCl_3$, δ): 3.66 (dd, J = 14.0, 3.6 Hz, 1H), 3.74 (s, 3H), 3.82–3.99 (m, 2H), 4.02–4.29 (m, 1H), 4.99–5.25 (m, 1H), 5.16 (q, J = 11.5 Hz, 2H), 5.83–6.15 (m, 2H), 7.27–7.40 (m, 5H). ^{13}C NMR (100 MHz, $CDCl_3$, δ): 43.0 (CH_2), 43.9 and 44.4 (a pair of CH_2), 54.6 (CH_3), 67.0 and 67.2 (a pair of CH_2), 68.6 (*C*H), 122.8 and 123.2 (a pair of *C*H), 127.5 and 127.8 (a pair of *C*H), 128.3 (*C*H), 129.5 (*C*H), 130.5 (*C*H), 136.3 (*C*), 136.4 (*C*), 155.0 (*C*). HRMS-ESI (*m*/*z*): $[M + Na]^+$ calcd for $C_{15}H_{17}NO_5Na$, 314.0999; found, 314.0996.

Fig. 2.26 Scheme of **1c**

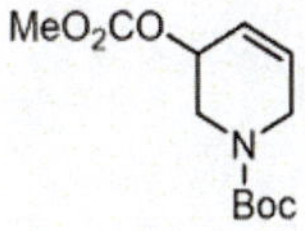

1c

Fig. 2.27 Scheme of **1d**

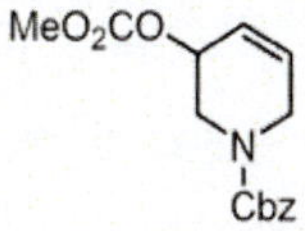

1d

78.2 (*C*H), 131.8 (*C*H), 132.9 (*C*H), 155.3 (*C*). HRMS-ESI (*m*/*z*): $[M + Na]^+$ calcd for $C_9H_{14}O_3Na$, 193.0835; found, 193.0836.

Preparation of *tert*-butyl 6-methoxy-3,6-dihydropyridine-1(2*H*)-carboxylate (1g, Fig. 2.31)

In a vacuum-dried 300 mL two-neck round-bottomed flask, *tert*-butyl 2-oxopiperidine-1-carboxylate (3.98 g, 20.0 mmol) was dissolved in dry THF (80 mL), and the mixture was cooled to −78 °C under a nitrogen atmosphere. Diisobutylaluminium hydride solution in hexane (1.0 M, 30.0 mL, 30.0 mmol) was added to the mixture dropwise, and the mixture was stirred for 2 h. The reaction mixture was quenched by saturated potassium sodium tartrate aqueous solution (40 mL), and the mixture was stirred for 3 h to turn into clear solution. Then, the mixture was extracted with ethyl acetate three times (40 mL × 3). The combined organic layer was dried over $MgSO_4$. After filtration, the solvents were removed by evaporation. The crude product, a mixture of **S3** and **S4**, was used to the next reaction without further purification.

In a vacuum-dried 100 mL two-neck round-bottomed flask equipped with reflux condenser and Dean–Stark apparatus, the crude mixture of **S3** and **S4** was dissolved in dry toluene (30 mL) under a nitrogen atmosphere. *p*-Toluenesulfonic acid monohydrate (10 mg, 0.05 mmol) was added to the mixture, and the mixture was warmed up to 135 °C and stirred for 1 h. The mixture was passed through a short silica gel (Φ: 15 mm, the height of the silica gel column: ca. 100 mm) eluting with ethyl acetate. After the solvents were removed by evaporation, the crude product **S4** was used to the next reaction without further purification.

In a vacuum-dried 200 mL two-neck round-bottomed flask, the crude **S4** was dissolved in dry DCM (20 mL), and the mixture was cooled to 0 °C under a nitrogen atmosphere. *N*-Iodosuccinimide (4.95 g, 22.0 mmol) solution in MeOH (44 mL) was added to the mixture dropwise until the color of the reaction mixture turn to yellow. After stirring for 16 h, 10% $Na_2S_2O_4$ aqueous solution was added to the mixture until the reaction mixture turn to colorless. Then, the mixture was extracted with dichloromethane three times (20 mL × 3). The combined organic layer was dried

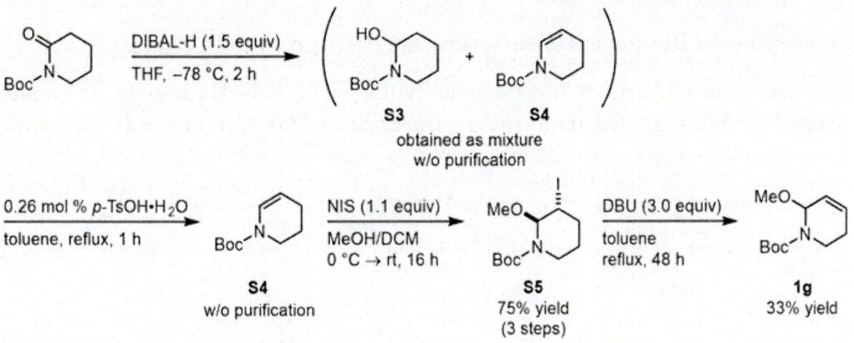

Fig. 2.31 Scheme of preparation procedure of **1g**

over $MgSO_4$. After filtration, the solvents were removed by evaporation. The crude product was purified by flash column chromatography to obtain the corresponding piperidine derivative **S5** (5.15 g, 15.1 mmol, 75% in three steps) as a colorless oil.

In a vacuum-dried 30 mL two-neck round-bottomed flask equipped with reflux condenser, **S5** (1.56 g, 5.0 mmol) and DBU (2.24 mL, 15.0 mmol) were dissolved in dry toluene (10 mL), and the mixture was warmed up to 135 °C under a nitrogen atmosphere. After stirring for 48 h, the solvents were removed by evaporation. The crude product was purified by flash column chromatography to obtain the corresponding allyl ether **1g** (353.3 mg, 1.66 mmol, 33%) as a colorless oil.

^{1}H NMR (400 MHz, $CDCl_3$, δ): 1.49 (s, 9H), 1.87–2.04 (br-m, 1H), 2.09–2.30 (br-m, 1H), 2.94–3.20 (br-m, 1H), 3.36 (s, 3H), 3.85–4.22 (br-m, 1H), 5.28–5.55 (br-m, 1H), 5.69–5.85 (br-m, 1H), 5.91–6.10 (br-m, 1H). ^{13}C NMR (100 MHz, $CDCl_3$, δ): 24.6 and 24.9 (a pair of CH_2), 28.3 (CH_3), 35.1 and 36.7 (a pair of CH_2), 55.3 and 55.4 (a pair of CH_3), 79.0 and 79.7 (a pair of *C*H), 79.9 and 80.3 (a pair of *C*), 124.9 and 125.4 (a pair of *C*H), 129.5 and 129.9 (a pair of *C*H), 154.3 and 154.9 (a pair of *C*). HRMS-ESI (*m*/*z*): $[M + Na]^+$ calcd for $C_{11}H_{19}O_3NNa$, 236.1257; found, 236.1262.

Preparation of 6-(benzyloxy)-3,6-dihydro-2*H*-pyran (1h, Fig. 2.32)

The dihydropyran **1h** was prepared according to the literature procedure [71]. In a vacuum-dried 300 mL two-neck round-bottomed flask, Grubbs catalyst® first generation (164.6 mg, 0.20 mmol) and a 1,7-diene **S6** (873.2 mg, 4.0 mmol) were dissolved in dry and degassed dichloromethane (133 mL) under an argon atmosphere. After stirring for 18 h, the solvents were removed by evaporation. The crude product was purified by flash column chromatography to obtain the corresponding dihydropyran derivative **1h** (563.9 mg, 2.97 mmol, 74%) as a colorless oil.

^{1}H NMR (400 MHz, $CDCl_3$, δ): 1.86–1.96 (m, 1H), 2.26–2.39 (m, 1H), 3.76 (dd, $J = 11.2, 6.3$ Hz, 1H), 3.99 (td, $J = 11.4, 3.6$ Hz, 1H), 4.57 (d, $J = 12.1$ Hz, 1H), 4.81 (d, $J = 11.7$ Hz, 1H), 5.01 (s, 1H), 5.75 (dtd, $J = 10.2, 2.8, 1.3$ Hz, 1H), 6.01–6.08 (m, 1H), 7.24–7.30 (m, 1H), 7.31–7.41 (m, 4H). ^{13}C NMR (100 MHz, $CDCl_3$, δ): 24.6 (CH_2), 57.3 (CH_2), 69.2 (CH_2), 92.9 (*C*H), 125.7 (*C*H), 127.5 (*C*H), 127.9 (*C*H), 128.3 (*C*H), 129.1 (*C*H), 138.1 (*C*). HRMS-ESI (*m*/*z*): $[M + Na]^+$ calcd for $C_{12}H_{14}O_2Na$, 213.0886; found, 213.0889.

Preparation of ((cyclopent-2-en-1-yloxy)methyl)benzene (1i, Fig. 2.33) [43]

In a vacuum-dried 50 mL two-neck round-bottomed flask, NaH (60% paraffin liquid dispersion, 648 mg, 16.2 mmol) was suspended in THF (20 mL) at 0 °C. To the

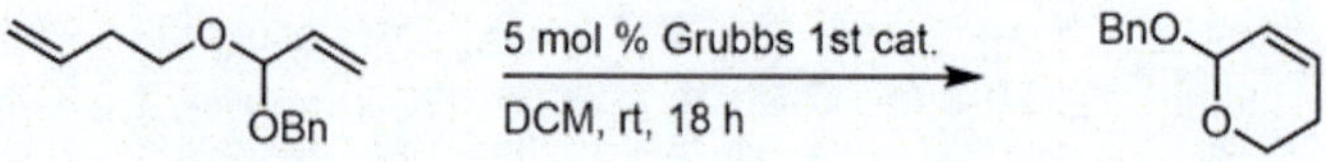

S6

1h
74% yield

Fig. 2.32 Scheme of preparation procedure of **1h**

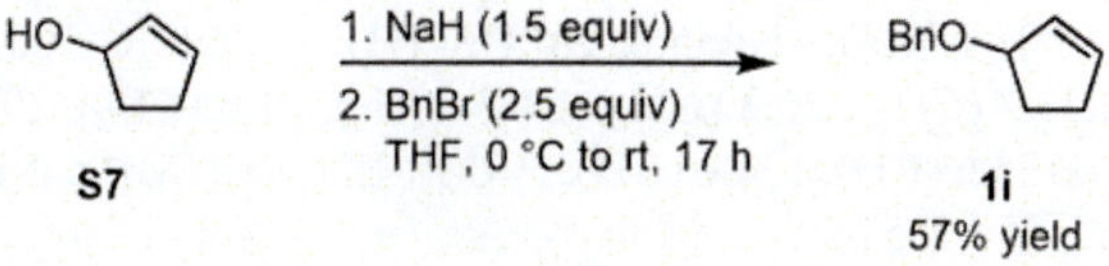

Fig. 2.33 Scheme of preparation procedure of **1i**

suspension, allyl alcohol **S7** (905 mg, 10.8 mmol) was slowly added and the mixture was warmed up to room temperature. After stirring for 30 min, the mixture was cooled to 0 °C again. Then, benzyl bromide (3.21 mL, 27.0 mmol) was added. After stirring for 17 h at room temperature, the reaction mixture was quenched by H_2O (20 mL) and extracted with hexane three times (20 mL × 3). The combined organic layer was then dried over $MgSO_4$. After filtration, the solvents were removed by evaporation. The crude product was purified by flash column chromatography to obtain the corresponding allyl ether (1.08 g, 6.21 mmol, 57%) as a colorless oil.

^{1}H NMR (400 MHz, $CDCl_3$, δ): 1.82–1.90 (m, 1H), 2.11–2.21 (m, 1H), 2.22–2.32 (m, 1H), 2.46–2.57 (m, 1H), 4.53 (q, J = 10.7 Hz, 2H), 4.65–4.70 (m, 1H), 5.89 (dq, J = 6.0, 1.9 Hz, 1H), 6.03 (dtd, J = 5.7, 2.3, 1.2 Hz, 1H), 7.24–7.29 (m, 1H), 7.30–7.38 (m, 4H). ^{13}C NMR (100 MHz, $CDCl_3$, δ): 29.7 (CH_2), 31.0 (CH_2), 70.4 (CH_2), 84.3 (CH), 127.3 (CH), 127.6 (CH), 128.2 (CH), 130.7 (CH), 135.5 (CH), 138.8 (C). HRMS-EI (m/z): $[M]^+$ calcd for $C_{12}H_{14}O$, 174.1045; found, 174.1045.

Preparation of 3-butyl-3-methoxycyclopent-1-ene (1j, Fig. 2.34)

In a vacuum-dried 50 mL two-neck round-bottomed flask, NaH (60% paraffin liquid dispersion, 264 mg, 6.6 mmol) was suspended in THF (13 mL) at 0 °C. To the suspension, allyl alcohol **S8** (616.5 mg, 4.4 mmol) was slowly added and the mixture was warmed up to room temperature. After stirring for 30 min, the mixture was cooled to 0 °C again. Then, methyl iodide (0.70 mL, 11.0 mmol) was added. After stirring for 17 h at room temperature, the reaction mixture was quenched by H_2O (10 mL) and extracted with hexane three times (10 mL × 3). The combined organic layer was then dried over $MgSO_4$. After filtration, the solvents were removed by evaporation. The crude product was purified by flash column chromatography to obtain the corresponding allyl ether (382.4 mg, 2.48 mmol, 56%) as a colorless oil.

^{1}H NMR (400 MHz, $CDCl_3$, δ): 0.90 (t, J = 6.9 Hz, 3H), 1.25–1.36 (m, 4H), 1.55–1.70 (m, 2H), 1.79 (ddd, J = 14.0, 8.9, 5.0 Hz, 1H), 1.95 (ddd, J = 13.9, 9.1, 4.7 Hz, 1H), 2.25–2.35 (m, 1H), 2.37–2.47 (m, 1H), 3.13 (s, 3H), 5.59 (dt, J

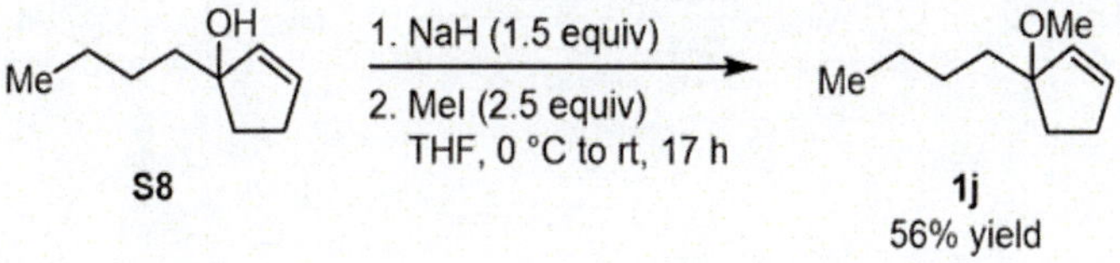

Fig. 2.34 Scheme of preparation procedure of **1j**

= 5.8, 2.2 Hz, 1H), 5.97 (dt, J = 5.6, 2.6 Hz, 1H). ^{13}C NMR (100 MHz, $CDCl_3$, δ): 14.1 (CH_3), 23.3 (CH_2), 26.3 (CH_2), 31.7 (CH_2), 31.9 (CH_2), 39.2 (CH_2), 50.0 (CH_3), 91.2 (C), 133.6 (CH), 134.6 (CH). HRMS-EI (m/z): $[M]^+$ calcd for $C_{10}H_{18}O$, 154.1358; found, 154.1357.

Preparation of (Z)-methyl (2-methyl-8-phenyloct-5-en-4-yl) carbonate [*rac*-(Z)-4a, Fig. 2.35]

An allyl alcohol (*Z*)-2-methyl-8-phenyloct-5-en-4-ol **S9** was synthesized according to the literature procedure [43]. In a vacuum-dried 20 mL two-neck round-bottomed flask, (*Z*)-2-methyl-8-phenyloct-5-en-4-ol **S9** (1.09 g, 5.0 mmol, *E*/*Z* < 5:95) and pyridine (1.81 mL, 22.5 mmol) were dissolved in dry DCM (5 mL), and the mixture was cooled to 0 °C under a nitrogen atmosphere. Methyl chloroformate (0.85 mL, 11 mmol) was added to the mixture dropwise, and the mixture was warmed to room temperature. After the mixture was stirred for 17 h, the reaction mixture was quenched by H_2O (5 mL) and extracted with hexane three times (5 mL × 3). The combined organic layer was then dried over $MgSO_4$. After filtration, the solvents were removed by evaporation. The crude product was purified by flash column chromatography to obtain the corresponding allyl carbonate *rac*-(Z)-**4a** (1.32 g, 4.78 mmol, 96%, *E*/*Z* < 5:95) as a colorless oil.

^{1}H NMR (400 MHz, $CDCl_3$, δ): 0.87 (d, J = 4.5 Hz, 3H), 0.89 (d, J = 4.5 Hz, 3H), 1.11–1.24 (m, 1H), 1.51–1.66 (m, 2H), 2.41–2.70 (m, 3H), 2.71–2.81 (m, 1H), 3.75 (s, 3H), 5.28–5.44 (m, 2H), 5.60 (dt, J = 13.3, 5.2 Hz, 1H), 7.13–7.23 (m, 3H), 7.24–7.32 (m, 2H). ^{13}C NMR (100 MHz, $CDCl_3$, δ): 22.3 (CH_3), 22.8 (CH_3), 24.2 (CH), 29.8 (CH_2), 35.7 (CH_2), 43.5 (CH_2), 54.4 (CH_3), 73.0 (CH), 125.9 (CH), 128.3 (CH), 128.5 (CH), 128.6 (CH), 133.1 (CH), 141.5 (C), 155.3 (C). HRMS-ESI (m/z): $[M + Na]^+$ calcd for $C_{17}H_{24}O_3Na$, 299.1618; found, 299.1717.

(Z)-Methyl (1-phenylhex-4-en-3-yl) carbonate [*rac*-(Z)-4b, Fig. 2.36]

rac-(Z)-**4b** was prepared from the corresponding alcohol according to the procedure described above. The product *rac*-(Z)-**4b** was obtained in 85% yield (274.8 mg, 1.17 mmol, *E*/*Z* < 5:95, colorless oil).

^{1}H NMR (400 MHz, $CDCl_3$, δ): 1.68–1.74 (m, 3H), 1.79–1.93 (m, 1H), 2.02–2.16 (m, 1H), 2.59–2.74 (m, 2H), 3.77 (s, 3H), 5.36–5.47 (m, 2H), 5.71 (dq, J = 15.4, 4.6 Hz, 1H), 7.15–7.23 (m, 3H), 7.24–7.33 (m, 2H). ^{13}C NMR (100 MHz, $CDCl_3$, δ):13.4 (CH_3), 31.2 (CH_2), 36.1 (CH_2), 54.4 (CH_3), 73.7 (CH), 125.9 (CH), 128.20

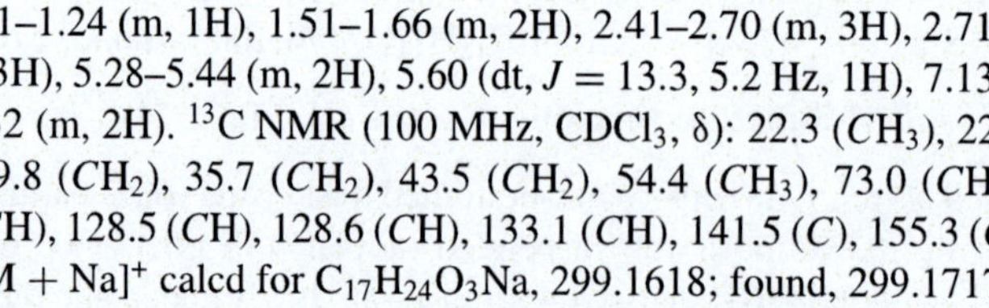

Fig. 2.35 Scheme of preparation procedure of *rac*-(Z)-**4a**

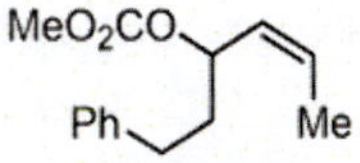

rac-(*Z*)-**4b**
E/*Z* < 5:95

Fig. 2.36 Scheme of *rac*-(Z)-**4b**

(*C*H), 128.23 (*C*H), 128.3 (*C*H), 129.2 (*C*H), 141.1 (*C*), 155.2 (*C*). HRMS-ESI (*m*/*z*): $[M + Na]^+$ calcd for $C_{14}H_{18}O_3Na$, 257.1148; found, 257.1151.

(Z)-Methyl (6-phenylhex-3-en-2-yl) carbonate [*rac*-(Z)-4c, Fig. 2.37]

rac-(Z)-**4c** was prepared from the corresponding alcohol according to the procedure described above. The product *rac*-(Z)-**4c** was obtained in 58% yield (808.9 mg, 3.45 mmol, *E*/*Z* < 5:95, colorless oil).

^1H NMR (400 MHz, $CDCl_3$, δ): 1.18 (d, J = 5.8 Hz, 3H), 2.40–2.57 (m, 2H), 2.58–2.69 (m, 1H), 2.70–2.82 (m, 1H), 3.75 (s, 3H), 5.34–5.46 (m, 2H), 5.49–5.62 (m, 1H), 7.12–7.22 (m, 3H), 7.24–7.33 (m, 2H). ^{13}C NMR (100 MHz, $CDCl_3$, δ): 20.6 (CH_3), 29.6 (CH_2), 35.5 (CH_2), 54.5 (CH_3), 71.0 (*C*H), 125.9 (*C*H), 128.3 (*C*H), 128.5 (*C*H), 129.4 (*C*H), 132.3 (*C*H), 141.4 (*C*), 155.1 (*C*). HRMS-ESI (*m*/*z*): $[M + Na]^+$ calcd for $C_{14}H_{18}O_3Na$, 257.1148; found, 257.1147.

(Z)-Methyl (1-phenylnon-4-en-3-yl) carbonate [*rac*-(Z)-4d, Fig. 2.38]

rac-(Z)-**4d** was prepared from the corresponding alcohol according to the procedure described above. The product *rac*-(Z)-**4d** was obtained in 87% yield (2.13 g, 7.72 mmol, *E*/*Z* < 5:95, colorless oil).

^1H NMR (400 MHz, $CDCl_3$, δ): 0.85–0.92 (m, 3H), 1.23–1.42 (m, 4H), 1.78–1.89 (m, 1H), 2.02–2.19 (m, 3H), 2.58–2.74 (m, 2H), 3.77 (s, 3H), 5.34–5.44 (m, 2H), 5.55–5.66 (m, 1H), 7.15–7.22 (m, 3H), 7.24–7.32 (m, 2H). ^{13}C NMR (100 MHz, $CDCl_3$, δ): 13.9 (CH_3), 22.3 (CH_2), 27.6 (CH_2), 31.2 (CH_2), 31.6 (CH_2), 36.3 (CH_2), 54.5 (CH_3), 74.1 (*C*H), 125.9 (*C*H), 127.1 (*C*H), 128.3 (*C*H), 128.4 (*C*H), 135.0 (*C*H), 141.1 (*C*), 155.2 (*C*). HRMS-FD (*m*/*z*): $[M]^+$ calcd for $C_{17}H_{24}O_3$, 276.1725; found, 276.1714.

(Z)-Methyl (2-methyl-7-phenylhept-4-en-3-yl) carbonate [*rac*-(Z)-4e, Fig. 2.39]

rac-(Z)-**4e** was prepared from the corresponding alcohol according to the procedure described above. The product *rac*-(Z)-**4e** was obtained in 92% yield (1.94 g, 7.40 mmol, *E*/*Z* < 5:95, colorless oil).

^1H NMR (400 MHz, $CDCl_3$, δ): 0.85 (d, J = 6.8 Hz, 3H), 0.92 (d, J = 6.8 Hz, 3H), 1.81 (octet, J = 6.7 Hz, 1H), 2.38–2.79 (m, 4H), 3.76 (s, 3H), 5.12 (dd, J = 9.5,

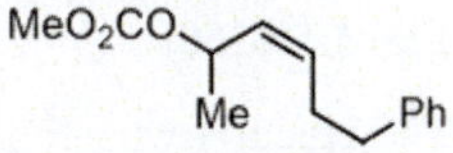

rac-(*Z*)-**4c**
E/*Z* < 5:95

Fig. 2.37 Scheme of *rac*-(Z)-**4c**

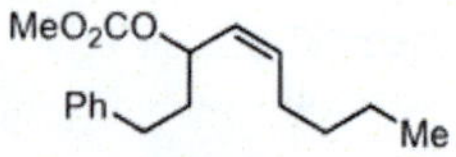

rac-(*Z*)-**4d**
E/*Z* < 5:95

Fig. 2.38 Scheme of *rac*-(Z)-**4d**

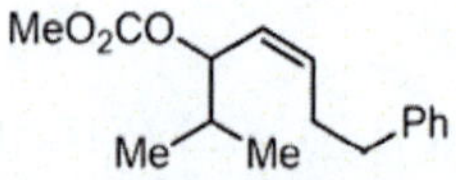

rac-(*Z*)-**4e**
E/*Z* < 5:95

Fig. 2.39 Scheme of *rac*-(Z)-**4e**

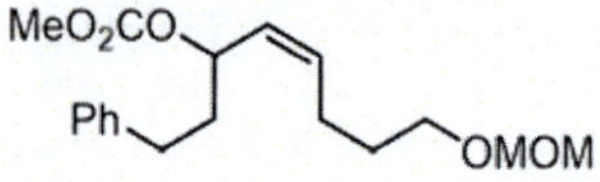

rac-(*Z*)-**4f**
E/*Z* < 5:95

Fig. 2.40 Scheme of *rac*-(Z)-**4f**

6.8 Hz, 1H), 5.32–5.40 (m, 1H), 5.68 (dt, $J = 13.4$, 5.5 Hz, 1H), 7.15–7.23 (m, 3H), 7.25–7.33 (m, 2H). ^{13}C NMR (100 MHz, $CDCl_3$, δ): 17.8 (CH_3), 18.0 (CH_3), 29.8 (CH_2), 32.1 (CH), 35.6 (CH_2), 54.5 (CH_3), 78.9 (CH), 125.9 (CH), 126.3 (CH), 128.3 (CH), 128.4 (CH), 134.2 (CH), 141.5 (C), 155.3 (C). HRMS-ESI (*m*/*z*): [M + Na]$^+$ calcd for $C_{16}H_{22}O_3Na$, 285.1461; found, 285.1455.

(Z)-8-(Methoxymethoxy)-1-phenyloct-4-en-3-yl methyl carbonate [*rac*-(Z)-4f, Fig. 2.40]

rac-(Z)-**4f** was prepared from the corresponding alcohol according to the procedure described above. The product *rac*-(Z)-**4f** was obtained in 56% yield (714.6 mg, 2.22 mmol, *E*/*Z* < 5:95, colorless oil).

^{1}H NMR (400 MHz, $CDCl_3$, δ): 1.58–1.74 (m, 2H), 1.79–1.91 (m, 1H), 2.03–2.14 (m, 1H), 2.15–2.29 (m, 2H), 2.59–2.75 (m, 2H), 3.36 (s, 3H), 3.51 (t, $J = 6.6$ Hz, 2H), 3.77 (s, 3H), 4.61 (s, 2H), 5.36–5.46 (m, 2H), 5.59–5.66 (m, 1H), 7.15–7.22 (m, 3H), 7.25–7.32 (m, 2H). ^{13}C NMR (100 MHz, $CDCl_3$, δ): 24.5 (CH_2), 29.4 (CH_2), 31.2 (CH_2), 36.2 (CH_2), 54.5 (CH_3), 55.0 (CH_3), 67.0 (CH_2), 73.9 (CH), 96.3 (CH_2), 125.9 (CH), 127.8 (CH), 128.2 (CH), 128.3 (CH), 133.9 (CH), 141.0 (C), 155.1 (C). HRMS-FD (*m*/*z*): [M]$^+$ calcd for $C_{18}H_{25}O_5$, 322.1780; found, 322.1787.

2.4.4 Borylation Product Characterization

(*S*)-2-(Cyclohex-2-en-1-yl)-4,4,5,5-tetramethyl-1,3,2-dioxaborolane [(*S*)-3a, Fig. 2.41] [50]

The reaction was conducted with 77.5 mg (0.50 mmol) of **1a** by using procedure A. The product (*S*)-**3a** was obtained as a colorless oil in 86% yield (88.9 mg). The

Fig. 2.41 Scheme of (*S*)-**3a**

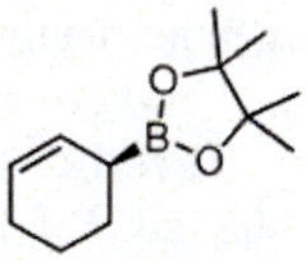

(*S*)-**3a**

enantioselectivity was determined by the following stereospecific derivatization of the boryl group.

^{1}H NMR (400 MHz, $CDCl_3$, δ): 1.25 (s, 12H), 1.53–1.71 (m, 3H), 1.72–1.85 (m, 2H), 1.95–2.03 (m, 2H), 5.63–5.75 (m, 2H). ^{13}C NMR (100 MHz, $CDCl_3$, δ): 21.7 (br, B–*C*H), 22.4 (CH_2), 24.0 (CH_2), 24.6 (CH_3), 24.7 (CH_3), 24.9 (CH_2), 83.0 (*C*), 125.9 (*C*H), 127.5 (*C*H). HRMS-EI (*m*/*z*): $[M]^+$ calcd for $C_{12}H_{21}{}^{11}BO_2$, 208.1637; found, 208.1637. $[\alpha]_D^{27}$ +1.2 (*c* 0.8, $CHCl_3$).

The allylboration reaction was conducted with 41.5 mg (0.20 mmol) of (*S*)-**3a**. The product (*S*,*R*)-**S1a** [43]. was obtained as a colorless oil in 81% yield (30.5 mg) with 93% ee (Fig. 2.42).

^{1}H NMR (400 MHz, $CDCl_3$, δ): 1.43–1.61 (m, 2H), 1.64–1.81 (m, 2H), 1.84 (d, *J* = 2.7 Hz, 1H), 1.95–2.04 (m, 2H), 2.46–2.55 (m, 1H), 4.59 (dd, *J* = 6.3, 2.7 Hz, 1H), 5.39 (dq, *J* = 10.1, 2.0 Hz, 1H), 5.82 (dq, *J* = 10.0, 3.3 Hz, 1H), 7.24–7.30 (m, 1H), 7.32–7.38 (m, 4H). ^{13}C NMR (100 MHz, $CDCl_3$, δ): 21.1 (CH_2), 23.8 (CH_2), 25.2 (CH_2), 43.0 (*C*H), 77.4 (*C*H), 126.5 (*C*H), 127.4 (*C*H), 128.0 (*C*H), 128.2 (*C*H), 130.3 (*C*H), 142.8 (*C*). HRMS-ESI (*m*/*z*): $[M + Na]^+$ calcd for $C_{13}H_{16}ONa$, 211.1099; found, 211.1091. $[\alpha]_D^{28}$ +4.7 (*c* 0.9, $CHCl_3$, 93% ee). Daicel CHIRALPAK® OD-3, 2-PrOH/hexane = 3/97, 0.50 mL/min, 40 °C, (*S*,*R*) isomer: t_R = 17.4 min., (*R*,*S*) isomer: t_R = 20.2 min.

(*R*)-3-(4,4,5,5-Tetramethyl-1,3,2-dioxaborolan-2-yl)-1-tosyl-1,2,3,6-tetrahydropyridine [(*R*)-3b, Fig. 2.43]

The reaction was conducted with 155.3 mg (0.50 mmol) of **1b** by using procedure A. The product (*R*)-**3b** was obtained as a white solid in 77% yield (140.1 mg). The

Fig. 2.42 Scheme of (*S*,*R*)-**S1a**

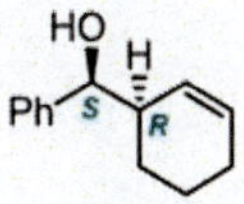

(*S*,*R*)-**S1a**

Fig. 2.43 Scheme of (*R*)-**3b**

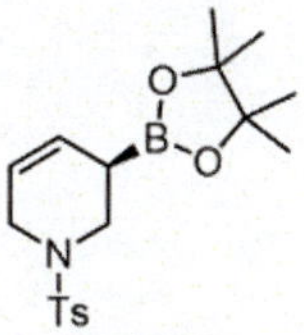

(*R*)-**3b**

enantioselectivity was determined by the following stereospecific derivatization of the boryl group.

^{1}H NMR (400 MHz, $CDCl_3$, δ): 1.24 (s, 12H), 2.05–2.14 (m, 1H), 2.43 (s, 3H), 3.02 (dd, J = 11.7, 8.1 Hz, 1H), 3.42 (dq, J = 16.3, 2.8 Hz 1H), 3.47 (dd, J = 11.6, 5.3 Hz, 1H), 3.68 (dq, J = 16.2, 2.8 Hz, 1H), 5.58 (dq, J = 10.1, 3.3 Hz, 1H), 5.81 (dq, J = 10.2, 2.6 Hz, 1H), 7.32 (d, J = 8.1 Hz, 2H), 7.69 (d, J = 8.5 Hz, 2H). ^{13}C NMR (100 MHz, $CDCl_3$, δ): 21.3 (CH_3), 22.9 (br, B–CH), 24.5 (CH_3), 24.6 (CH_3), 44.0 (CH_2), 44.7 (CH_2), 83.6 (C), 121.2 (CH), 126.1 (CH), 127.5 (CH), 129.4 (CH), 133.2 (C), 143.2 (C). HRMS-ESI (m/z): $[M + Na]^+$ calcd for $C_{18}H_{26}{}^{11}BNO_4SNa$, 386.1568; found, 386.1571. $[\alpha]_D^{27}$ −19.2 (c 0.8, $CHCl_3$).

The oxidation reaction was conducted with 78.5 mg (0.22 mmol) of (R)-**3b**. The product (R)-**S1b** was obtained as a white solid in 74% yield (40.4 mg) with 95% ee (Fig. 2.44).

^{1}H NMR (400 MHz, $CDCl_3$, δ): 1.97 (d, J = 9.4 Hz, 1H), 2.44 (s, 3H), 3.06 (dd, J = 11.7, 3.6 Hz, 1H), 3.31–3.42 (m, 2H), 3.77 (d, J = 16.6 Hz, 1H), 4.15–4.26 (br-m, 1H), 5.77–5.84 (m, 1H), 5.87–5.94 (m, 1H), 7.34 (d, J = 8.1 Hz, 2H), 7.69 (d, J = 8.1 Hz, 2H). ^{13}C NMR (100 MHz, $CDCl_3$, δ): 21.5 (CH_3), 44.8 (CH_2), 50.0 (CH_2), 63.4 (CH), 125.7 (CH), 127.7 (CH), 128.1 (CH), 129.8 (CH), 132.9 (C), 143.9 (C). HRMS-ESI (m/z): $[M + Na]^+$ calcd for $C_{12}H_{15}NO_3SNa$, 276.0665; found, 276.0664. $[\alpha]_D^{28}$ −92.9 (c 0.9, $CHCl_3$, 95% ee). Daicel CHIRALPAK® OD-3, 2-PrOH/hexane = 10/90, 0.50 mL/min, 40 °C, (R) isomer: t_R = 25.3 min., (S) isomer: t_R = 23.8 min.

tert-Butyl (*R*)-3-(4,4,5,5-tetramethyl-1,3,2-dioxaborolan-2-yl)-3,6-dihydropyridine-1(2*H*)-carboxylate [(*R*)-3c, Fig. 2.45]

The reaction was conducted with 127.5 mg (0.50 mmol) of **1c** by using procedure A. The product (R)-**3c** was obtained as a colorless oil in 52% yield (79.9 mg). The enantioselectivity was determined by the following stereospecific derivatization of the boryl group.

Fig. 2.44 Scheme of (R)-**S1b**

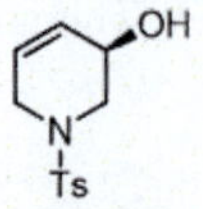

(R)-**S1b**

Fig. 2.45 Scheme of (R)-**3c**

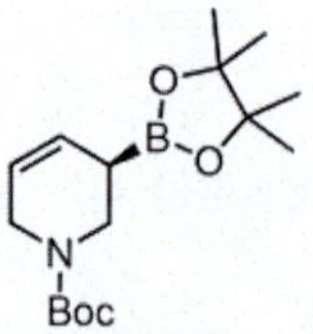

(R)-**3c**

^{1}H NMR (400 MHz, $CDCl_3$, δ): 1.24 (s, 12H), 1.47 (s, 9H), 1.89–2.05 (m, 1H), 3.37–3.52 (m, 1H), 3.57–3.75 (m, 1H), 3.76–3.86 (m, 1H), 3.88–3.99 (m, 1H), 5.55–5.71 (m, 1H), 5.79–5.92 (m, 1H). ^{13}C NMR (100 MHz, $CDCl_3$, δ): 22.8 (br, B–*C*H), 24.6 (CH_3), 24.7 (CH_3), 28.4 (CH_3), 41.0 and 41.9 (a pair of CH_2), 42.7 and 43.5 (a pair of CH_2), 79.1 (*C*), 83.4 (*C*), 123.1 (*C*H), 125.9 and 126.3 (a pair of *C*H), 154.7 (*C*). HRMS-ESI (*m*/*z*): [M + Na]$^+$ calcd for $C_{16}H_{28}{}^{11}BNO_4Na$, 332.2007; found, 332.2004. $[\alpha]_D{}^{27}$ −48.6 (*c* 1.0, $CHCl_3$).

The allylboration reaction was conducted with 26.4 mg (0.09 mmol) of (*R*)-**3c**. The product (*S*,*S*)-**S1c** was obtained as a colorless oil in 73% yield (18.1 mg) with 90% ee (Fig. 2.46).

^{1}H NMR (400 MHz, $CDCl_3$, δ): 1.50 (s, 9H), 2.21–2.55 (br-m, 1H), 3.08–3.59 (br-m, 1H), 3.73 and 3.77 (a pair pf q, $J = 2.5$ Hz, 1H), 3.91–4.25 (br-m, 2H), 4.27–4.55 (br-m, 1H), 5.24–5.50 (br-m, 1H), 5.56–5.80 (br-m, 1H), 7.21–7.26 (m, 2H), 7.32–7.43 (m, 4H). ^{13}C NMR (100 MHz, $CDCl_3$, δ): 28.4 (CH_3), 41.6 (CH_2), 43.2 (*C*H), 44.4 (CH_2), 75.3 (*C*H), 80.1 (*C*), 125.7 and 126.3 (a pair of *C*H), 126.3 and 127.0 (a pair of *C*H), 126.6 (*C*H), 127.6 (*C*H), 128.3 (*C*H), 142.1 (*C*), 154.9 and 155.8 (a pair of *C*). HRMS-ESI (*m*/*z*): [M + Na]$^+$ calcd for $C_{17}H_{23}NO_3Na$, 312.1570; found, 312.1569. $[\alpha]_D{}^{28}$ +52.1 (*c* 0.8, $CHCl_3$, 90% ee). Daicel CHIRALPAK® OD-3, 2-PrOH/hexane = 5/95, 0.50 mL/min, 40 °C, (*S*,*S*) isomer: $t_R = 13.7$ min., (*R*,*R*) isomer: $t_R = 15.0$ min.

Benzyl (*R*)-3-(4,4,5,5-tetramethyl-1,3,2-dioxaborolan-2-yl)-3,6-dihydropyridine-1(2*H*)-carboxylate [(*R*)-3d, Fig. 2.47]

The reaction was conducted with 147.0 mg (0.50 mmol) of **1d** by using procedure A. The product (*R*)-**3d** was obtained as a colorless oil in 79% yield (122.6 mg). The enantioselectivity was determined by the following stereospecific derivatization of the boryl group.

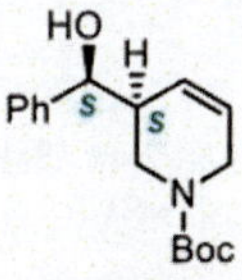

(*S*,*S*)-**S1c**

Fig. 2.46 Scheme of (*S*,*S*)-**S1c**

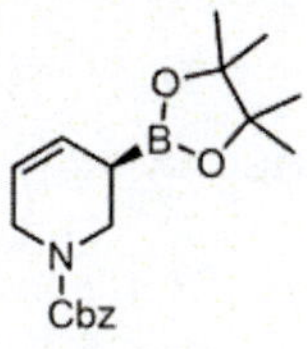

(*R*)-**3d**

Fig. 2.47 Scheme of (*R*)-**3d**

^{1}H NMR (400 MHz, $CDCl_3$, δ): 1.22 (s, 12H), 1.93–2.08 (m, 1H), 3.36–3.94 (m, 2H), 3.81–4.10 (m, 2H), 5.15 (s, 2H), 5.56–5.72 (m, 1H), 5.83–5.92 (m, 1H), 7.28–7.44 (m, 5H). ^{13}C NMR (100 MHz, $CDCl_3$, δ): 22.5 (br, B–*C*H), 24.5 (*C*H_3), 24.6 (*C*H_3), 41.6 and 41.7 (a pair of *C*H_2), 43.2 (*C*H_2), 66.7 (*C*H_2), 83.4 (*C*), 122.5 and 122.8 (a pair of *C*H), 126.0 and 126.4 (a pair of *C*H), 127.7 (*C*H), 128.2 (*C*H), 136.8 (*C*), 155.2 (*C*). HRMS-ESI (*m/z*): $[M + Na]^+$ calcd for $C_{19}H_{26}{}^{11}BNO_4Na$, 366.1851; found, 366.1844. $[\alpha]_D{}^{27}$ −38.9 (*c* 0.9, $CHCl_3$).

The allylboration reaction was conducted with 41.7 mg (0.14 mmol) of (*R*)-**3d**. The product (*S*,*S*)-**S1d** was obtained as a colorless oil in 73% yield (32.0 mg) with 90% ee (Fig. 2.48).

^{1}H NMR (400 MHz, $CDCl_3$, δ): 1.90–2.14 and 2.93–3.09 (a pair of br-s, 1H), 2.46–2.57 (br-m, 1H), 3.27–3.57 (br-m, 1H), 3.82 and 3.86 (a pair of q, $J = 2.6$ Hz, 1H), 3.99–4.25 (br-m, 2H), 4.31–4.48 (br-m, 1H), 5.09–5.26 (br-s, 2H), 5.28–5.46 (br-m, 1H), 5.57–5.80 (br-m, 1H), 7.23–7.43 (m, 10H). ^{13}C NMR (100 MHz, $CDCl_3$, δ): 42.2 (*C*H_2), 42.9 (*C*H), 43.9 (*C*H_2), 67.3 (*C*H_2), 75.1 (*C*H), 125.3 and 126.1 (a pair of *C*H), 126.1 and 126.9 (a pair of *C*H), 126.6 (*C*H), 127.7 (*C*H), 127.9 (*C*H), 128.0 (*C*H), 128.3 (*C*H), 128.5 (*C*H), 136.6 (*C*), 142.1 (*C*). HRMS-ESI (*m/z*): $[M + Na]^+$ calcd for $C_{20}H_{21}NO_3Na$, 346.1414; found, 346.1408. $[\alpha]_D{}^{28}$ +83.0 (*c* 0.8, $CHCl_3$, 90% ee). Daicel CHIRALPAK® OD-3, 2-PrOH/hexane = 3/97, 0.50 mL/min, 40 °C, (*S*,*S*) isomer: $t_R = 61.9$ min., (*R*,*R*) isomer: $t_R = 71.8$ min.

(*R*)-2-(3,6-Dihydro-2*H*-pyran-3-yl)-4,4,5,5-tetramethyl-1,3,2-dioxaborolane [(*R*)-3e, Fig. 2.49] [72]

The reaction was conducted with 79.0 mg (0.50 mmol) of **1e** by using procedure A. The product (*R*)-**3e** was obtained as a colorless oil in 57% yield (60.3 mg). The enantioselectivity was determined by the following stereospecific derivatization of the boryl group.

^{1}H NMR (400 MHz, $CDCl_3$, δ): 1.25 (s, 12H), 2.02–2.12 (m, 1H), 3.76 (dd, $J = 11.1, 7.9$ Hz, 1H), 3.96 (dd, $J = 11.1, 5.2$ Hz, 1H), 4.13 (q, $J = 2.7$ Hz, 2H), 5.70 (dq, $J = 10.2, 2.6$ Hz, 1H), 5.90 (dq, $J = 10.2, 2.5$ Hz, 1H). ^{13}C NMR (100 MHz, $CDCl_3$,

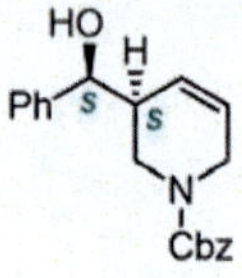

(*S*,*S*)-**S1d**

Fig. 2.48 Scheme of (*S*,*S*)-**S1d**

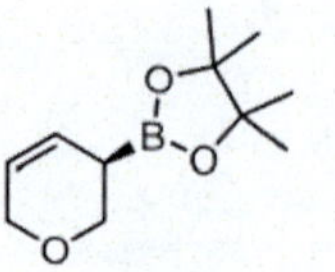

(*R*)-**3e**

Fig. 2.49 Scheme of (*R*)-**3e**

δ): 23.1 (br, B–*C*H), 24.6 (CH_3), 24.7 (CH_3), 65.3 (CH_2), 65.6 (CH_2), 83.4 (*C*), 125.1 (*C*H), 125.2 (*C*H). HRMS-EI (*m*/*z*): $[M]^+$ calcd for $C_{11}H_{19}{}^{11}BO_3$, 210.1429; found, 210.1438. $[\alpha]_D{}^{27}$ −24.3 (*c* 0.9, $CHCl_3$).

The allylboration reaction was conducted with 27.5 mg (0.13 mmol) of (*R*)-**3e**. The product (*S*,*S*)-**S1e** was obtained as a colorless oil in 82% yield (20.5 mg) with 96% ee (Fig. 2.50).

^1H NMR (400 MHz, $CDCl_3$, δ): 2.38–2.46 (m, 1H), 2.48 (d, $J = 4.0$ Hz, 1H), 3.75 (dd, $J = 11.7, 4.0$ Hz, 1H), 4.06 (dd, $J = 11.4, 3.4$ Hz, 1H), 4.08–4.21 (m, 2H), 4.72 (dd, $J = 7.2, 4.0$ Hz, 1H), 5.45–5.53 (m, 1H), 5.84 (dq, $J = 10.4, 2.2$ Hz, 1H), 7.27–7.33 (m, 1H), 7.33–7.41 (m, 4H). ^{13}C NMR (100 MHz, $CDCl_3$, δ): 41.8 (*C*H), 65.6 (CH_2), 65.6 (CH_2), 75.2 (*C*H), 125.4 (*C*H), 126.4 (*C*H), 127.6 (*C*H), 128.4 (*C*H), 142.9 (*C*). HRMS-ESI (*m*/*z*): $[M + Na]^+$ calcd for $C_{12}H_{14}O_2Na$, 213.0886; found, 213.0889. $[\alpha]_D{}^{28}$ +79.7 (*c* 0.9, $CHCl_3$, 96% ee). Daicel CHIRALPAK® OD-3, 2-PrOH/hexane = 3/97, 0.50 mL/min, 40 °C, (*S*,*S*) isomer: $t_R = 27.0$ min., (*R*,*R*) isomer: $t_R = 30.6$ min.

(*S*)-2-(5,5-Dimethylcyclohex-2-en-1-yl)-4,4,5,5-tetramethyl-1,3,2-dioxaborolane [(*S*)-3f, Fig. 2.51] [49]

The reaction was conducted with 92.6 mg (0.50 mmol) of **1f** by using procedure A. The product (*S*)-**3f** was obtained as a colorless oil in 58% yield (68.5 mg). The enantioselectivity was determined by the following stereospecific derivatization of the boryl group.

^1H NMR (400 MHz, $CDCl_3$, δ): 0.87 (s, 3H), 0.92 (s, 3H), 1.25 (s, 12H), 1.35–1.49 (m, 2H), 1.66–1.75 (m, 1H), 1.79–1.89 (m, 2H), 5.56–5.63 (m, 1H), 5.65–5.72 (m, 1H). ^{13}C NMR (100 MHz, $CDCl_3$, δ): 19.6 (br, B–*C*H), 24.6 (CH_3), 24.7 (CH_3), 25.9 (CH_3), 28.4 (*C*), 30.6 (CH_3), 36.6 (CH_2), 38.7 (CH_2), 83.1 (*C*), 124.9 (*C*H), 126.1 (*C*H). HRMS-EI (*m*/*z*): $[M]^+$ calcd for $C_{14}H_{25}{}^{11}BO_2$, 236.1950; found, 236.1953. $[\alpha]_D{}^{27}$ −38.8 (*c* 0.8, $CHCl_3$).

Fig. 2.50 Scheme of (*S*,*S*)-**S1e**

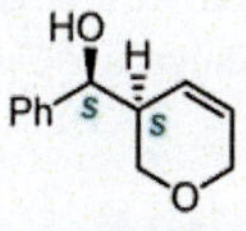

(*S*,*S*)-**S1e**

Fig. 2.51 Scheme of (*S*)-**3f**

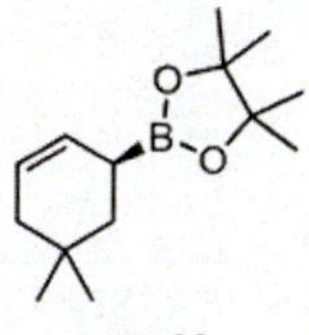

(*S*)-**3f**

The allylboration reaction was conducted with 47.6 mg (0.20 mmol) of (*S*)-**3f**. The product (*S*,*R*)-**S1f** was obtained as a colorless oil in 87% yield (38.1 mg) with 84% ee (Fig. 2.52).

^{1}H NMR (400 MHz, $CDCl_3$, δ): 0.85 (s, 3H), 0.95 (s, 3H), 1.28 (d, J = 11.2 Hz, 1H), 1.31–1.40 (m, 1H), 1.66–1.76 (m, 1H), 1.81–1.92 (m, 1H), 1.83 (d, J = 2.7 Hz, 1H), 2.47–2.59 (m, 1H), 4.65 (dd, J = 5.8, 2.2 Hz, 1H), 5.40–5.47 (m, 1H), 5.75 (ddt, J = 10.2, 5.2, 2.5 Hz, 1H), 7.23–7.32 (m, 1H), 7.33–7.40 (m, 4H). ^{13}C NMR (100 MHz, $CDCl_3$, δ): 25.1 (CH_3), 29.3 (*C*), 32.1 (CH_3), 36.8 (CH_2), 39.1 (CH_2), 41.7 (*C*H), 77.6 (*C*H), 126.5 (*C*H), 126.7 (*C*H), 127.3 (*C*H), 128.2 (*C*H), 129.1 (*C*H), 142.8 (*C*). HRMS-ESI (*m/z*): [M + Na]$^+$ calcd for $C_{15}H_{20}ONa$, 239.1412; found, 239.1421. $[\alpha]_D^{28}$ −22.8 (*c* 1.1, $CHCl_3$, 84% ee). Daicel CHIRALPAK® OD-3, 2-PrOH/hexane = 3/97, 0.50 mL/min, 40 °C, (*S*,*R*) isomer: t_R = 15.3 min., (*R*,*S*) isomer: t_R = 20.6 min.

tert-Butyl (_R_)-6-[(_R_)-hydroxy(phenyl)methyl]-3,6-dihydropyridine-1(2H)-carboxylate [(_R_,_R_)-S1g, Fig. 2.53] [53]

The reaction was conducted with 106.5 mg (0.50 mmol) of **1g** by using procedure B. Although the borylation product (*S*)-**3g** was detected in ^{1}H NMR analysis (90% ^{1}H NMR yield), the product could not be isolated due to its instability to silica gel. After the allylboration of benzaldehyde using the crude material including (*S*)-**3g**, the product (*R*,*R*)-**S1g** was obtained as a colorless oil with a small amount of pinacol in 41% yield (59.4 mg) with 79% ee.

^{1}H NMR (400 MHz, $CDCl_3$, δ): 1.35–1.60 (br-s, 9H), 1.87–2.01 (br-m, 1H), 2.09–2.30 (br-m, 1H), 2.72–3.10 (br-m, 1H), 3.81–4.29 (br-m, 1H), 4.46–4.76 (br-m, 2H), 5.04–5.42 (br-m, 1H), 5.75–5.99 (br-m, 1H), 7.27–7.42 (m, 5H). ^{13}C NMR (100 MHz, $CDCl_3$, δ): 24.4 and 24.7 (a pair of CH_2), 28.3 (CH_3), 36.7 and 38.1 (a pair of CH_2), 58.1 (*C*H), 75.7 and 76.7 (a pair of *C*H), 80.3 and 80.5 (a pair of *C*), 124.4 and 124.7 (a pair of *C*H), 127.0 (*C*H), 127.8 (*C*H), 128.2 (*C*H), 130.0 and 133.1 (a pair of *C*H), 140.9 and 141.4 (a pair of *C*), 155.1 and 157.1 (a pair of *C*). The NMR charts contain inseparable pinacol impurities. HRMS-ESI (*m/z*): [M + Na]$^+$ calcd for $C_{17}H_{23}O_3NNa$, 312.1570; found, 312.1573. $[\alpha]_D^{27}$ +124.1 (*c* 0.8, $CHCl_3$,

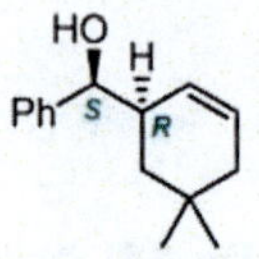

(*S*,*R*)-**S1f**

Fig. 2.52 Scheme of (*S*,*R*)-**S1f**

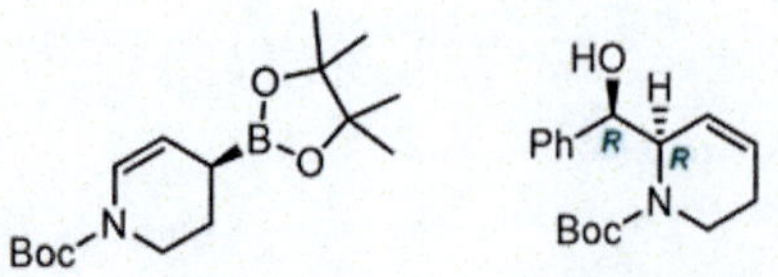

(*S*)-**3g** (*R*,*R*)-**S1g**

Fig. 2.53 Scheme of (*S*)-**3g** and (*R*,*R*)-**S1g**

79% ee). Daicel CHIRALPAK® IA-3, 2-PrOH/hexane = 3/97, 0.50 mL/min, 40 °C, (*R*,*R*) isomer: t_R = 36.3 min., (*S*,*S*) isomer: t_R = 44.2 min.

(*R*)-[(*R*)-5,6-Dihydro-2H-pyran-2-yl](phenyl)methanol [(*R*,*R*)-S1h, Fig. 2.54] [53]

The reaction was conducted with 95.2 mg (0.50 mmol) of **1h** by using procedure A. The borylation product (*S*)-**3h** was detected in ^{1}H NMR analysis (40% ^{1}H NMR yield). After the allylboration of benzaldehyde using the crude material including (*S*)-**3h**, the product (*R*,*R*)-**S1h** was obtained as a colorless oil with a small amount of pinacol in 31% yield (29.3 mg) with 56% ee.

^{1}H NMR (400 MHz, $CDCl_3$, δ): 2.00–2.10 (m, 1H), 2.20–2.31 (m, 1H), 3.14 (s, 1H), 3.74 (ddd, *J* = 15.3, 7.7, 3.9 Hz, 1H), 4.05 (dt, *J* = 11.0, 4.8 Hz, 1H), 4.14 (doublet of quintet, *J* = 7.7, 2.5 Hz, 1H), 4.55 (d, *J* = 8.2 Hz, 1H), 5.35 (dq, *J* = 10.4, 1.8 Hz, 1H), 5.91 (doublet of quintet, *J* = 10.3, 2.5 Hz, 1H), 7.28–7.42 (m, 5H). ^{13}C NMR (100 MHz, $CDCl_3$, δ): 25.1 (CH_2), 62.5 (CH_2), 76.3 (*C*H), 78.0 (*C*H), 125.7 (*C*H), 126.7 (*C*H), 127.3 (*C*H), 128.0 (*C*H), 128.3 (*C*H), 139.7 (*C*). The NMR charts contain inseparable pinacol impurities. HRMS-ESI (*m*/*z*): $[M + Na]^+$ calcd for $C_{12}H_{14}O_2Na$, 213.0886; found, 213.0887. $[\alpha]_D^{27}$ +7.9 (*c* 0.9, $CHCl_3$, 56% ee). Daicel CHIRALPAK® OZ-3, 2-PrOH/hexane = 3/97, 0.50 mL/min, 40 °C, (*R*,*R*) isomer: t_R = 26.7 min., (*S*,*S*) isomer: t_R = 29.1 min.

(*S*)-2-(Cyclopent-2-en-1-yl)-4,4,5,5-tetramethyl-1,3,2-dioxaborolane [(*S*)-3i, Fig. 2.55] [70]

The reaction was conducted with 87.2 mg (0.50 mmol) of **1i** by using procedure C. The product (*S*)-**3i** was obtained as a colorless oil in 51% yield (50.0 mg). The enantioselectivity was determined by the following stereospecific derivatization of the boryl group.

^{1}H NMR (400 MHz, $CDCl_3$, δ): 1.24 (s, 12H), 1.81–1.91 (m, 1H), 1.97–2.08 (m, 1H), 2.12–2.21 (m, 1H), 2.30–2.40 (m, 2H), 5.69–5.75 (m, 2H). ^{13}C NMR (100 MHz, $CDCl_3$, δ): 24.7 (CH_3), 25.5 (CH_2), 29.4 (br, B–*C*H) 32.8 (CH_2), 83.1 (*C*), 129.7 (*C*H), 131.3 (*C*H). HRMS-EI (*m*/*z*): $[M]^+$ calcd for $C_{11}H_{19}{}^{11}BO_2$, 194.1478; found, 194.1484. $[\alpha]_D^{26}$ −97.9 (*c* 1.2, $CHCl_3$).

Fig. 2.54 Scheme of (*S*)-**3h** and (*R*,*R*)-**S1h**

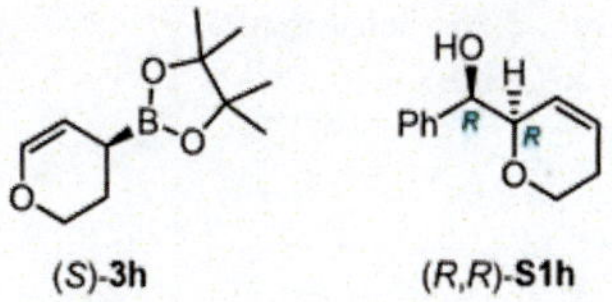

Fig. 2.55 Scheme of (*S*)-**3i**

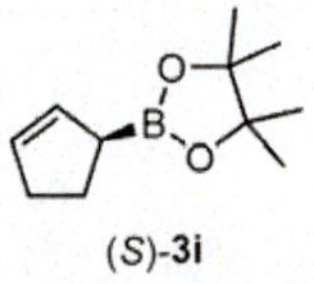

The allylboration reaction was conducted with 21.9 mg (0.11 mmol) of (*S*)-**3i**. The product (*S*,*R*)-**S1i** was obtained as a colorless oil in 78% yield (15.3 mg) with 92% ee (Fig. 2.56).

^{1}H NMR (400 MHz, $CDCl_3$, δ): 1.80–1.99 (m, 2H), 1.85 (d, J = 2.7 Hz, 1H), 2.25–2.45 (m, 2H), 3.07–3.16 (m, 1H), 4.59 (dd, J = 6.5, 2.5 Hz, 1H), 5.39–5.45 (m, 1H), 5.84–5.90 (m, 1H), 7.24–7.31 (m, 1H), 7.32–7.39 (m, 4H). ^{13}C NMR (100 MHz, $CDCl_3$, δ): 25.0 (CH_2), 32.2 (CH_2), 53.9 (CH), 76.9 (CH), 126.3 (CH), 127.3 (CH), 128.2 (CH), 131.2 (CH), 133.7 (CH), 143.4 (C). HRMS-ESI (m/z): $[M + Na]^+$ calcd for $C_{12}H_{14}ONa$, 197.0942; found, 197.0939. $[\alpha]_D^{26}$ +21.6 (c 1.5, $CHCl_3$, 92% ee). Daicel CHIRALPAK® OD-3 + OD-3, 2-PrOH/hexane = 1/99, 0.50 mL/min, 40 °C, (*S*,*R*) isomer: t_R = 76.2 min., (*R*,*S*) isomer: t_R = 67.7 min.

(*S*)-2-(3-Butylcyclopent-2-en-1-yl)-4,4,5,5-tetramethyl-1,3,2-dioxaborolane [(*S*)-3j, Fig. 2.57]

The reaction was conducted with 76.8 mg (0.50 mmol) of **1j** by using procedure D. The product (*S*)-**3j** was obtained as a colorless oil in 79% yield (98.1 mg). The enantioselectivity was determined by the following stereospecific derivatization of the boryl group.

^{1}H NMR (400 MHz, $CDCl_3$, δ): 0.89 (t, J = 7.2 Hz, 3H), 1.18–1.34 (m, 2H), 1.24 (s, 12H), 1.36–1.46 (m, 2H), 1.87 (dq, J = 18.0, 5.1 Hz, 1H), 1.97–2.09 (m, 3H), 2.11–2.18 (m, 1H), 2.21–2.29 (m, 2H), 5.27–5.31 (m, 1H). ^{13}C NMR (100 MHz, $CDCl_3$, δ): 13.9 (CH_3), 22.5 (CH_2), 24.67 (CH_3), 24.69 (CH_3), 26.0 (CH_2), 29.3 (br, B–CH), 30.0 (CH_2), 30.8 (CH_2), 35.6 (CH_2), 82.9 (C), 123.4 (CH), 144.2 (C). HRMS-EI (m/z): $[M]^+$ calcd for $C_{15}H_{27}{}^{11}BO_2$, 250.2104; found, 250.2114. $[\alpha]_D^{23}$ −66.4 (c 1.1, $CHCl_3$).

The allylboration reaction was conducted with 50.9 mg (0.20 mmol) of (*S*)-**3j**. The product (*S*,*R*)-**S1j** was obtained as a colorless oil in 46% yield (21.3 mg) with 94% ee (Fig. 2.58).

^{1}H NMR (400 MHz, $CDCl_3$, δ): 0.86 (t, J = 6.7 Hz, 3H), 1.15–1.29 (m, 4H), 1.33–1.50 (m, 3H), 1.91 (d, J = 1.8 Hz, 1H), 2.04–2.31 (m, 3H), 4.64 (d, J = 1.8 Hz, 1H), 5.46–5.52 (m, 1H), 5.83–5.90 (m, 1H), 7.23–7.38 (m, 5H). ^{13}C NMR (100 MHz, $CDCl_3$, δ): 14.2 (CH_3), 23.5 (CH_2), 26.7 (CH_2), 28.6 (CH_2), 32.7 (CH_2), 36.0 (CH_2),

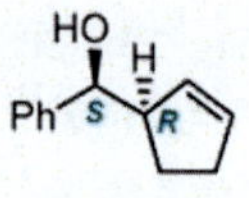

(*S*,*R*)-**S1i**

Fig. 2.56 Scheme of (*S*,*R*)-**S1i**

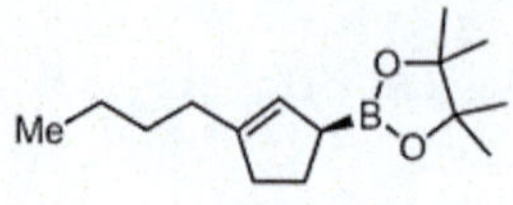

(*S*)-**3j**

Fig. 2.57 Scheme of (*S*)-**3j**

Fig. 2.58 Scheme of (*S*,*R*)-**S1j**

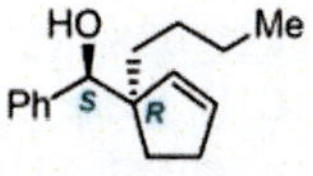

(S,R)-**S1j**

58.7 (*C*), 78.9 (*C*H), 127.3 (*C*H), 127.6 (*C*H), 127.7 (*C*H), 133.3 (*C*H), 135.2 (*C*H), 141.6 (*C*). HRMS-ESI (*m*/*z*): [M + Na]$^+$ calcd for $C_{16}H_{22}ONa$, 253.1574; found, 253.1562. $[\alpha]_D^{23}$ −34.1 (*c* 1.1, $CHCl_3$, 94% ee). Daicel CHIRALPAK® OD-3, 2-PrOH/hexane = 5/95, 0.50 mL/min, 40 °C, (*S*,*R*) isomer: t_R = 10.6 min., (*R*,*S*) isomer: t_R = 12.8 min.

(*S*)-2-(Cyclohept-2-en-1-yl)-4,4,5,5-tetramethyl-1,3,2-dioxaborolane [(*S*)-3 k, Fig. 2.59] [73]

The reaction was conducted with 85.2 mg (0.50 mmol) of **1k** by using procedure A. The product (*S*)-**3k** was obtained as a colorless oil in 72% yield (79.7 mg). The enantioselectivity was determined by the following stereospecific derivatization of the boryl group.

^{1}H NMR (400 MHz, $CDCl_3$, δ): 1.25 (s, 12H), 1.39–1.58 (m, 2H), 1.63–1.84 (m, 4H), 1.91–2.01 (m, 1H), 2.04–2.18 (m, 2H), 5.75–5.88 (m, 2H). ^{13}C NMR (100 MHz, $CDCl_3$, δ): 24.7 (CH_3), 24.8 (CH_3), 25.9 (br, B–CH), 27.3 (CH_2), 28.80 (CH_2), 28.85 (CH_2), 31.3 (CH_2), 83.1 (*C*), 131.5 (*C*H), 133.0 (*C*H). HRMS-EI (*m*/*z*): [M]$^+$ calcd for $C_{13}H_{23}{}^{11}BO_2$, 222.1791; found, 222.1788. $[\alpha]_D^{23}$ −51.4 (*c* 1.0, $CHCl_3$).

The allylboration reaction was conducted with 37.0 mg (0.17 mmol) of (*S*)-**3k**. The product (*S*,*R*)-**S1k** was obtained as a colorless oil in 75% yield (25.5 mg) with 68% ee (Fig. 2.60).

^{1}H NMR (400 MHz, $CDCl_3$, δ): 1.24–1.41 (m, 2H), 1.48–1.61 (m, 1H), 1.63–1.73 (m, 1H), 1.84 (d, *J* = 3.6 Hz, 1H), 1.85–1.93 (m, 1H), 1.94–2.03 (m, 1H), 2.04–2.21 (m, 2H), 2.62–2.71 (m, 1H), 4.63–4.70 (m, 1H), 5.50–5.58 (m, 1H), 5.74–5.83 (m, 1H), 7.24–7.30 (m, 1H), 7.32–7.38 (m, 4H). ^{13}C NMR (100 MHz, $CDCl_3$, δ): 26.7

Fig. 2.59 Scheme of (*S*)-**3k**

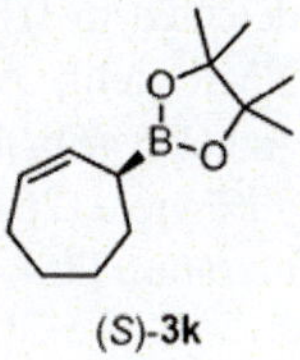

(S)-**3k**

Fig. 2.60 Scheme of (*S*,*R*)-**S1k**

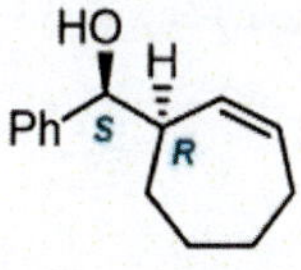

(S,R)-**S1k**

(*C*H_2), 28.1 (*C*H_2), 28.6 (*C*H_2), 30.0 (*C*H_2), 46.9 (*C*H), 77.4 (*C*H), 126.7 (*C*H), 127.4 (*C*H), 128.2 (*C*H), 132.4 (*C*H), 133.5 (*C*H), 143.2 (*C*). HRMS-ESI (*m*/*z*): $[M + Na]^+$ calcd for $C_{14}H_{18}ONa$, 225.1255; found, 225.1266. $[\alpha]_D^{23}$ −14.4 (*c* 1.0, $CHCl_3$, 68% ee). Daicel CHIRALPAK® OD-3, 2-PrOH/hexane = 5/95, 0.50 mL/min, 40 °C, (*S*,*R*) isomer: t_R = 14.1 min., (*R*,*S*) isomer: t_R = 15.8 min.

(*S*,*E*)-7-Methyl-1-phenyloct-4-en-3-ol [(*S*,*E*)-S2a, Fig. 2.61]

The reaction was conducted with 138.1 mg (0.50 mmol) of *rac*-(*Z*)-**4a** by using procedure E. The borylation product (*S*,*E*)-**5a** and the carbamate (*S*,*Z*)-**4a** were detected in ^{1}H NMR analysis [(*S*,*E*)-**5a**: 42% ^{1}H NMR yield; (*S*,*Z*)-**4a**: 58% ^{1}H NMR yield]. After the oxidation of the roughly purified material including (*S*,*Z*)-**4a** and (*S*,*E*)-**5a**, the corresponding alcohol (*S*,*E*)-**S2a** was obtained as a colorless oil in 33% yield (36.1 mg) with 86% ee.

^{1}H NMR (400 MHz, $CDCl_3$, δ): 0.89 (d, *J* = 1.8 Hz, 3H), 0.90 (d, *J* = 1.8 Hz, 3H), 1.43 (d, *J* = 3.6 Hz, 1H), 1.63 (nonet, *J* = 6.8 Hz, 1H), 1.75–2.00 (m, 4H), 2.62–2.77 (m, 2H), 4.09 (dq, *J* = 6.5, 5.5 Hz, 1H), 5.44–5.54 (m, 1H), 5.59–5.70 (m, 1H), 7.15–7.23 (m, 3H), 7.24–7.39 (m, 2H). ^{13}C NMR (100 MHz, $CDCl_3$, δ): 22.2 (*C*H_3), 22.3 (*C*H_3), 28.2 (*C*H), 31.8 (*C*H_2), 38.8 (*C*H_2), 41.5 (*C*H_2), 72.4 (*C*H), 125.7 (*C*H), 128.3 (*C*H), 128.4 (*C*H), 131.2 (*C*H), 133.9 (*C*H), 142.0 (*C*). HRMS-ESI (*m*/*z*): $[M + Na]^+$ calcd for $C_{15}H_{22}ONa$, 241.1563; found, 241.1563. $[\alpha]_D^{27}$ −5.9 (*c* 1.0, $CHCl_3$, 86% ee). Daicel CHIRALPAK® OD-3, 2-PrOH/hexane = 3/97, 0.50 mL/min, 40 °C, (*S*,*E*) isomer: t_R = 19.8 min., (*R*,*E*) isomer: t_R = 28.0 min.

After oxidation of the mixture of (*S*,*Z*)-**4a** and (*S*,*E*)-**5a**, the substrate (*S*,*Z*)-**4a** was recovered in 49% (67.3 mg) with 63% ee (Fig. 2.62). $[\alpha]_D^{27}$ +18.7 (*c* 1.1, $CHCl_3$, 63% ee). Daicel CHIRALPAK® OD-3 + OD-3, 2-PrOH/hexane = 0.25/99.75, 0.50 mL/min, 40 °C, (*S*,*Z*) isomer: t_R = 31.7 min., (*R*,*Z*) isomer: t_R = 38.2 min.

(*S*,*E*)-6-Phenylhex-3-en-2-ol [(*S*,*E*)-S2b, Fig. 2.63] [74]

The reaction was conducted with 116.6 mg (0.50 mmol) of *rac*-(*Z*)-**4b** by using procedure E. The borylation product (*S*,*E*)-**5b** and the carbamate (*S*,*Z*)-**4b** were detected in ^{1}H NMR analysis [(*S*,*E*)-**5b**: 45% ^{1}H NMR yield; (*S*,*Z*)-**4b**: 51% ^{1}H NMR yield]. After the oxidation of the roughly purified material including (*S*,*Z*)-**4b** and (*S*,*E*)-**5b**, the corresponding alcohol (*S*,*E*)-**S2b** was obtained as a colorless oil in 30% yield (26.0 mg) with 89% ee. The absolute configuration was confirmed by the literature [75].

Fig. 2.61 Scheme of (*S*,*E*)-**5a** and (*S*,*E*)-**S2a**

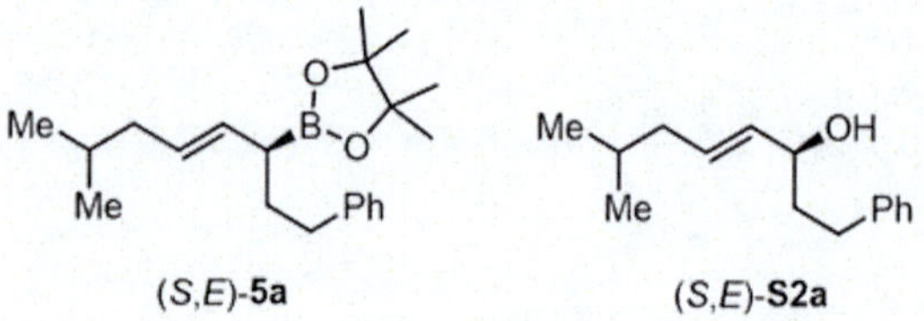

Fig. 2.62 Scheme of (*S*,*Z*)-**4a**

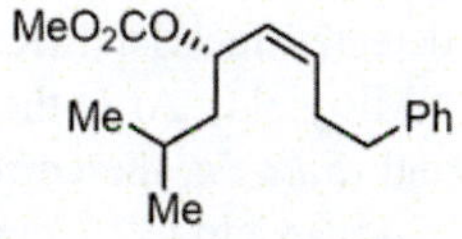

(*S*,*Z*)-**4a**

Fig. 2.63 Scheme of (*S*,*E*)-**5b** and (*S*,*E*)-**S2b**

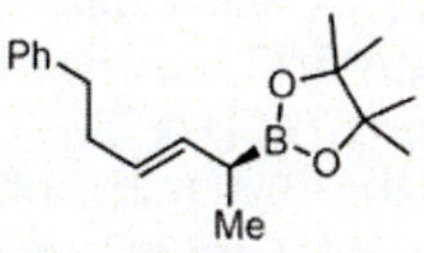

(*S*,*E*)-**5b**

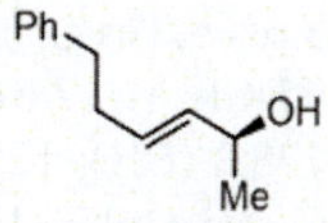

(*S*,*E*)-**S2b**

^{1}H NMR (400 MHz, $CDCl_3$, δ): 1.24 (d, *J* = 6.3 Hz, 3H), 1.38 (s, 1H), 2.34 (q, *J* = 7.3 Hz, 2H), 2.69 (t, *J* = 7.9 Hz, 2H), 4.21–4.31 (m, 1H), 5.49–5.58 (m, 1H), 5.62–5.73 (m, 1H), 7.13–7.23 (m, 3H), 7.24–7.34 (m, 2H). ^{13}C NMR (100 MHz, $CDCl_3$, δ): 23.3 (CH_3), 33.9 (CH_2), 35.5 (CH_2), 68.8 (*C*H), 125.8 (*C*H), 128.3 (*C*H), 128.4 (*C*H), 129.9 (*C*H), 134.8 (*C*H), 141.7 (*C*). HRMS-EI (*m*/*z*): $[M]^+$ calcd for $C_{12}H_{16}O$, 176.1201; found, 176.1202. $[\alpha]_D^{27}$ −7.5 (*c* 0.9, $CHCl_3$, 89% ee). Daicel CHIRALPAK® IA-3, 2-PrOH/hexane = 3/97, 0.75 mL/min, 40 °C, (*S*,*E*) isomer: t_R = 23.9 min., (*R*,*E*) isomer: t_R = 25.5 min.

After oxidation of the mixture of (*S*,*Z*)-**4b** and (*S*,*E*)-**5b**, the substrate (*S*,*Z*)-**4b** was recovered in 45% (52.8 mg) with 74% ee (Fig. 2.64). $[\alpha]_D^{27}$ −5.9 (*c* 1.2, $CHCl_3$, 74% ee). Daicel CHIRALPAK® OD-3 + OD-3, 2-PrOH/hexane = 0.25/99.75, 0.50 mL/min, 40 °C, (*S*,*Z*) isomer: t_R = 64.5 min., (*R*,*Z*) isomer: t_R = 51.9 min.

(*S*,*E*)-1-Phenylhex-4-en-3-ol [(*S*,*E*)-S2c, Fig. 2.65] [75]

The reaction was conducted with 116.9 mg (0.50 mmol) of *rac*-(*Z*)-**4c** by using procedure E. The borylation product (*S*,*E*)-**5c** and the carbamate (*S*,*Z*)-**4c** were

Fig. 2.64 Scheme of (*S*,*Z*)-**4b**

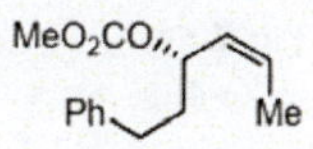

(*S*,*Z*)-**4b**

Fig. 2.65 Scheme of (*S*,*E*)-**5c** and (*S*,*E*)-**S2c**

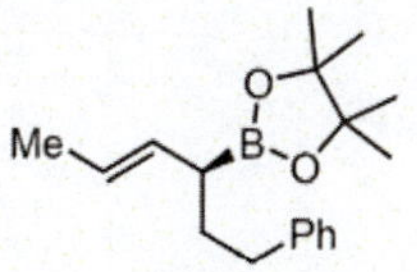

(*S*,*E*)-**5c**

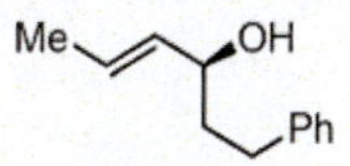

(*S*,*E*)-**S2c**

detected in ^{1}H NMR analysis [(*S*,*E*)-**5c**: 51% ^{1}H NMR yield; (*S*,*Z*)-**4c**: 44% ^{1}H NMR yield]. After the oxidation of the roughly purified material including (*S*,*Z*)-**4c** and (*S*,*E*)-**5c**, the corresponding alcohol (*S*,*E*)-**S2c** was obtained as a colorless oil in 45% yield (39.9 mg) with 83% ee. The absolute configuration was confirmed by the literature [74].

^{1}H NMR (400 MHz, $CDCl_3$, δ): 1.43 (d, $J = 4.0$ Hz, 1H), 1.71 (dd, $J = 7.2, 0.9$ Hz, 3H), 1.75–1.93 (m, 2H), 2.61–2.77 (m, 2H), 4.02–4.12 (m, 1H), 5.48–5.57 (m, 1H), 5.62–5.73 (m, 1H), 7.14–7.23 (m, 3H), 7.24–7.33 (m, 2H). ^{13}C NMR (100 MHz, $CDCl_3$, δ): 17.6 (CH_3), 31.7 (CH_2), 38.7 (CH_2), 72.3 (*C*H), 125.7 (*C*H), 127.1 (*C*H), 128.3 (*C*H), 128.4 (*C*H), 134.0 (*C*H), 142.0 (*C*). HRMS-EI (*m/z*): $[M]^+$ calcd for $C_{12}H_{16}O$, 176.1201; found, 176.1207. $[\alpha]_D^{28}$ −15.6 (*c* 1.2, $CHCl_3$, 83% ee). Daicel CHIRALPAK® OZ-3, 2-PrOH/hexane = 3/97, 0.50 mL/min, 40 °C, (*S*,*E*) isomer: $t_R = 15.6$ min., (*R*,*E*) isomer: $t_R = 14.5$ min.

After oxidation of the mixture of (*S*,*Z*)-**4c** and (*S*,*E*)-**5c**, the substrate (*S*,*Z*)-**4c** was recovered in 38% (44.8 mg) with 92% ee (Fig. 2.66). $[\alpha]_D^{28}$ +36.2 (*c* 1.2, $CHCl_3$, 92% ee). Daicel CHIRALPAK® OD-3 + OD-3, 2-PrOH/hexane = 0.25/99.75, 0.50 mL/min, 40 °C, (*S*,*Z*) isomer: $t_R = 51.1$ min., (*R*,*Z*) isomer: $t_R = 49.4$ min. After deprotection of the carbonate group, the absolute configuration was confirmed by comparing the optical rotation of the corresponding alcohol with that in the literature [76].

(*S*,*E*)-1-Phenylnon-3-en-5-ol [(*S*,*E*)-S2d, Fig. 2.67]

The reaction was conducted with 138.4 mg (0.50 mmol) of *rac*-(*Z*)-**4d** by using procedure E. The borylation product (*S*,*E*)-**5d** and the carbamate (*S*,*Z*)-**4d** were detected in ^{1}H NMR analysis [(*S*,*E*)-**5d**: 45% ^{1}H NMR yield; (*S*,*Z*)-**4d**: 45% ^{1}H NMR yield]. After the oxidation of the roughly purified material including (*S*,*Z*)-**4d** and (*S*,*E*)-**5d**, the corresponding alcohol (*S*,*E*)-**S2d** was obtained as a colorless oil in 38% yield (41.6 mg) with 88% ee.

^{1}H NMR (400 MHz, $CDCl_3$, δ): 0.90 (t, $J = 7.0$ Hz, 3H), 1.17–1.37 (m, 5H), 1.39–1.59 (m, 2H), 2.36 (q, $J = 7.3$ Hz, 2H), 2.70 (t, $J = 7.9$ Hz, 2H), 4.02 (qd, $J = 6.6, 3.3$ Hz, 1H), 5.47 (ddt, $J = 15.3, 7.1, 1.3$ Hz, 1H), 5.67 (dt, $J = 15.3$, 6.8 Hz, 1H), 7.14–7.21 (m, 3H), 7.24–7.31 (m, 2H). ^{13}C NMR (100 MHz, $CDCl_3$,

Fig. 2.66 Scheme of (*S*,*Z*)-**4c**

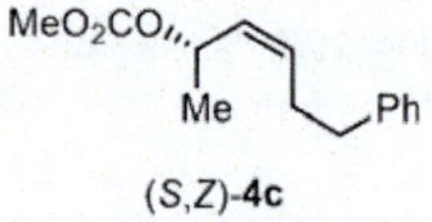

Fig. 2.67 Scheme of (*S*,*E*)-**5d** and (*S*,*E*)-**S2d**

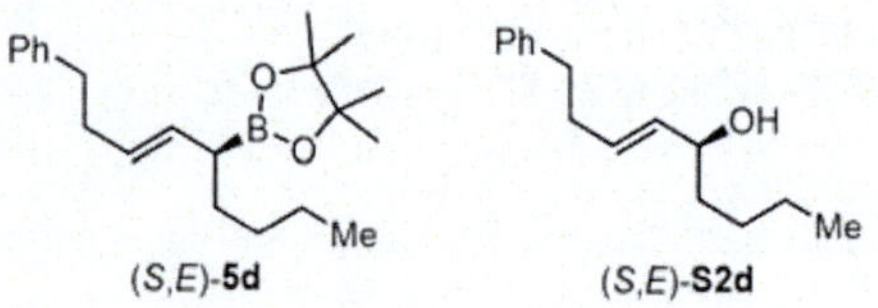

δ): 14.0 (CH_3), 22.6 (CH_2), 27.5 (CH_2), 33.9 (CH_2), 35.6 (CH_2), 36.9 (CH_2), 73.0 (*C*H), 125.8 (*C*H), 128.2 (*C*H), 128.4 (*C*H), 130.9 (*C*H), 133.8 (*C*H), 141.6 (*C*). HRMS-FD (*m*/*z*): $[M]^+$ calcd for $C_{15}H_{22}O$, 218.1671; found, 218.1679. $[\alpha]_D^{25}$ −23.9 (*c* 1.0, $CHCl_3$, 88% ee). Daicel CHIRALPAK® IA-3, 2-PrOH/hexane = 3/97, 0.50 mL/min, 40 °C, (*S*,*E*) isomer: t_R = 23.1 min., (*R*,*E*) isomer: t_R = 25.3 min.

After oxidation of the mixture of (*S*,*Z*)-**4d** and (*S*,*E*)-**5d**, the substrate (*S*,*Z*)-**4d** was recovered in 38% (53.0 mg) with 81% ee (Fig. 2.68). $[\alpha]_D^{25}$ +9.7 (*c* 1.1, $CHCl_3$, 81% ee). Daicel CHIRALPAK® OD-3 + OD-3, 2-PrOH/hexane = 0.25/99.75, 0.50 mL/min, 40 °C, (*S*,*Z*) isomer: t_R = 45.4 min., (*R*,*Z*) isomer: t_R = 37.0 min.

(*S*,*E*)-6-Methyl-1-phenylhept-4-en-3-ol [(*S*,*E*)-S2e, Fig. 2.69]

The reaction was conducted with 131.3 mg (0.50 mmol) of *rac*-(*Z*)-**4e** by using procedure E. The borylation product (*S*,*E*)-**5e** and the carbamate (*S*,*Z*)-**4e** were detected in ^{1}H NMR analysis [(*S*,*E*)-**5e**: 36% ^{1}H NMR yield; (*S*,*Z*)-**4e**: 51% ^{1}H NMR yield]. After the oxidation of the roughly purified material including (*S*,*Z*)-**4e** and (*S*,*E*)-**5e**, the corresponding alcohol (*S*,*E*)-**S2e** was obtained as a colorless oil in 27% yield (28.1 mg) with 88% ee.

^{1}H NMR (400 MHz, $CDCl_3$, δ): 0.99 (s, 3H), 1.00 (s, 3H), 1.43 (d, *J* = 4.1 Hz, 1H), 1.75–1.94 (m, 2H), 2.30 (octet, *J* = 6.6 Hz, 1H), 2.61–2.76 (m, 2H), 4.07 (qd, *J* = 6.5, 4.0 Hz, 1H), 5.44 (dd, *J* = 15.4, 7.2 Hz, 1H), 5.63 (dd, *J* = 15.4, 6.3 Hz, 1H), 7.15–7.23 (m, 3H), 7.24–7.32 (m, 2H). ^{13}C NMR (100 MHz, $CDCl_3$, δ): 22.26 (CH_3), 22.30 (CH_3), 30.7 (*C*H), 31.8 (CH_2), 38.8 (CH_2), 72.5 (*C*H), 125.7 (*C*H), 128.3 (*C*H), 128.4 (*C*H), 129.7 (*C*H), 139.5 (*C*H), 142.0 (*C*). HRMS-EI (*m*/*z*): $[M]^+$ calcd for $C_{14}H_{20}O$, 204.1514; found, 204.1517. $[\alpha]_D^{27}$ −6.5 (*c* 1.0, $CHCl_3$, 88% ee). Daicel CHIRALPAK® ID-3, 2-PrOH/hexane = 3/97, 0.50 mL/min, 40 °C, (*S*,*E*) isomer: t_R = 18.3 min., (*R*,*E*) isomer: t_R = 17.6 min.

After oxidation of the mixture of (*S*,*Z*)-**4e** and (*S*,*E*)-**5e**, the substrate (*S*,*Z*)-**4e** was recovered in 47% (23.7 mg) with 55% ee (Fig. 2.70). $[\alpha]_D^{26}$ +25.9 (*c* 0.9, $CHCl_3$, 55% ee). Daicel CHIRALPAK® OD-3 + OD-3, 2-PrOH/hexane = 0.25/99.75, 0.50 mL/min, 40 °C, (*S*,*Z*) isomer: t_R = 29.6 min., (*R*,*Z*) isomer: t_R = 36.0 min.

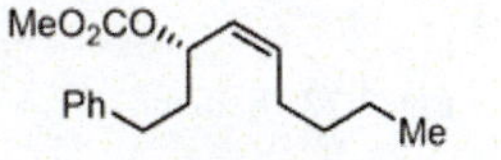

(*S*,*Z*)-**4d**
E/*Z* < 5:95

Fig. 2.68 Scheme of (*S*,*Z*)-**4d**

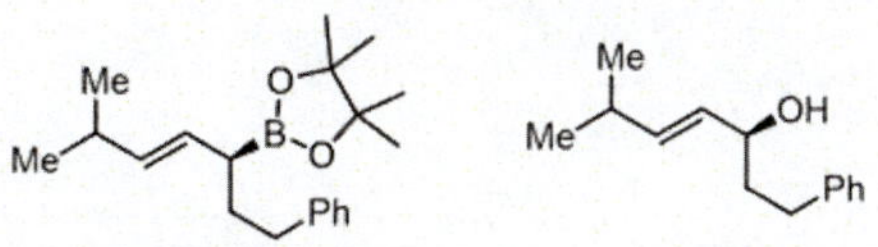

(*S*,*E*)-**5e** (*S*,*E*)-**S2e**

Fig. 2.69 Scheme of (*S*,*E*)-**5e** and (*S*,*E*)-**S2e**

Fig. 2.70 Scheme of (*S*,*Z*)-**4e**

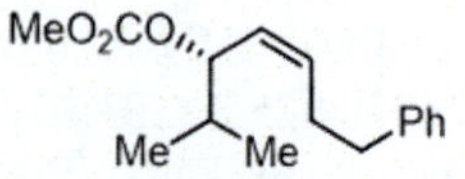

(*S*,*Z*)-**4e**
E/*Z* < 5:95

(*S*,*E*)-1-(Methoxymethoxy)-8-phenyloct-5-en-4-ol [(*S*,*E*)-S2f, Fig. 2.71]

The reaction was conducted with 161.3 mg (0.50 mmol) of *rac*-(*Z*)-**4f** by using procedure E. The borylation product (*S*,*E*)-**5f** and the carbamate (*S*,*Z*)-**4f** were detected in ^{1}H NMR analysis [(*S*,*E*)-**5f**: 42% ^{1}H NMR yield; (*S*,*Z*)-**4f**: 57% ^{1}H NMR yield]. After the oxidation of the roughly purified material including (*S*,*Z*)-**4f** and (*S*,*E*)-**5f**, the corresponding alcohol (*S*,*E*)-**S2f** was obtained as a colorless oil with a small amount of pinacol in 34% yield (44.9 mg) with 90% ee.

^{1}H NMR (400 MHz, $CDCl_3$, δ): 1.52–1.70 (m, 4H), 1.81 (d, $J = 3.6$ Hz, 1H), 2.36 (q, $J = 7.3$ Hz, 2H), 2.70 (t, $J = 7.9$ Hz, 2H), 3.32–3.40 (m, 3H), 3.48–3.58 (m, 2H), 4.04–4.11 and 4.43–4.48 (a pair of m, 1H), 4.63 (s, 2H), 5.48 (dd, J = 15.5, 7.0 Hz, 1H), 5.69 (dt, J = 15.1, 6.8 Hz, 1H), 7.12–7.22 (m, 3H), 7.24–7.33 (m, 2H). ^{13}C NMR (100 MHz, $CDCl_3$, δ): 25.7 (CH_2), 33.9 (CH_2), 34.1 (CH_2), 35.5 (CH_2), 55.1 (CH_3), 67.6 (CH_2), 72.5 (CH), 96.3 (CH_2), 125.8 (CH), 128.2 (CH), 128.4 (CH), 130.9 (CH), 133.5 (CH), 141.6 (C). The NMR charts contain inseparable pinacol impurities. HRMS-FD (*m/z*): $[M]^+$ calcd for $C_{16}H_{24}O_3$, 264.1725; found, 264.1733. $[\alpha]_D^{26}$ −2.8 (*c* 0.9, $CHCl_3$, 90% ee). Daicel CHIRALPAK® IA-3, 2-PrOH/hexane = 5/95, 0.50 mL/min, 40 °C, (*S*,*E*) isomer: $t_R = 33.4$ min., (*R*,*E*) isomer: $t_R = 39.8$ min.

After oxidation of the mixture of (*S*,*Z*)-**4f** and (*S*,*E*)-**5f**, the substrate (*S*,*Z*)-**4f** was recovered in 52% (84.1 mg) with 69% ee (Fig. 2.72). $[\alpha]_D^{26}$ −4.4 (*c* 1.0, $CHCl_3$, 69% ee). Daicel CHIRALPAK® IB-3, 2-PrOH/hexane = 1/99, 0.50 mL/min, 40 °C, (*S*,*E*) isomer: t_R = 30.7 min., (*R*,*E*) isomer: t_R = 29.6 min.

Fig. 2.71 Scheme of (*S*,*E*)-**5f** and (*S*,*E*)-**S2f**

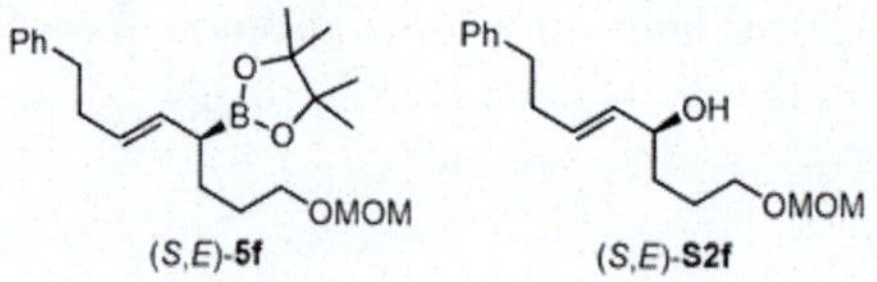

(*S*,*E*)-**5f** (*S*,*E*)-**S2f**

Fig. 2.72 Scheme of (*S*,*Z*)-**4f**

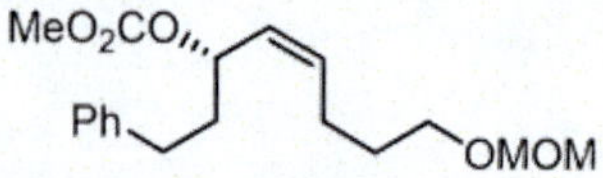

(*S*,*Z*)-**4f**
E/*Z* < 5:95

2.4.5 One-Pot Borylation/Allylboration Procedure

[Cu(MeCN)$_4$]BF$_4$ (7.9 mg, 0.025 mmol) and bis(pinacolato)diboron (**2**) (253.9 mg, 1.00 mmol), (*R*,*R*)-5,8-TMS-QuinoxP* (12.0 mg, 0.025 mmol) were placed in an oven-dried reaction vial in an argon-filled glove box (Fig. 2.73). After the vial was sealed with a screw cap containing a Teflon®-coated rubber septum, the vial was taken from the glove box. Dry THF (0.25 mL) and K(O-*t*-Bu)/THF (1.0 M, 0.25 mL, 0.25 mmol) were added in the vial through the rubber septum using a syringe. After stirring for 15 min, **1c** (128.1 mg, 0.50 mmol) was added to the mixture at 50 °C. After the reaction was completed, piperonal (225.2 mg, 1.50 mmol) solution in THF (0.50 mL) was added to the reaction mixture. After the mixture was stirred additionally for 24 h at the same temperature, the reaction mixture was passed through a short silica gel column (*Φ*: 10 mm, the height of the silica gel column: ca. 30 mm) eluting with ethyl acetate/hexane 50:50. The crude material was purified by flash column chromatography (SiO$_2$, ethyl acetate/hexane, typically 5:95–20:80) to give the corresponding alcohol product (*S*,*S*)-**6** (144.5 mg, 0.433 mmol, 87%) as a colorless oil. The enantioselectivity was determined by HPLC (87% ee).

^{1}H NMR (392 MHz, CDCl$_3$, δ): 1.50 (s, 9H), 2.36–2.46 (br-m, 1H), 3.09–3.47 (br-m, 1H), 3.75 (br-dd, *J* = 18.6, 2.5 Hz, 1H), 3.88–4.43 (br-m, 2H), 4.13 (br-d, *J* = 4.7 Hz, 1H), 5.23–5.52 (br-m, 1H), 5.55–5.79 (br-m, 1H), 5.95 (s, 2H), 6.77 (s, 2H), 6.87 (s, 1H). ^{13}C NMR (99 MHz, CDCl$_3$, δ): 28.3 (*C*H$_3$), 41.5 (br, *C*H$_2$), 43.1 (*C*H), 44.3 (br, *C*H$_2$), 74.9 (*C*H), 80.0 (*C*), 100.8 (*C*H$_2$), 106.9 (*C*H), 107.8 (*C*H), 120.0 (*C*H), 125.6 and 126.1 (a pair of *C*H), 126.1 and 126.9 (a pair of *C*H), 136.2 (*C*), 146.8 (*C*), 147.5 (*C*), 154.8 and 155.7 (a pair of *C*). HRMS-ESI (*m*/*z*): [M + Na]$^+$ calcd for C$_{18}$H$_{23}$O$_5$NNa, 356.1468; found, 356.1471. $[\alpha]_D^{28}$ +122.7 (*c* 1.2, CHCl$_3$, 87% ee). Daicel CHIRALPAK® OD-3, 2-PrOH/hexane = 5/95, 0.50 mL/min, 40 °C, (*S*,*S*) isomer: t_R = 32.5 min., (*R*,*R*) isomer: t_R = 27.3 min.

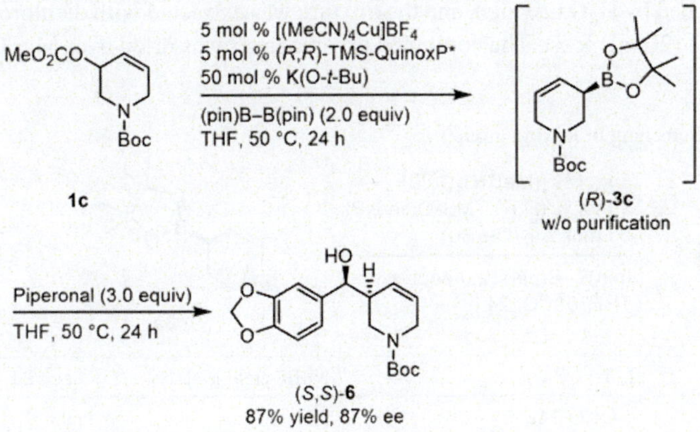

Fig. 2.73 Scheme of preparation procedure of (*S*,*S*)-**6**

2.4.6 Leaving Group Effect

The leaving group effect on the reactivity and enantioselectivity of the direct enantioconvergent borylation reaction was investigated (Table 2.4). Methyl carbonate group showed high reactivity and enantioselectivity (entry 1). Similarly, Cbz group also showed good reactivity and almost the same enantioselectivity (entry 2). In contrast, the reactivity was dropped when Boc group was used for the leaving group, while the enantioselectivity retained (entry 3).

2.4.7 Large-Scale Borylation Reaction

The direct enantioconvergent borylation of allyl carbonate **1a** was conducted in large-scale (Fig. 2.74). The enantioselectivity remained almost the same as the standard conditions; however, the substrate conversion and the yield were decreased probably because the reduced efficiency of carbon dioxide evolution would affect the reactivity.

2.4.8 Deuterium Labeling Study

Preparation of cyclohex-2-en-1-yl-1-*d* methyl carbonate (1a-D, Fig. 2.75) [77]

In a vacuum-dried 100 mL two-neck round-bottomed flask, cyclohex-2-en-1-one (0.97 mL, 10.0 mmol) was dissolved in dry MeOH (30 mL), and the mixture was cooled to 0 °C under a nitrogen atmosphere. Cerium(III) chloride heptahydrate (3.73 g, 10.0 mmol) and deuterated sodium borohydride (418.8 mg, 10.0 mmol, 98%-D) were added to the mixture, and the mixture was stirred for 1 h. The reaction mixture was quenched by H_2O (20 mL), and the mixture was extracted with dichloromethane three times (20 mL × 3). The combined organic layer was dried over $MgSO_4$. After

Table 2.4 Screening of leaving group

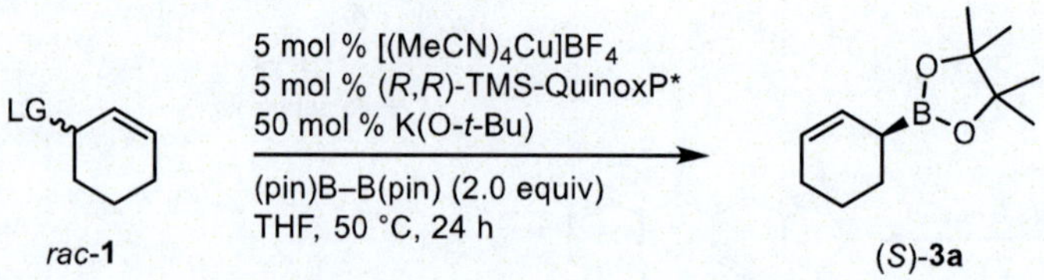

Entry	LG	NMR yield [%]	ee [%]
1	$-OCO_2Me$	92	93
2	$-OCO_2Bn$ (Cbz)	92	94
3	$-OCO_2(t\text{-}Bu)$ (Boc)	82	93

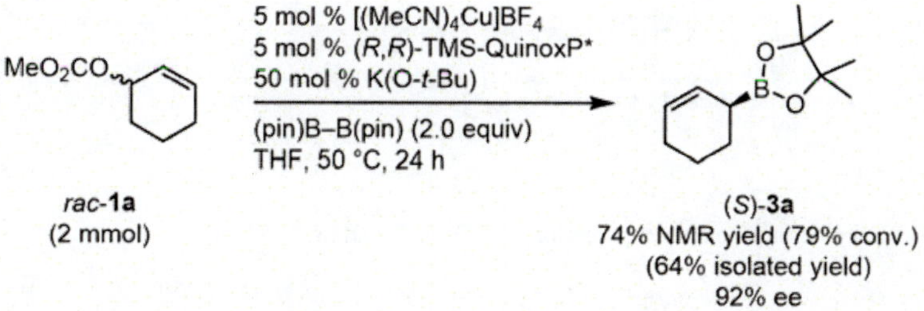

Fig. 2.74 Large-scale borylation reaction

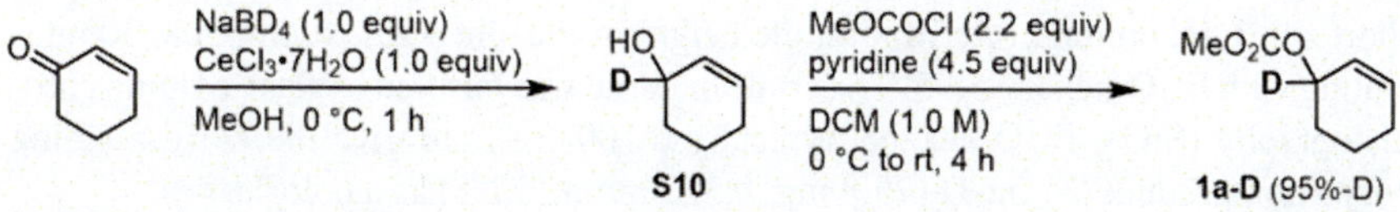

Fig. 2.75 Scheme of preparation procedure of **1a-D**

filtration, the solvents were removed by evaporation. The crude product was purified by flash column chromatography to obtain cyclohex-2-en-1-*d*-1-ol (**S10**) (708.3 mg, 7.1 mmol, 71%) as a colorless oil.

In a vacuum-dried 30 mL two-neck round-bottomed flask, cyclohex-2-en-1-*d*-1-ol (**S10**) (708.3 mg, 7.1 mmol) and pyridine (2.59 mL, 32.1 mmol) were dissolved in dry DCM (7 mL), and the mixture was cooled to 0 °C under a nitrogen atmosphere. Methyl chloroformate (1.21 mL, 15.7 mmol) was added to the mixture dropwise, and the mixture was warmed to room temperature. After the mixture was stirred for 4 h, the reaction mixture was quenched by H_2O (7 mL) and extracted with hexane three times. (7 mL × 3) The combined organic layer was then dried over $MgSO_4$. After filtration, the solvents were removed by evaporation. The crude product was purified by flash column chromatography to obtain cyclohex-2-en-1-yl-1-*d* methyl carbonate (**1a-D**) (643.6 mg, 4.1 mmol, 57%, 95%-D) as a colorless oil.

^{1}H NMR (392 MHz, $CDCl_3$, δ): 1.58–1.69 (m, 1H), 1.70–1.93 (m, 3H), 1.94–2.15 (m, 2H), 3.78 (s, 3H), 5.09–5.15 (m, 0.05H), 5.77 (dt, $J = 10.1, 2.0$ Hz, 1H), 5.99 (dt, $J = 10.1, 3.7$ Hz, 1H). ^{13}C NMR (99 MHz, $CDCl_3$, δ): 18.6 (CH_2), 24.9 (CH_2), 28.1 (CH_2), 54.5 (CH_3), 71.4 (t, $J = 23.0$ Hz, *C*D), 124.9 (*C*H), 133.4 (*C*H), 155.5 (*C*). HRMS-EI (*m*/*z*): $[M]^+$ calcd for $C_8H_{11}DO_3$, 157.0849; found, 157.0855.

Copper(I)-Catalyzed Direct Enantioconvergent Borylation of *rac*-1a-D (Fig. 2.76)

$[Cu(MeCN)_4]BF_4$ (7.9 mg, 0.025 mmol), bis(pinacolato)diboron (**2**) (253.9 mg, 1.00 mmol), and (*R*,*R*)-5,8-TMS-QuinoxP* (12.0 mg, 0.025 mmol) were placed in an oven-dried reaction vial in an argon-filled glove box. After the vial was sealed with a screw cap containing a Teflon®-coated rubber septum, the vial was taken from the glove box. Dry THF (0.25 mL) and K(O-*t*-Bu)/THF (1.0 M, 0.25 mL,

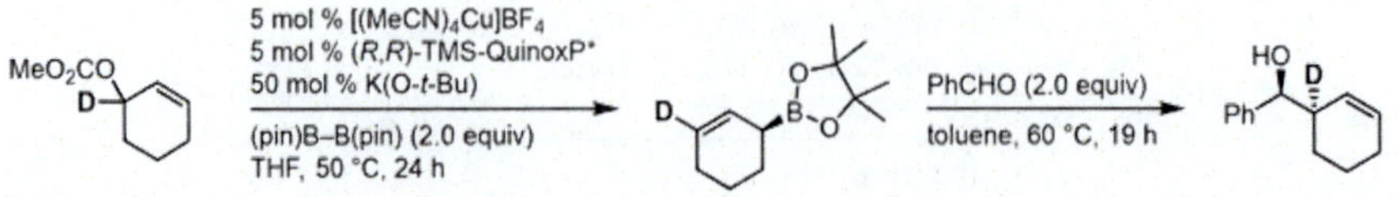

Fig. 2.76 Scheme of preparation procedure of (*S*)-**3a-D** and (*S*,*R*)-**S1a-D**

0.25 mmol) were added in the vial through the rubber septum using a syringe. After stirring for 15 min, *rac*-**1a-D** (78.4 mg, 0.50 mmol) was added to the mixture at 50 °C. After the reaction was completed, the reaction mixture was passed through a short silica gel column (*Φ*: 10 mm, the height of the silica gel column: ca. 30 mm) eluting with Et_2O/hexane 5:95. The crude material was purified by flash column chromatography (SiO_2, Et_2O/hexane, typically 0:100–3:97) to give the corresponding borylation product (*S*)-**3a-D** (79.8 mg, 0.382 mmol, 77%) as a colorless oil.

In an oven-dried reaction vial, (*S*)-**3a-D** (37.3 mg, 0.178 mmol) and benzaldehyde (36.4 μL, 0.356 mmol) were dissolved in dry toluene (350 μL) under a nitrogen atmosphere. The reaction mixture was warmed up to 60 °C and stirred for 19 h. After the reaction was complete, the reaction mixture was quenched by triethanolamine solution in dichloromethane (10 vol%, 2 mL) and H_2O (1.0 mL), and extracted with Et_2O three times (2.0 mL × 3). The combined organic layer was dried over $MgSO_4$. After filtration, the solvents were removed by evaporation. The crude product was purified by flash column chromatography to obtain (*S*)-[(*R*)-cyclohex-2-en-1-yl-1-*d*](phenyl)methanol [(*S*,*R*)-**S1a-D**] (28.2 mg, 0.149 mmol, 84%) as a colorless oil. The deuterated position and the deuteration ratio were determined by ^{1}H NMR analysis (95%-D). The enantioselectivity was determined by HPLC (93% ee).

^{1}H NMR (392 MHz, $CDCl_3$, δ): 1.42–1.59 (m, 2H), 1.63–1.82 (m, 2H), 1.85 (d, *J* = 2.7 Hz, 1H), 1.94–2.05 (m, 2H), 2.46–2.55 (m, 0.05H), 4.59 (d, *J* = 2.1 Hz, 1H), 5.38 (d, *J* = 10.2 Hz, 1H), 5.82 (dt, *J* = 10.2, 3.7 Hz, 1H), 7.23–7.31 (m, 1H), 7.32–7.40 (m, 4H).^{13}C NMR (99 MHz, $CDCl_3$, δ): 21.0 (CH_2), 23.7 (CH_2), 25.2 (CH_2), 42.4 (t, *J* = 19.7 Hz, *C*D), 77.3 (*C*H), 126.5 (*C*H), 127.3 (*C*H), 127.9 (*C*H), 128.2 (*C*H), 130.4 (*C*H), 142.8 (*C*). HRMS-ESI (*m*/*z*): $[M + Na]^+$ calcd for $C_{13}H_{15}DONa$, 212.1162; found, 212.1152. $[\alpha]_D^{28}$ +6.2 (*c* 1.0, $CHCl_3$, 93% ee). Daicel CHIRALPAK® OD-3, 2-PrOH/hexane = 3/97, 0.50 mL/min, 40 °C, (*S*,*R*) isomer: t_R = 17.5 min., (*R*,*S*) isomer: t_R = 20.4 min.

2.4.9 *Asymmetric Allylic Boryl Substitution of an Achiral Allyl Carbonate*

For the initial performance evaluation of the new ligand (*R*,*R*)-5,8-TMS-QuinoxP*, the ligand was applied to a reported asymmetric borylation reaction (Table 2.5) [78].

Table 2.5 Enantioselective allylic borylation of (*Z*)-allyl carbonate

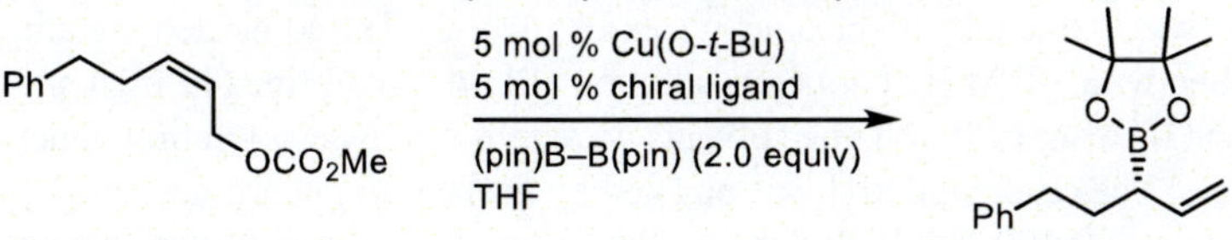

S11   (*R*)-**S12** isolated yield

Entry	Chiral ligand	Temp. [°C]	Time [h]	Yield [%]	ee [%]
1	(*R*,*R*)-QuinoxP*	0	18	57	96
2	(*R*,*R*)-5,8-TMS-QuinoxP*	0	2	81	97
3	(*R*,*R*)-5,8-TMS-QuinoxP*	−20	3	82	98

In the enantioselective allylic borylation of an achiral (*Z*)-allyl carbonate **S11**, (*R*,*R*)-5,8-TMS-QuinoxP* showed higher reactivity and enantioselectivity than the parent ligand (*R*,*R*)-QuinoxP*.

2.4.9.1 Ligand Synthesis and Characterization

(*R*,*R*)-2,3-Bis(*tert*-butyl(methyl)phosphino)-5,8-bis(trimethylsilyl)quinoxaline [(*R*,*R*)-5,8-TMS-QuinoxP*, Fig. 2.77]

The bisphosphine was synthesized according to the literature procedure [35]. A 2,3-dichloroquinoxaline derivative **S11** [44] and a phosphine borane **S12** [79] were prepared according to the literature. A hexane solution of *n*-BuLi (1.55 M, 6.0 mmol, 3.9 mL) was added to a solution of the chiral phosphine–borane complex **S12** (6.0 mmol, 707.8 mg, >99% ee) in dry THF (12 mL) at −78 °C under nitrogen atmosphere. After stirring for 15 min, the solution of the bis(trimethylsilyl) dichloroquinoxaline **S11** (2.0 mmol, 686.8 mg) in THF (12 mL) was added to the reaction mixture in one portion. The mixture was warmed up to room temperature over 1 h and stirred for additional 3 h. Then, TMEDA (3 mL) was added to the mixture. After

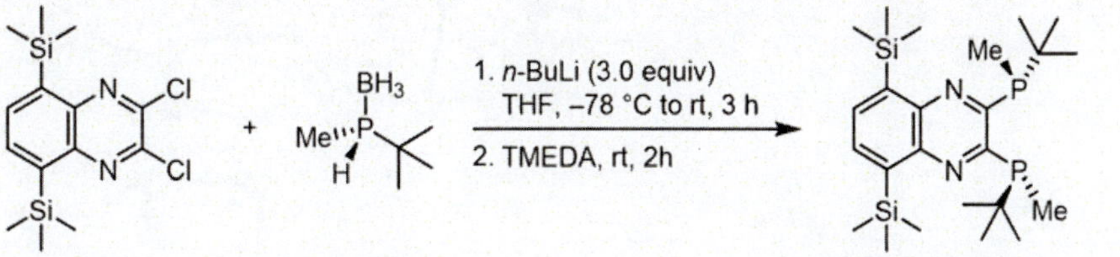

S11   **S12** (3.0 equiv)   **(*R*,*R*)-5,8-TMS-QuinoxP*** 47% yield

Fig. 2.77 Scheme of preparation procedure of (*R*,*R*)-5,8-TMS-QuinoxP*

stirring for 2 h, the reaction was quenched by 1.0 M HCl aqueous solution (24 mL) and the mixture was extracted with hexane twice (24 mL × 2). The combined organic layer was washed with 1.0 M HCl aqueous solution (12 mL) and brine (12 mL) and dried over Na_2SO_4. After filtration, the solvents were removed by evaporation under reduced pressure. The crude material was purified by flash column chromatography (SiO_2, Et_2O/hexane, 0:100–2:98). Recrystallization of the purified material with MeOH/hexane (1:1) gave the chiral bisphosphine as an orange crystal (452.6 mg, 0.945 mmol, 47% yield) [Fig. 2.78 for (*R*,*R*)-QuinoxP* and Fig. 2.79 for (*R*,*R*)-5,8-TMS-QuinoxP*, Table 2.6]. The single-crystal suitable for X-ray structural analysis was obtained by liquid–liquid diffusion (MeOH/hexane = 1:1).

^{1}H NMR (400 MHz, $CDCl_3$, δ): 0.45 (s, 18H), 0.94–1.01 (m, 18H), 1.44–1.48 (m, 6H), 7.85 (s, 2H). ^{13}C NMR (100 MHz, $CDCl_3$, δ): −0.1 (CH_3), 4.6 (*C*), 27.8–28.1 (m, CH_3), 32.3–32.5 (m, CH_3), 135.9 (*C*H), 143.3 (*C*), 145.3 (*C*), 162.6–162.8 (m, *C*). ^{31}P NMR (160 MHz, $CDCl_3$, δ): −12.9 (s). HRMS-ESI (*m*/*z*): $[M + H]^+$ calcd for $C_{24}H_{45}N_2P_2Si_2$, 479.2591; found, 479.2595. $[\alpha]_D^{27}$ +112.3 (*c* 1.0, $CHCl_3$).

(*R*,*R*)-2,3-Bis(*tert*-butyl(methyl)phosphino)-5,8-bis(triethylsilyl)quinoxaline [(*R*,*R*)-5,8-TES-QuinoxP*, Fig. 2.80]

A 2,3-dichloroquinoxaline derivative **S13** [44] and a phosphine borane **S12** [79] were prepared according to the literature. In a vacuum-dried 30 mL two-neck round-bottomed flask, the 2,3-dichloroquinoxaline derivative **S13** (2.0 mmol, 855 mg) and the chiral phosphine–borane complex **S12** (4.4 mmol, 519 mg, >99% ee) were dissolved in dry and degassed THF (4 mL) and TMEDA (10 mmol, 1.49 mL), and the mixture was cooled to −20 °C under argon atmosphere. Then, K(O-*t*-Bu)/THF (1.0 M, 4.8 mL, 4.8 mmol) was added dropwise to the mixture over 10 min. The mixture was allowed to warm to room temperature over 1.5 h to form orange precipitates and stirred overnight. The reaction was quenched by 2.0 M HCl aqueous solution (8.6 mL) at 0 °C. The resulting yellow precipitate was filtered and washed with water and MeOH. Recrystallization of the precipitate with EtOAc under argon atmosphere

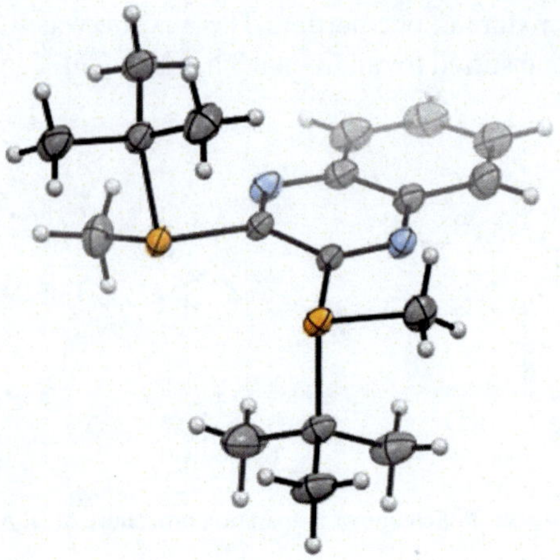

Fig. 2.78 ORTEP structure of (*R*,*R*)-QuinoxP*. Thermal ellipsoids are drawn at the 50% probability level

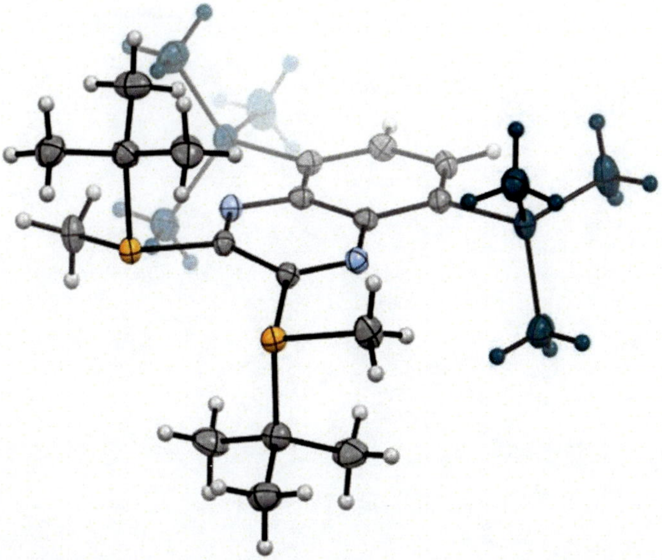

Fig. 2.79 ORTEP structure of (*R*,*R*)-5,8-TMS-QuinoxP*. Thermal ellipsoids are drawn at the 50% probability level

Table 2.6 Summary of X-ray crystallographic data

Compound	(*R*,*R*)-QuinoxP*	(*R*,*R*)-5,8-TMS-QuinoxP*
CCDC name	2,013,923	1,949,894
Empirical formula	$C_{18}H_{28}N_2P_2$	$C_{24}H_{44}N_2P_2Si_2$
Formula weight	334.36	478.73
Crystal system	Monoclinic	Monoclinic
Crystal size (mm)	0.200 × 0.200 × 0.400	0.100 × 0.100 × 0.100
a (Å)	10.1381(2)	13.7047(2)
b (Å)	17.5752(4)	10.18090(10)
c (Å)	11.1531(2)	11.7274(2)
α (°)	90.0000	90.0000
β (°)	93.175(2)	113.970(2)
γ (°)	90.0000	90.0000
V (Å^3)	1984.20(7)	1495.16(4)
Space group	$C2$ (#5)	$C2$ (#5)
Z value	4	2
Dcalc (g cm^{-3})	1.119	1.063
Temperature (K)	123	123
$2\theta_{max}$ (°)	147.7	147.062
μ (CuKα) (mm^{-1})	1.964	2.171

(continued)

Table 2.6 (continued)

Compound	(*R*,*R*)-QuinoxP*	(*R*,*R*)-5,8-TMS-QuinoxP*
μ (MoKα) (mm^{-1})	–	–
No. of reflections measured	Total: 7669 Unique: 7669 (R_{int} = N/A[a])	Total: 7908 Unique: 2784 (R_{int} = 0.0204)
No. of observations (all reflections)	7669	2784
Residuals: R_1 (I > 2.00σ (I))	0.0346	0.0316
Residuals: wR_2 (all reflections)	0.1038	0.0851
Goodness of fit indicator (GOF)	1.049	1.157
Maximum peak in final diff. map (Å^3)	0.25	0.37
Minimum peak in final diff. map (Å^3)	−0.25	−0.23
Flack parameter	0.001(15)	−0.009(16)

[a] Not available because of the twin analyses

Compound	(*R*,*R*)-5,8-TES-QuinoxP*	(*R*,*R*)-5,8-TIPS-QuinoxP*
CCDC name	2,012,612	2,012,613
Empirical formula	$C_{30}H_{56}N_2P_2Si_2$	$C_{36}H_{68}N_2P_2Si_2$
Formula weight	562.88	647.04
Crystal system	Monoclinic	Orthorhombic
Crystal size (mm)	0.400 × 0.200 × 0.200	0.200 × 0.100 × 0.100
a (Å)	14.1804(2)	8.51892(5)
b (Å)	10.19430(10)	23.66109(13)
c (Å)	12.3812(2)	10.24320(6)
α (°)	90.0000	90.0000
β (°)	112.342(2)	90.0000
γ (°)	90.0000	90.0000
V (Å^3)	1655.46(4)	2064.69(2)
Space group	*C*2 (#5)	$P2_12_12$ (#18)
Z value	2	2
*D*calc (g cm^{-3})	1.129	1.041
Temperature (K)	123	123
$2\theta_{max}$ (°)	147.458	147.466
μ (CuKα) (mm^{-1})	2.027	1.678
μ (MoKα) (mm^{-1})	–	–
No. of reflections measured	Total: 8772 Unique: 3249 (R_{int} = 0.0202)	Total: 12,846 Unique: 4091 (R_{int} = 0.0230)
No. of observations (all reflections)	3249	4091

(continued)

Table 2.6 (continued)

Compound	(*R*,*R*)-5,8-TES-QuinoxP*	(*R*,*R*)-5,8-TIPS-QuinoxP*
Residuals: R_1 ($I > 2.00\sigma$ (I))	0.0333	0.0261
Residuals: wR_2 (all reflections)	0.0928	0.0707
Goodness of fit indicator (GOF)	1.086	1.049
Maximum peak in final diff. map (Å^3)	0.32	0.19
Minimum peak in final diff. map (Å^3)	−0.29	−0.23
Flack parameter	−0.007(19)	−0.001(7)

Compound	[(*R*,*R*)-5,8-TMS-QuinoxP*]$PdCl_2$·acetone
CCDC name	1,949,893
Empirical formula	$C_{27}H_{50}Cl_2N_2OP_2PdSi_2$
Formula weight	714.11
Crystal system	Orthorhombic
Crystal size (mm)	0.200 × 0.050 × 0.050
a (Å)	8.0306(6)
b (Å)	19.5519(13)
c (Å)	22.4572(13)
α (°)	90.0000
β (°)	90.0000
γ (°)	90.0000
V (Å^3)	3526.1(4)
Space group	$P2_12_12_1$ (#19)
Z value	4
Dcalc (g cm^{-3})	1.345
Temperature (K)	123
$2\theta_{max}$ (°)	50.696
μ (CuKα) (mm^{-1})	–
μ (MoKα) (mm^{-1})	0.859
No. of reflections measured	Total: 27,383 Unique: 6460 ($R_{int} = 0.0457$)
No. of observations (all reflections)	6460
Residuals: R_1 ($I > 2.00\sigma$ (I))	0.0355
Residuals: wR_2 (all reflections)	0.0714
Goodness of fit indicator (GOF)	1.112

(continued)

Table 2.6 (continued)

Compound	[(*R*,*R*)-5,8-TMS-QuinoxP*]$PdCl_2$·acetone
Maximum peak in final diff. map (Å^3)	0.92
Minimum peak in final diff. map (Å^3)	−0.38
Flack parameter	0.013(11)

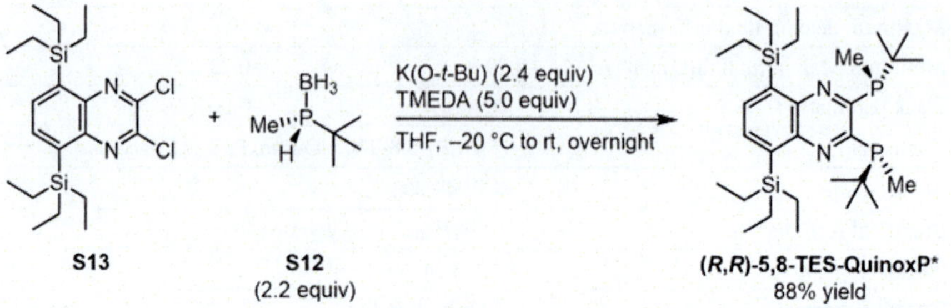

Fig. 2.80 Scheme of preparation procedure of (*R*,*R*)-5,8-TES-QuinoxP*

gave the chiral bisphosphine as an orange crystal (995 mg, 1.77 mmol, 88% yield) (Fig. 2.81, Table 2.6). The single-crystal suitable for X-ray structural analysis was obtained by slow evaporation from the solution in $CDCl_3$.

^{1}H NMR (400 MHz, $CDCl_3$, δ): 0.87–0.98 (m, 36H), 0.99–1.15 (m, 12H), 1.43–1.49 (m, 6H), 7.81 (s, 2H). ^{13}C NMR (100 MHz, $CDCl_3$, δ): 3.9 (CH_2), 4.4–4.5 (m,

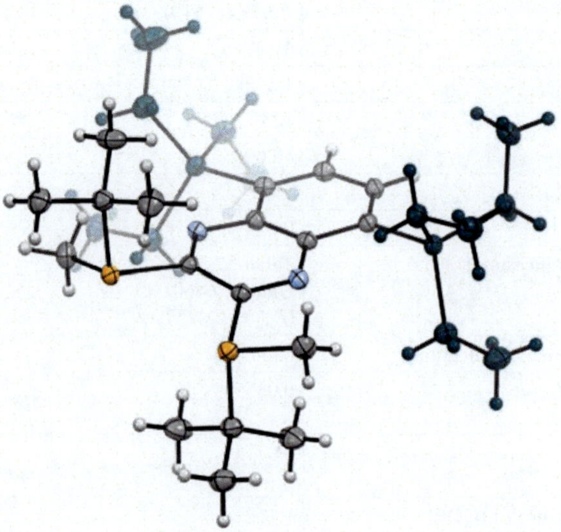

Fig. 2.81 ORTEP structure of (*R*,*R*)-5,8-TES-QuinoxP*. Thermal ellipsoids are drawn at the 50% probability level

C), 7.7 (CH_3), 27.6–28.0 (m, CH_3), 32.1–32.4 (m, CH_3), 137.1 (*C*H), 140.2 (*C*), 145.5 (*C*), 162.0–162.2 (m, *C*). ^{31}P NMR (160 MHz, $CDCl_3$, δ): −12.6 (s). HRMS-EI (*m/z*): $[M]^+$ calcd for $C_{30}H_{56}N_2P_2Si_2$, 562.3457; found, 562.3441. $[\alpha]_D^{22}$ +195.8 (*c* 1.0, $CHCl_3$).

(*R,R*)-2,3-Bis(*tert*-butyl(methyl)phosphino)-5,8-bis(triisopropylsilyl)quinoxaline [(*R,R*)-5,8-TIPS-QuinoxP*, Fig. 2.82]

(*R,R*)-5,8-TIPS-QuinoxP* was prepared from the corresponding 2,3-dichloroquinoxaline derivative according to the procedure described above. The product (*R,R*)-5,8-TIPS-QuinoxP* was obtained in 54% yield (246 mg, 0.38 mmol, yellow powder). The recrystallization was conducted with MeOH/$CHCl_3$ (1:1) (Fig. 2.83, Table 2.6). The single-crystal suitable for X-ray structural analysis was obtained by slow evaporation from the solution in $CDCl_3$.

^{1}H NMR (400 MHz, $CDCl_3$, δ): 0.82–0.89 (m, 18H), 1.08 (d, $J = 7.6$ Hz, 18H), 1.12 (d, $J = 7.6$ Hz, 18H), 1.40–1.45 (m, 6H), 1.86 (septet, $J = 7.6$ Hz, 6H), 7.86 (s, 2H). ^{13}C NMR (100 MHz, $CDCl_3$, δ): 3.9–4.1 (m, *C*), 12.8 (*C*H), 19.2 (CH_3), 19.3 (CH_3), 27.5–27.8 (m, CH_3), 32.1–32.3 (m, CH_3), 137.4 (*C*H), 138.8 (*C*), 145.6 (*C*), 160.5–160.7 (m, *C*). ^{31}P NMR (160 MHz, $CDCl_3$, δ): −11.0 (s). HRMS-ESI (*m/z*): $[M + H]^+$ calcd for $C_{36}H_{69}N_2P_2Si_2$, 647.4469; found, 647.4467. $[\alpha]_D^{23}$ +386.7 (*c* 1.0, $CHCl_3$).

[(*R,R*)-5,8-TMS-QuinoxP*]$PdCl_2$ complex (Fig. 2.84)

In order to investigate a structure of the ligand–metal complex with (*R,R*)-5,8-TMS-QuinoxP*, I attempted to prepare a [(*R,R*)-5,8-TMS-QuinoxP*]CuCl complex. However, I failed to obtain crystals which are suitable to the single-crystal X-ray diffraction analysis. Instead, a preparation of [(*R,R*)-5,8-TMS-QuinoxP*]$PdCl_2$ complex was conducted.

In the reaction vial, (COD)$PdCl_2$ complex (28.6 mg, 0.10 mmol) was dissolved in Et_2O (1.0 mL) under nitrogen atmosphere. To the reaction mixture, a solution of (*R,R*)-5,8-TMS-QuinoxP* (47.9 mg, 0.10 mmol) in Et_2O (1.0 mL) was added at room temperature. After stirred for 21 h, the mixture was filtered and washed with cooled Et_2O to give the corresponding palladium complex (53.7 mg, 0.082 mmol, 82% yield). The recrystallization was conducted with acetone (Fig. 2.85, Table 2.6). The

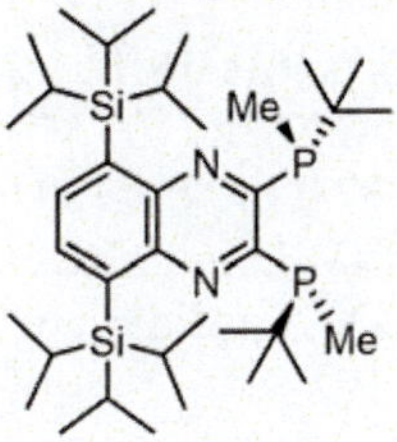

(*R,R*)-5,8-TIPS-QuinoxP*

Fig. 2.82 Scheme of (*R,R*)-5,8-TIPS-QuinoxP*

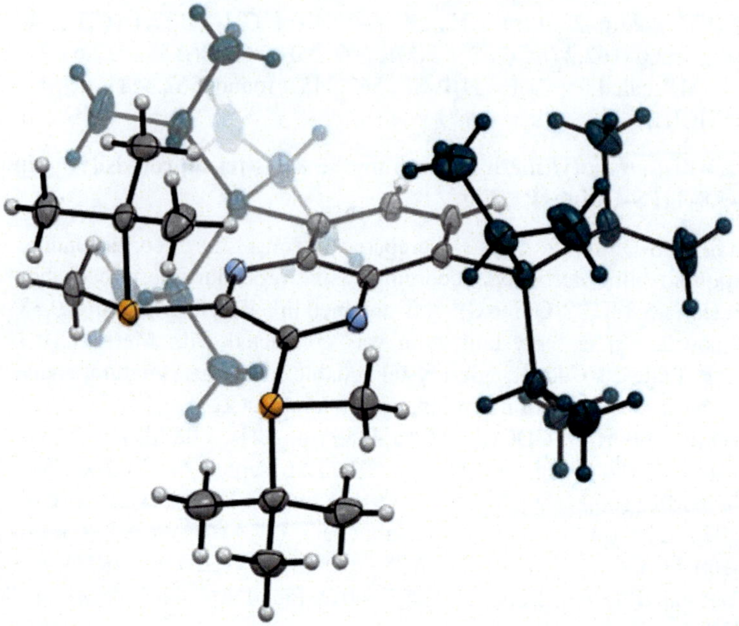

Fig. 2.83 ORTEP structure of (*R*,*R*)-5,8-TIPS-QuinoxP*. Thermal ellipsoids are drawn at the 50% probability level

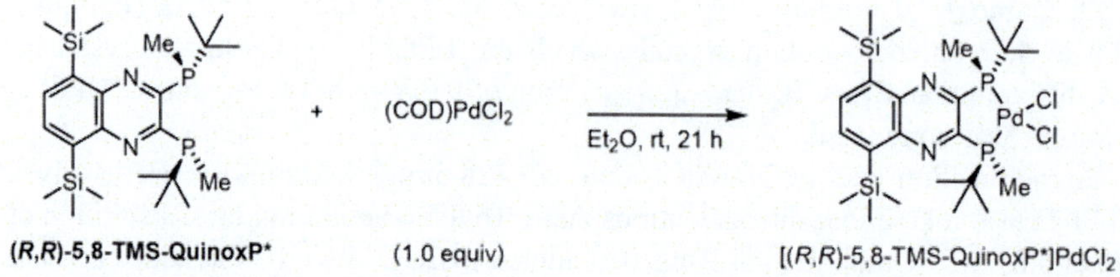

Fig. 2.84 Scheme of preparation procedure of [(*R*,*R*)-5,8-TMS-QuinoxP*]$PdCl_2$

single-crystal suitable for X-ray structural analysis was obtained by slow evaporation from the solution in acetone.

^{1}H NMR (400 MHz, $CDCl_3$, δ): 0.46 (s, 18H), 1.20 (d, $J = 16.6$ Hz, 18H), 2.28 (d, $J = 11.7$ Hz, 6H), 8.13 (s, 2H). ^{13}C NMR (100 MHz, $CDCl_3$, δ): −0.1 (CH_3), 5.5–6.1 (m, CH_3), 28.0 (CH_3), 36.5–37.0 (m, *C*), 139.6 (*C*H), 144.5 (*C*), 146.4–146.7 (m, *C*), 150.4–151.6 (m, *C*). ^{31}P NMR (160 MHz, $CDCl_3$, δ): 55.7 (s). HRMS-ESI (*m*/*z*): $[M + Na]^+$ calcd for $C_{24}H_{44}N_2NaP_2PdSi_2$, 679.0821; found, 679.0810. $[\alpha]_D^{27}$ +97.9 (*c* 1.0, $CHCl_3$).

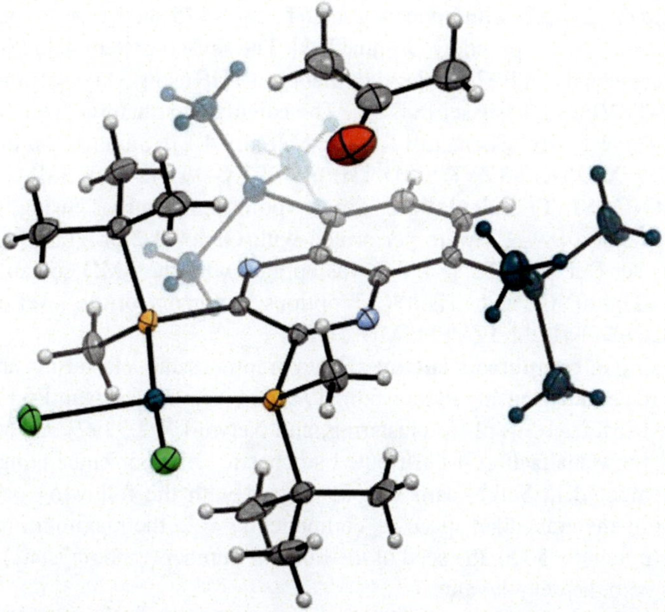

Fig. 2.85 ORTEP structure of [(*R*,*R*)-5,8-TMS-QuinoxP*]$PdCl_2$·acetone. Thermal ellipsoids are drawn at the 50% probability level

2.5 Computational Details

2.5.1 Computational Methods and Software Libraries

Quantum chemical calculations: All geometry optimizations and thermal energy correction calculations (frequency analyses) using density functional theory (DFT) were performed with Gaussian 09 (revision D.01) [80] and Gaussian 16 (revision B.01) [81] suite of programs. A substrate **1a**, bisphosphine ligands (*R*,*R*)-QuinoxP* and (*R*,*R*)-5,8-TMS-QuinoxP*, and a pinacol boronate ester were chosen for the computational model without simplification. The thresholds defined in Gaussian 09 were used for all the computations otherwise noted. The geometry optimizations were carried out at ωB97X-D [82] level of theory in THF using SMD solvation model [83] with Def2-SVP as a basis set [84, 85]. Harmonic frequency calculations were conducted at the same level of theory on the optimized geometries to check all the stationary points as either minima or first-order saddle points. Intrinsic reaction coordinate (IRC) [86] calculations were carried out to confirm the transition states connecting the correct reactants and products on the potential energy surface. The zero-point energy (ZPE) and thermal energy corrections were calculated using vibrational frequencies with *GoodVibes* python script 2.0.2 [87], employing a quasi-harmonic approximation for entropy calculation at the solution of 0.05 M and temperature of 323.15 K with frequency scale factor 0.975 and free-rotor description below 100 cm^{-1}, as proposed by Grimme [88]. The self-consistent field (SCF) energies were corrected at ωB97X-D level of theory in THF using SMD solvation model with Def2-TZVP as a basis set [84, 85]. The calculated structures were visualized with GaussView 6.0.16, Chemcraft 1.8, and Mercury 4.1.0. Summary of the level of theory: ωB97X-D/Def2-TZVP/SMD(THF)//ωB97X-D/Def2-SVP/SMD (THF).

DLPNO-CCSD(T) calculation: Single point calculations using DLPNO-CCSD(T) method [89–91] were performed with Orca 4.1.2 program [92, 93]. A large basis set Def2-TZVPP [84, 85] was applied with the SMD solvation model [83] with "TightPNO" and "TightSCF" options. Summary of the level of theory: DLPNO-CCSD(T)/Def2-TZVPP/SMD(THF).

Clustering of computed structures: Data preprocessing, clustering, and visualization were conducted using Python 3.6.5 [94] and its modules; NumPy 1.17.0 [95] and Scikit-learn 0.21.3 [96] for clustering, and NetworkX 2.3 [97] and Matplotlib 3.1.1 [98] for visualization. t-Distributed stochastic neighbor embedding (t-SNE) [99] implemented in Scikit-learn was conducted with the following parameters; dimension of the embedded space (n_components) = 2, the maximum number of iterations (n_iter) = 5000, the seed of the random number (random_state) = 0, and the others with the default value.

Electron density analysis: Qualitative non-covalent interactions (NCIs) in the computed structures were visualized with the non-covalent interaction index [100] using the optimized electron density at the same level of theory as the SCF energy correction. The wave function files in a wfn file format were obtained from the corresponding formatted Gaussian checkpoint files in a fchk file format using Multiwfn program [101]. The following thresholds were applied to generate the NCI isosurface with NCIPLOT program [102]; sign(λ_2)ρ ranging from −0.2 to 0.2 au and reduced density gradient (RDG) = 0.45 au. The surfaces were colored on a blue-green–red (BGR) scale using VMD program [103] according to values of sign(λ_2)ρ ranging from −0.02 to 0.02 au. The blue region indicates strong attractive interactions, and the red region indicates strong repulsive interactions.

Interfunctional group interaction analysis: Quantitative interaction analysis between pairs of functional groups in the computed structures was conducted with the intramolecular symmetry-adapted perturbation theory (I-SAPT) [104] implemented in the PSI4 1.3 program package [105]. The reference wave function was recalculated at SAPT0 level of theory. The basis set recommended by the developers was applied; Def2-SVP, Def2-SVP-JKFIT, Def2-SVP-RI, and cc-pVTZ-MINAO for Cu atom, and jun-cc-pVDZ, jun-cc-pVDZ-JKFIT, jun-cc-pVDZ-RI, and cc-pVTZ-MINAO for the other atoms. The multiplicities of the two functional groups were set to doublet, that of the other link structure was set to singlet, and that of the overall molecule was set to singlet. Finally, it was confirmed that two link bonds were assigned into the link structure as a result of automatic IBO localization and link bond selection in the I-SAPT calculations.

2.5.2 *Detailed Computational Investigation on Boryl Copper(I) Dimers*

2.5.2.1 Distortion–Interaction Analysis (DIA)

A distortion–interaction analysis was performed on the dimerization reaction of the boryl copper(I) complexes (Fig. 2.86 and Table 2.7). As shown in Fig. 2.86, for both (*R*,*R*)-QuinoxP* and (*R*,*R*)-5,8-TMS-QuinoxP* complex, two different patterns should be present when the dimer is split (split pattern 1: **SP1**, split pattern 2: **SP2**) into the monomers (left: **Part A**, right: **Part B**), depending on which boryl group is assigned in which part. In each split pattern, **Part A** and **Part B** have almost the same geometry because the dimer has C_2 or near-C_2 symmetry. Conversely, the geometries of **Part** $\mathbf{A_{SP1}}$ and **Part** $\mathbf{A_{SP2}}$ are different (**Part** $\mathbf{B_{SP1}}$ and **Part** $\mathbf{B_{SP2}}$ as well). Subsequently, the electronic energies of these parts were calculated and converted to the distortion and interaction energies (E^{dist} and E^{int} in Table 2.7). Here, the relative dimerization electronic energy (ΔE) is sum of the distortion energies and the interaction energy. As for the (*R*,*R*)-QuinoxP* complex, the distortion energy of **Part** $\mathbf{A_{SP1}}$ [$E^{\mathrm{dist}}(\mathbf{A_{SP1}})$ = 13.5 kcal/mol] and **Part** $\mathbf{B_{SP1}}$ [$E^{\mathrm{dist}}(\mathbf{B_{SP1}})$ = 15.9 kcal/mol] were different because the dimer does not have strict C_2 symmetry [**Part** $\mathbf{A_{SP2}}$: $E^{\mathrm{dist}}(\mathbf{A_{SP2}})$ = 18.4 kcal/mol and **Part** $\mathbf{B_{SP2}}$: $E^{\mathrm{dist}}(\mathbf{B_{SP2}})$ = 14.9 kcal/mol as well]. On the other hand, the (*R*,*R*)-5,8-TMS-QuinoxP* complex has more strict C_2 symmetry, and thus, the distortion energies in each part were identical [$E^{\mathrm{dist}}(\mathbf{A_{SP1}})$ = $E^{\mathrm{dist}}(\mathbf{B_{SP1}})$ = 14.8 kcal/mol; $E^{\mathrm{dist}}(\mathbf{A_{SP2}})$ = $E^{\mathrm{dist}}(\mathbf{B_{SP2}})$ = 16.0 kcal/mol]. Then, I focused on the relative values between the two ligands, (summarized in the bottom of Table 2.7). For both split patterns, one relative value of the distortion energy (ΔE^{dist}) was positive, and the other was negative [**SP1**: $\Delta E^{\mathrm{dist}}(\mathbf{A_{SP1}})$ = 1.3 kcal/mol, $\Delta E^{\mathrm{dist}}(\mathbf{B_{SP1}})$ = −1.1 kcal/mol; **SP2**: $\Delta E^{\mathrm{dist}}(\mathbf{A_{SP2}})$ = −2.4 kcal/mol, $\Delta E^{\mathrm{dist}}(\mathbf{B_{SP2}})$ = 1.1 kcal/mol]. Nevertheless, the difference of the relative dimerization electronic energies ($\Delta\Delta E$) is smaller than the relative value of the entropy terms ($|\Delta\Delta E|$ = 1.6 kcal/mol, $|\Delta(-T\Delta S)|$ = 6.0 kcal/mol, Table 2.7). Hence, as I conclude in the previous Sect. 2.2, the formation of the boryl copper(I) dimer should be suppressed by the entropy effect.

2.5.2.2 Structure Analysis

Then, to reveal the intramolecular interaction in the boryl copper(I) dimer, the structure analysis and non-covalent interaction (NCI) plot were performed (Figs. 2.87 and 2.88). Some short H–H contacts were found in the (*R*,*R*)-QuinoxP* complex (Fig. 2.87d, g, j). The shortest distance was 2.18 Å. Besides, the green region found in the NCI plot indicated the intramolecular van der Waals interaction (Fig. 2.87e, h, k). On the other hand, the shortest H–H contacts in the (*R*,*R*)-5,8-TMS-QuinoxP* complex was shorter than that in (*R*,*R*)-QuinoxP* complex (2.06 Å, Fig. 2.88d, g, j). However, the additional green regions were observed around the TMS groups

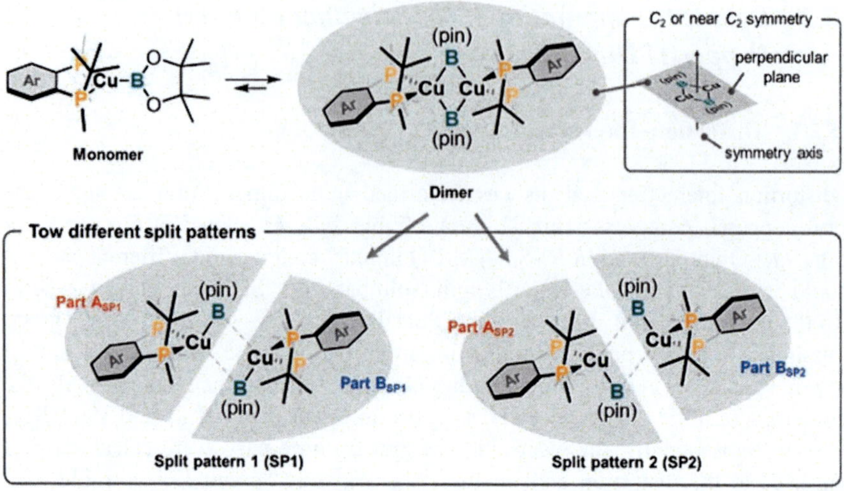

Fig. 2.86 Schematic illustration of distortion–interaction analysis (DIA) of the boryl copper(I) dimers

Table 2.7 Distortion–interaction analysis (DIA) of the boryl copper(I) dimers

	Distortion–interaction analysis	
(*R*,*R*)-QuinoxP* ligand	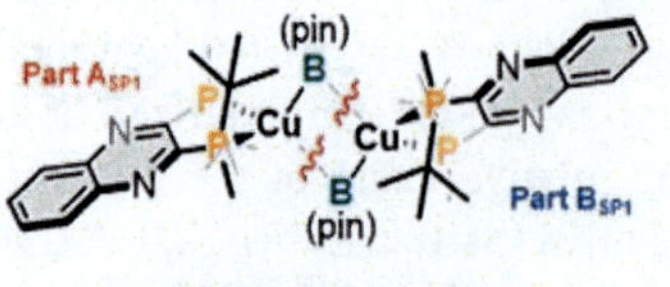 **Split pattern 1 (SP1)** E^{dist} (A_{SP1}) = 13.5 E^{dist} (B_{SP1}) = 15.9 E^{int} (A_{SP1}, B_{SP1}) = −59.4 ΔE = −30.0	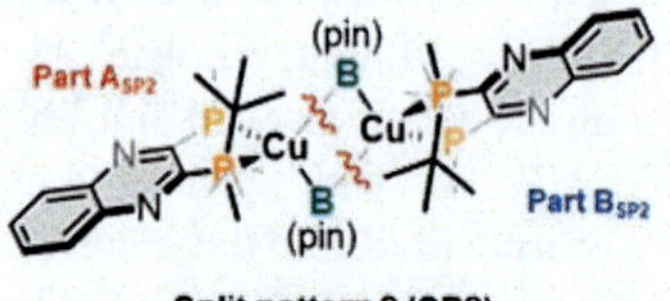 **Split pattern 2 (SP2)** E^{dist} (A_{SP2}) = 18.4 E^{dist} (B_{SP2}) = 14.9 E^{int} (A_{SP2}, B_{SP2}) = −63.4 ΔE = −30.0
(*R*,*R*)-5,8-TMS-QuinoxP* ligand	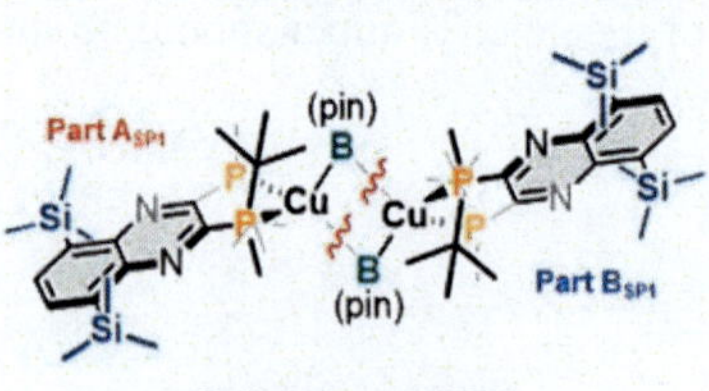 **Split pattern 1 (SP1)** E^{dist} (A_{SP1}) = 14.8 E^{dist} (B_{SP1}) = 14.8 E^{int} (A_{SP1}, B_{SP1}) = −61.1 ΔE = −31.6	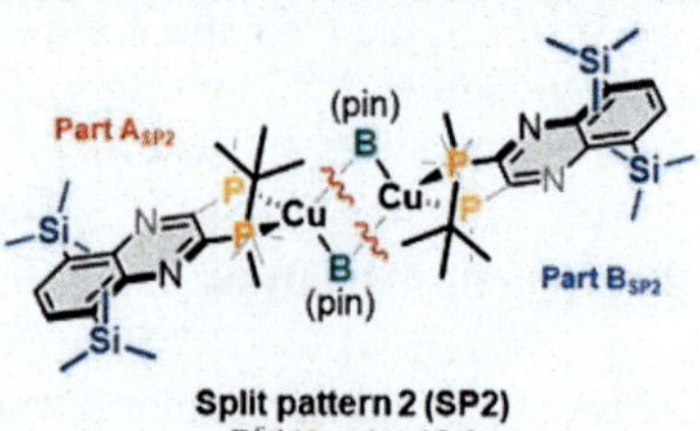 **Split pattern 2 (SP2)** E^{dist} (A_{SP2}) = 16.0 E^{dist} (B_{SP2}) = 16.0 E^{int} (A_{SP2}, B_{SP2}) = −63.6 ΔE = −31.6
Relative values	ΔE^{dist} (A_{SP1}) = 1.3 ΔE^{dist} (B_{SP1}) = −1.1 ΔE^{int} (A_{SP1}, B_{SP1}) = −1.7 $\Delta\Delta E$ = −1.6	ΔE^{dist} (A_{SP2}) = −2.4 ΔE^{dist} (B_{SP2}) = 1.1 ΔE^{int} (A_{SP2}, B_{SP2}) = 0.1 $\Delta\Delta E$ = −1.6

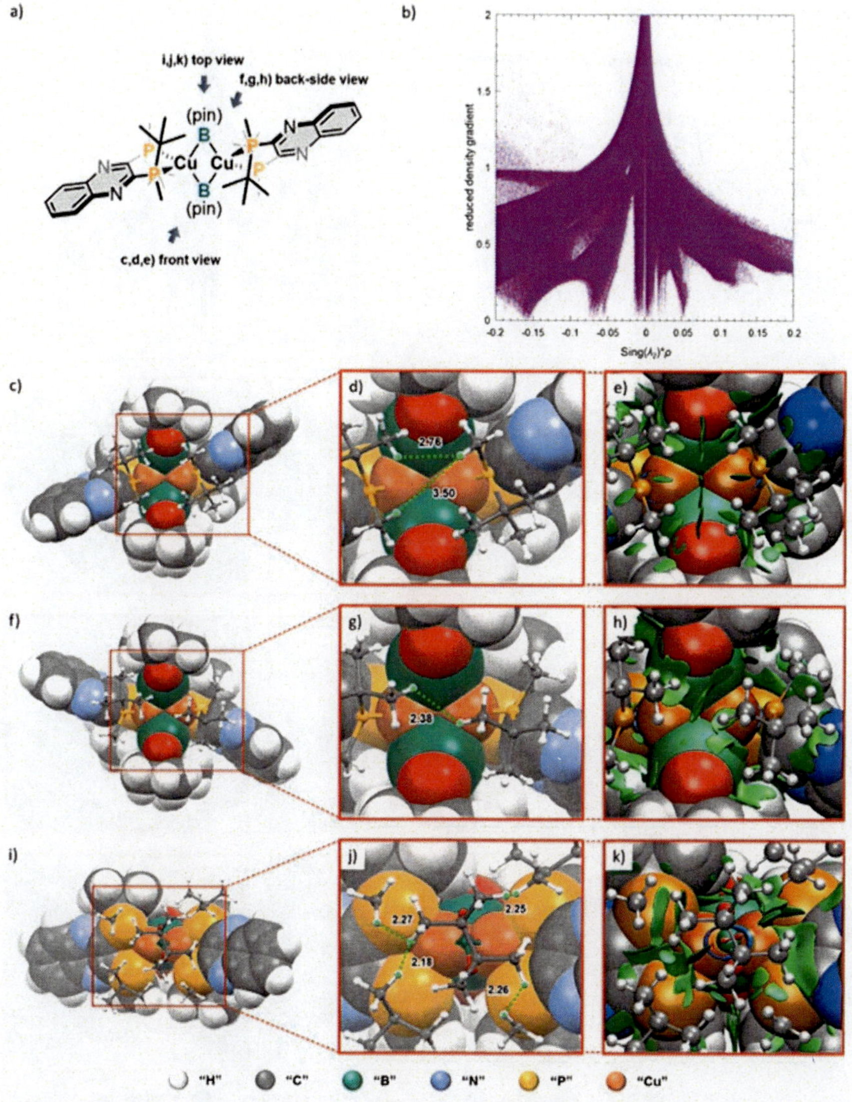

Fig. 2.87 **a** Structure of boryl copper(I) dimer with (*R*,*R*)-QuinoxP* **V**. **b** Plots of the non-covalent interaction index within the dimer **V**. **c**, **f**, **i** Ball-and-stick and space-filling models of the dimer **V**: a ball-and-stick model for the front-side structure, and a space-filling model for the back-side structure. Color code: white = H; green = B; gray = C; light blue = N; red = O; orange = P; brown = Cu. **d**, **g**, **j** Short H–H contacts in the dimer. **e**, **h**, **k** Non-covalent interaction (NCI) plot in the dimer

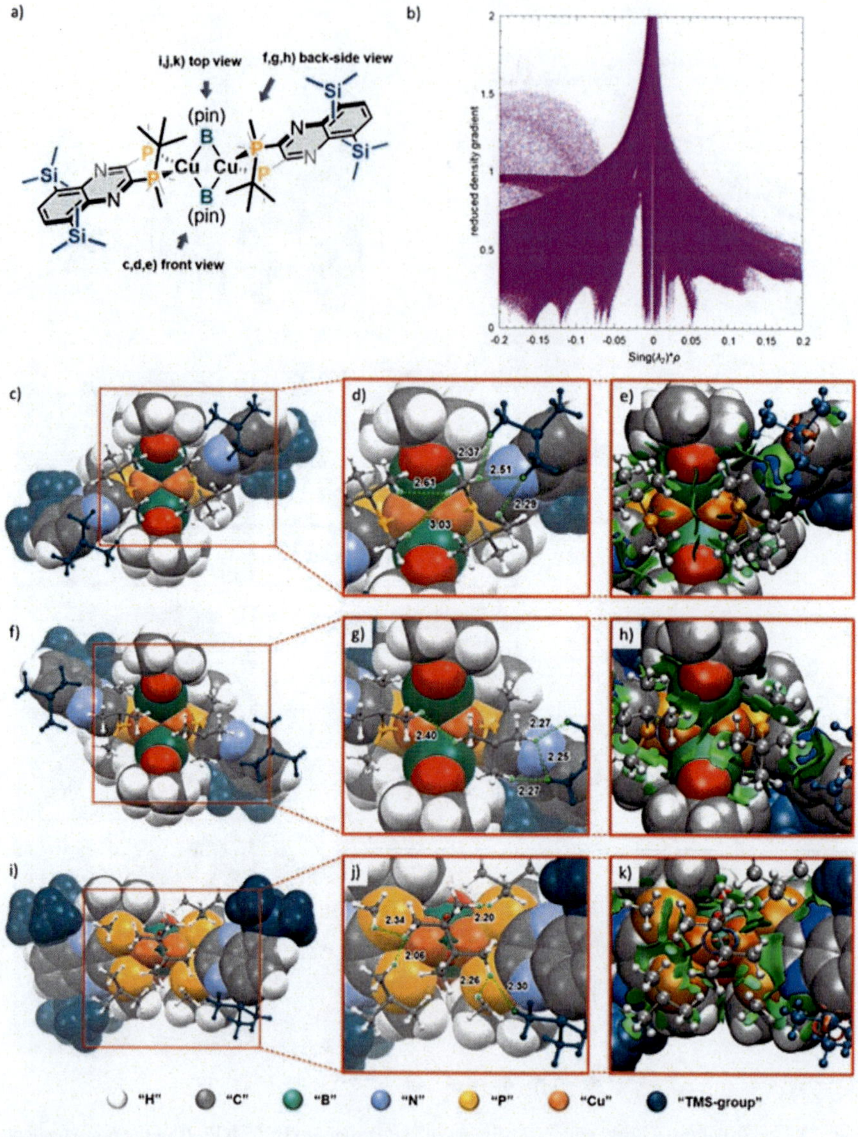

Fig. 2.88 **a** Structure of boryl copper(I) dimer with (*R*,*R*)-5,8-TMS-QuinoxP* **V′**. **b** Plots of the non-covalent interaction index within the dimer **V′**. **c**, **f**, **i** Ball-and-stick and space-filling models of the dimer **V′**: a ball-and-stick model for the front-side structure, and a space-filling model for the back-side structure. Color code: white = H; green = B; gray = C; light blue = N; red = O; orange = P; brown = Cu; dark blue = TMS group. **d**, **g**, **j** Short H–H contacts in the dimer. **e**, **h**, **k** Non-covalent interaction (NCI) plot in the dimer

(Fig. 2.88e, h, k). These results suggested that the steric repulsions enhanced by the TMS group were canceled out by the intramolecular weak attractive interactions. I assumed that the short H–H contacts around TMS were not the result of the strong steric repulsion, but these interactions can inhibit the rotation motion of the TMS group, which can contribute to the entropy loss upon the dimer formation.

2.5.2.3 Screening of Computational Method

Abnormally large entropy loss was found in the dimerization of a boryl copper(I) complex with (*R*,*R*)-5,8-TMS-QuinoxP* ligand ($-T\Delta S = 26.4$ kcal/mol) compared to (*R*,*R*)-QuinoxP* ligand ($-T\Delta S = 20.4$ kcal/mol) (Fig. 2.13). In order to check the dependency of computational methods on the balance between the dimerization enthalpy and entropy, DFT functionals and quasi-harmonic (QH) approximations were systematically examined (Table 2.8). As for DFT functionals, ωB97XD showed larger differences of the relative dimerization free energy and entropy [entries 1–3, $\Delta\Delta G$: 4.6–8.6 kcal/mol, $\Delta(-T\Delta S)$: 4.7–8.8 kcal/mol], while M06 showed smaller $\Delta\Delta G$ and $\Delta(-T\Delta S)$ [entries 7–9, $\Delta\Delta G$: 0.5–0.8 kcal/mol, $\Delta(-T\Delta S)$: 0.8–1.1 kcal/mol]. Next, as for QH approximations, the cutoff of low frequencies decreased $\Delta(-T\Delta S)$ value by 0.1–4.1 kcal/mol compared to uncorrected results. Truhlar's method with ωB97XD or M06 functional tended to suppress the entropy effect more largely than Grimme's method, whereas the difference of $\Delta(-T\Delta S)$ value of those QH approximations with B3LYP-D3 functional was negligible. Although there is such a tendency depended on the DFT functionals and QH approximations, the dimerization of the boryl copper(I) complex was suppressed by entropy

Table 2.8 Thermochemical properties for the dimerization of a boryl copper(I) complexes

Entry	Functional	Quasi-harmonic approximations	Ligand	Dimerization		
				ΔG	ΔH	$-T\Delta S$
1	ωB97XD	Uncorrected	(*R*,*R*)-QuinoxP*	−8.8	−28.7	19.9
			(*R*,*R*)-5,8-TMS-QuinoxP*	−0.1	−28.8	28.7
			$\Delta\Delta X$ ($\Delta X = \Delta G$, ΔH or $-T\Delta S$)	**8.6**	**−0.2**	**8.8**
2	ωB97XD	Grimme's	(*R*,*R*)-QuinoxP*	−8.3	−28.7	20.4
			(*R*,*R*)-5,8-TMS-QuinoxP*	−2.4	−28.8	26.4
			$\Delta\Delta X$ ($\Delta X = \Delta G$, ΔH or $-T\Delta S$)	**5.9**	**−0.2**	**6.0**
3	ωB97XD	Truhlar's	(*R*,*R*)-QuinoxP*	−9.3	−28.7	19.3
			(*R*,*R*)-5,8-TMS-QuinoxP*	−4.8	−28.8	24.1
			$\Delta\Delta X$ ($\Delta X = \Delta G$, ΔH or $-T\Delta S$)	**4.6**	**−0.2**	**4.7**

(continued)

Table 2.8 (continued)

Entry	Functional	Quasi-harmonic approximations	Ligand	Dimerization		
				ΔG	ΔH	$-T\Delta S$
4	B3LYP-D3	Uncorrected	(*R*,*R*)-QuinoxP*	−5.7	−25.6	19.9
			(*R*,*R*)-5,8-TMS-QuinoxP*	−2.8	−25.4	22.6
			$\boldsymbol{\Delta\Delta X}$ ($\Delta X = \Delta G$, ΔH or $-T\Delta S$)	**3.0**	**0.2**	**2.7**
5	B3LYP-D3	Grimme's	(*R*,*R*)-QuinoxP*	−5.0	−25.5	20.5
			(*R*,*R*)-5,8-TMS-QuinoxP*	−2.6	−25.3	22.7
			$\boldsymbol{\Delta\Delta X}$ ($\Delta X = \Delta G$, ΔH or $-T\Delta S$)	**2.4**	**0.2**	**2.2**
6	B3LYP-D3	Truhlar's	(*R*,*R*)-QuinoxP*	−6.3	−25.5	19.2
			(*R*,*R*)-5,8-TMS-QuinoxP*	−3.9	−25.3	21.5
			$\boldsymbol{\Delta\Delta X}$ ($\Delta X = \Delta G$, ΔH or $-T\Delta S$)	**2.5**	**0.2**	**2.3**
7	M06	Uncorrected	(*R*,*R*)-QuinoxP*	−3.3	−25.8	22.5
			(*R*,*R*)-5,8-TMS-QuinoxP*	−2.5	−26.1	23.6
			$\boldsymbol{\Delta\Delta X}$ ($\Delta X = \Delta G$, ΔH or $-T\Delta S$)	**0.8**	**−0.3**	**1.1**
8	M06	Grimme's	(*R*,*R*)-QuinoxP*	−3.7	−25.8	22.1
			(*R*,*R*)-5,8-TMS-QuinoxP*	−2.9	−26.1	23.1
			$\boldsymbol{\Delta\Delta X}$ ($\Delta X = \Delta G$, ΔH or $-T\Delta S$)	**0.8**	**−0.3**	**1.0**
9	M06	Truhlar's	(*R*,*R*)-QuinoxP*	−5.6	−25.8	20.2
			(*R*,*R*)-5,8-TMS-QuinoxP*	−5.1	−26.1	21.0
			$\boldsymbol{\Delta\Delta X}$ ($\Delta X = \Delta G$, ΔH or $-T\Delta S$)	**0.5**	**−0.3**	**0.8**

effect because the difference of the enthalpy term ($\Delta\Delta H$) was smaller than the $\Delta(-T\Delta S)$ value in all cases.

2.5.3 *Computational Analysis on Enantiodetermining Transition State*

2.5.3.1 Conformation Search of Transition States

To investigate the variety of the ligand structures on the potential energy surface determining the enantioselectivity, conformation search of transition states (TSs) on the borylcupration of the cyclic allyl carbonate **1a** was conducted (Figs. 2.89 and 2.90). The conformations based on the following structures were considered in this

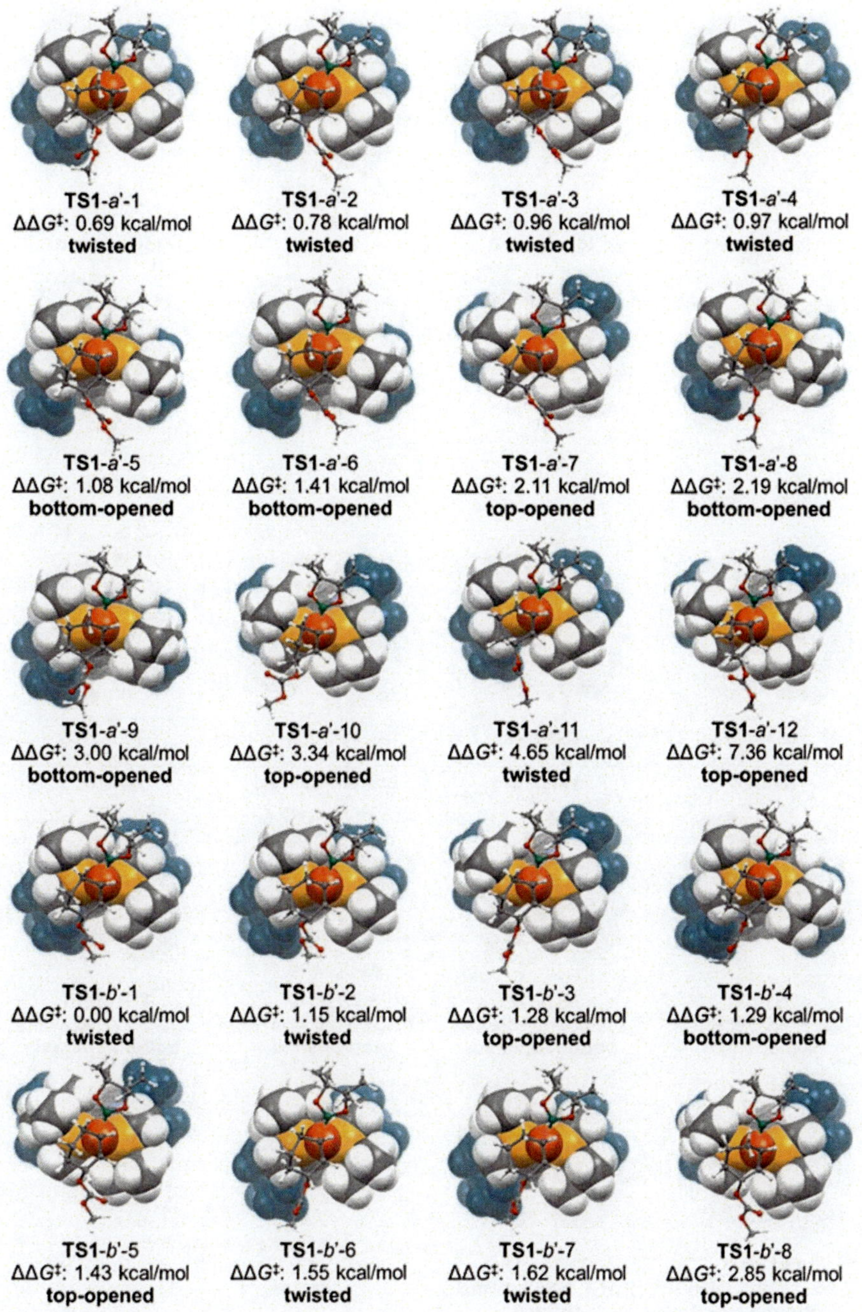

Fig. 2.89 Ball-and-stick and space-filling models of transition states for the borylcupration of the substrate **1a** with (*R*,*R*)-5,8-TMS-QuinoxP*: a ball-and-stick model for the substrate and B(pin) structure, and space-filling model for the ligand and copper atom. Color code: white = H; green = B; gray = C; light blue = N; red = O; orange = P; brown = Cu; dark blue = TMS group

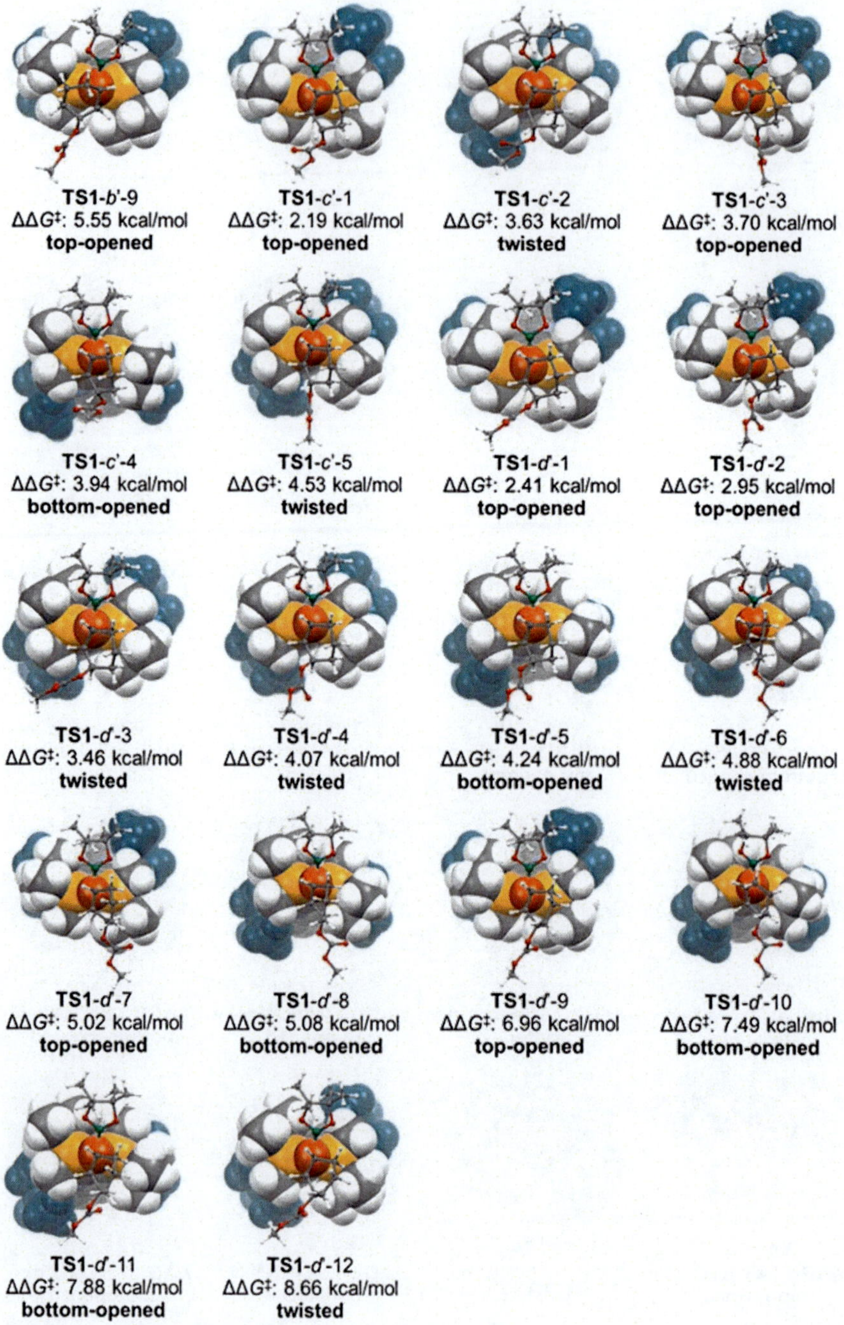

Fig. 2.89 (continued)

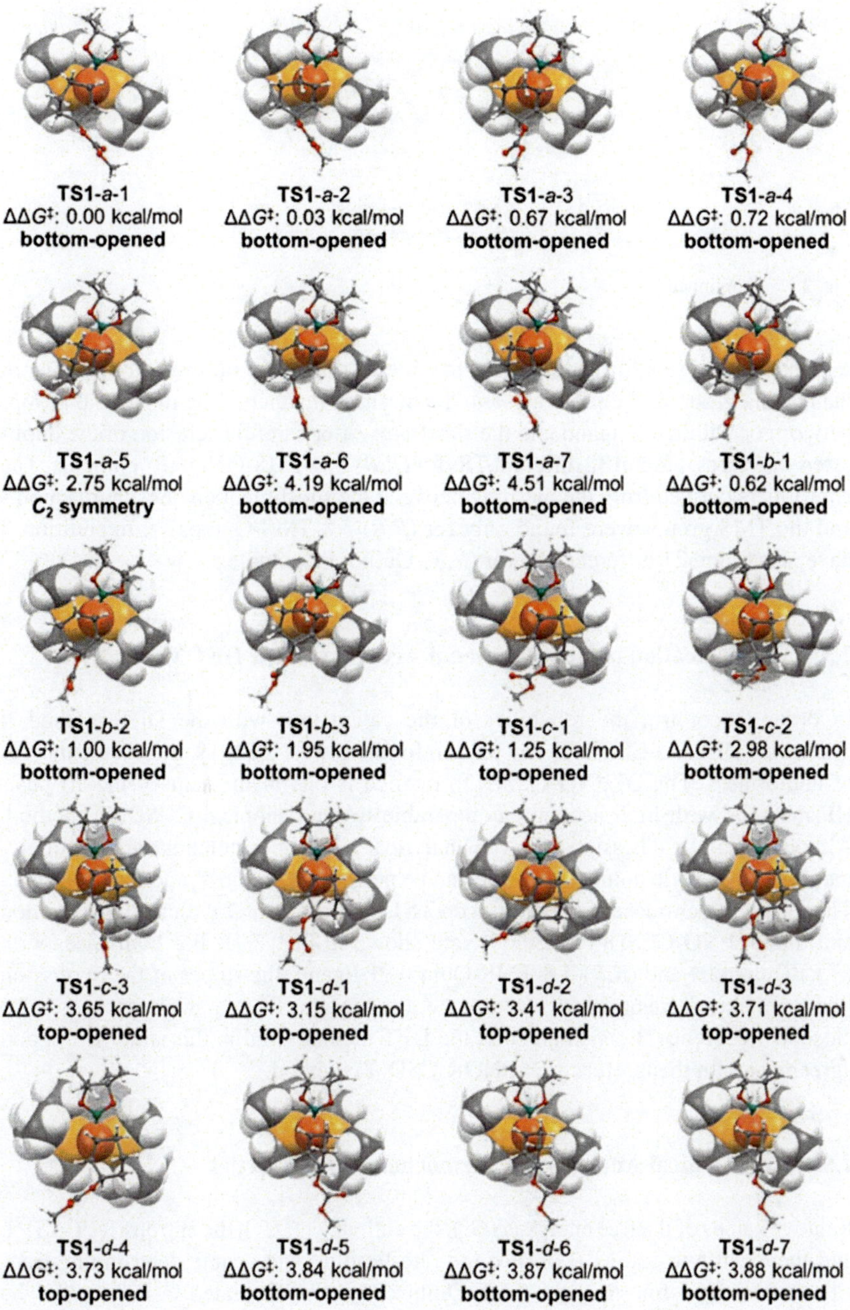

Fig. 2.90 Ball-and-stick and space-filling models of transition states for the borylcupration of the substrate **1a** with (*R*,*R*)-QuinoxP*: a ball-and-stick model for the substrate and B(pin) structure, and a space-filling model for the ligand and copper atom. Color code: white = H; green = B; gray = C; light blue = N; red = O; orange = P; brown = Cu

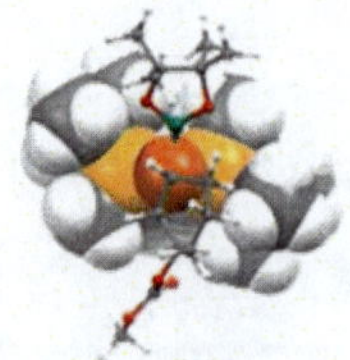

Fig. 2.90 (continued)

search; the cyclic conformer of the six-membered ring of the substrate, the rotamer of the leaving group of the substrate, and the meshing manner of the functional groups between or within the ligand and the substrate. After careful exclusion of the duplicated structures, I found 38 different TSs for (*R*,*R*)-5,8-TMS-QuinoxP* complex. The conformers derived from the meshing between the substituent on phosphorus atoms and the TMS groups were found only for (*R*,*R*)-5,8-TMS-QuinoxP*. In addition, I have also found 21 different TSs for (*R*,*R*)-QuinoxP* complex.

2.5.3.2 Verification of Computational Accuracy of the DFT Method

In order to confirm the reliability of the calculation with the DFT method, I performed DLPNO-CCSD(T) calculation for the lowest **TS1**s as the representatives of conformers. The DLPNO-CCSD(T) method is one of the state-of-the-art post-HF methods with high accuracy comparable to the canonical CCSD(T) method. Although analytical hessian and even analytical gradient calculations are not implemented yet, a single point calculation can be performed within a practical wall time. The relative electronic energies based on **TS1**-*a*-1 or **TS1**-*a*′-1 with the DFT method and the DLPNO-CCSD(T) method were shown in Fig. 2.91. For both cases with (*R*,*R*)-QuinoxP* and (*R*,*R*)-5,8-TMS-QuinoxP* ligand, the slopes of the regression line were almost identical to 1.0, and the goodness of fit R^2 parameters were also close to 1.0. Hence, I concluded that the DFT method used in this study is in good agreement with the reference DLPNO-CCSD(T) method.

2.5.3.3 Statistical Analysis of Thermochemical Properties

Relative values of the free energy ($\Delta G^‡$), the enthalpy ($\Delta H^‡$), the entropy [$(-T\Delta S)^‡$], and the electronic energy ($\Delta E^‡$) of the rate-limiting and enantiodetermining transition states **TS1** for (*R*,*R*)-5,8-TMS-QuinoxP* were illustrated in Fig. 2.92. The TS types in the figures represent the suffix defined in Figs. 2.14 and 2.15. The Boltzmann-weighted average enantioselectivity of **TS1** is 95% ee; this predicted value is in excellent agreement with the experimental value (93% ee, Fig. 2.92a).

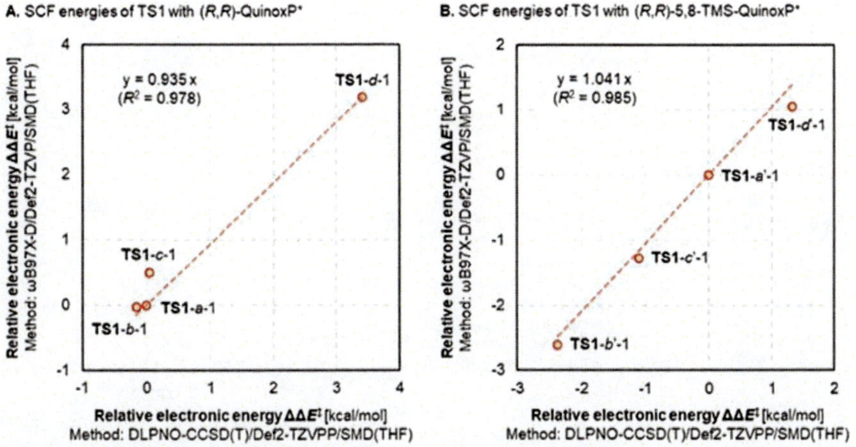

Fig. 2.91 Comparison of SCF energies of the lowest transition states **TS1** with (*R*,*R*)-QuinoxP* and (*R*,*R*)-5,8-TMS-QuinoxP* ligand

Furthermore, a pairwise correlation plot between those thermochemical properties was presented to illustrate the relationship between those values (Fig. 2.93). The diagonal plots show the distribution of each value, while the off-diagonal plots show the relationship between the two values.

Focused on correlations of the thermochemical properties, there is no definite correlation between the entropy $[(-T\Delta S)^{\ddagger}]$ and the other properties ($\Delta G^{\ddagger}$, $\Delta H^{\ddagger}$ and $\Delta E^{\ddagger}$, Fig. 2.93d, e, i, respectively). On the other hand, there is moderate positive correlation between the free energy ($\Delta G^{\ddagger}$) and the enthalpy ($\Delta H^{\ddagger}$) (Fig. 2.93b). Therefore, it is assumed that the free energy is determined mainly under enthalpy control. Furthermore, there is strong positive correlation between the enthalpy ($\Delta H^{\ddagger}$) and the electronic energy ($\Delta E^{\ddagger}$) (Fig. 2.93h). This enables the discussion of the enantioselection using the electronic energy instead of the Gibbs free energy and the enthalpy.

2.5.3.4 Distortion–Interaction Analysis (DIA)

Next, I attempted distortion–interaction analysis (DIA) for enantiodetermining **TS1** (Figs. 2.94 and 2.95, Eq. S1). Based on the standard protocol for the distortion–interaction analysis, the **TS1** was split into the boryl copper(I) species (catalyst) and the substrate **1a**. Then, the electronic energies of those species were calculated and converted to the distortion and interaction energies. As shown in Fig. 2.95, the catalyst and the substrate on the major paths (**TS1**-*a′* and **TS1**-*b′*) tend to exhibit a smaller distortion compared to those on the minor paths (**TS1**-*c′* and **TS1**-*d′*). Contrary, the interaction energies are almost identical in those TSs. Therefore, it is assumed that the enantioselectivity is determined by steric match/mismatch effect.

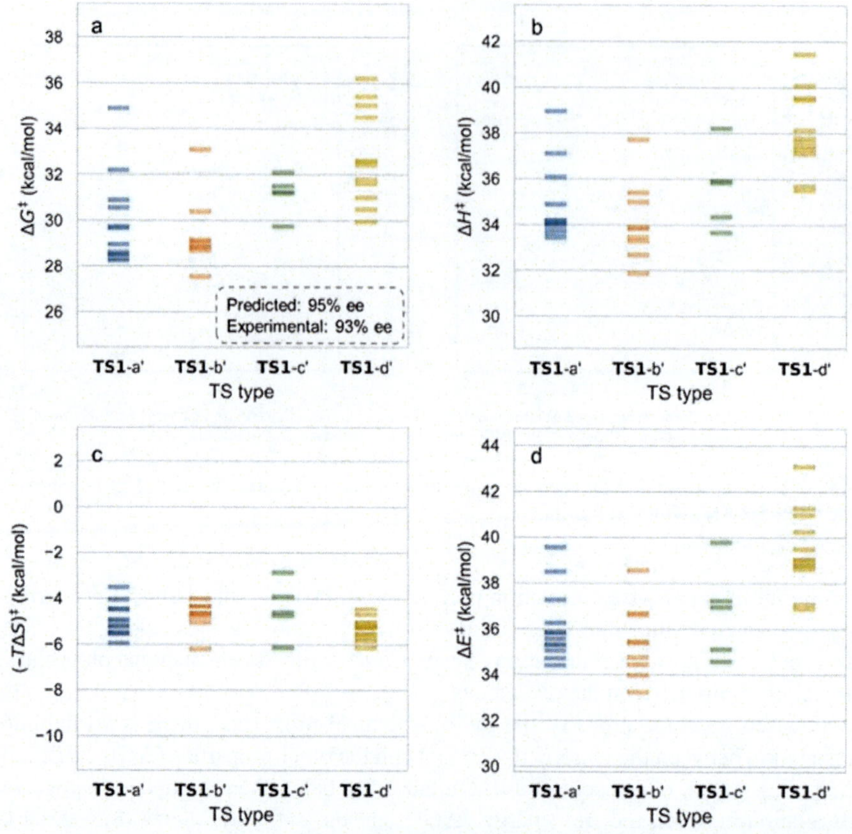

Fig. 2.92 Diagram of Gibbs free energy ($\Delta G^{\ddagger}$), enthalpy ($\Delta H^{\ddagger}$), entropy [$-(T\Delta S^{\ddagger})$], and electronic energy ($\Delta E^{\ddagger}$) for the transition state **TS1**

$$\Delta E^{\ddagger} = E_{\text{dist}}^{\text{catalyst}} + E_{\text{dist}}^{\text{substrate}} - E_{\text{int}} + \Delta E(\text{dimerization}) \qquad \text{(S1)}$$

2.5.3.5 Transition State Analysis of Acetal Substrate

According to the conformation search using substrate **1a**, the lowest transition states on borylcupration of the simplified acetal substrate **1h** were calculated (Fig. 2.96). The leaving group OBn in the substrate **1h** was simplified to OMe. The estimated ee value from the four transition states (48% ee) was good agreement with the experimental value (56% ee). The differences of the free energy [$\Delta(\Delta\Delta G^{\ddagger})$] for both (*S*)-**1h** [**TS1**-*a*′ vs. **TS1**-*c*′: $\Delta(\Delta\Delta G^{\ddagger}) = 2.84$ kcal/mol] and (*R*)-**1h** [**TS1**-*b*′ vs.

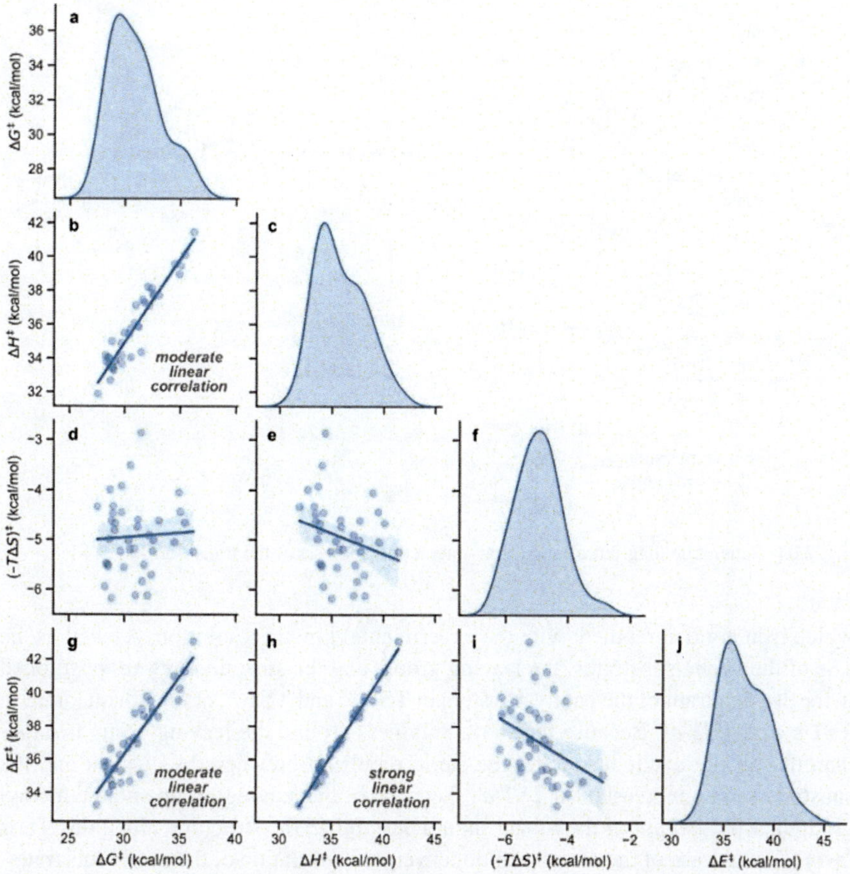

Fig. 2.93 Pairwise correlation plot of thermochemical properties for the transition state **TS1**

TS1-*d*′:$\Delta(\Delta\Delta G^{\ddagger}) = 0.01$ kcal/mol] was decreased compared to the substrate **1a**. The drop of enantioselectivity was attributable to the diminish of steric repulsion between the substrate and the 4th quadrant substituent (*t*-Bu group) of the ligand due to the substitution of $-CH_2-$ group to $-O-$ in the substrate structure.

2.5.3.6 Transition State Analysis of Linear Substrate

The transition states of the borylcupration of simplified linear substrates were calculated with (*R*,*R*)-5,8-TMS-QuinoxP* as the ligand (Fig. 2.97). After the conformational search, the respective lowest transition states of the four paths were found [Fig. 2.97, **TS1**-*a*′ to *d*′ (linear)]. In contrast to the cyclic substrates, the *anti*-insertion paths (**TS1**-*b*′ and **TS1**-*c*′) are more favored than the syn-insertion paths (**TS1**-*a*′ and **TS1**-*d*′). However, **TS1**-*b*′ has the prominently low reaction barrier in the four paths,

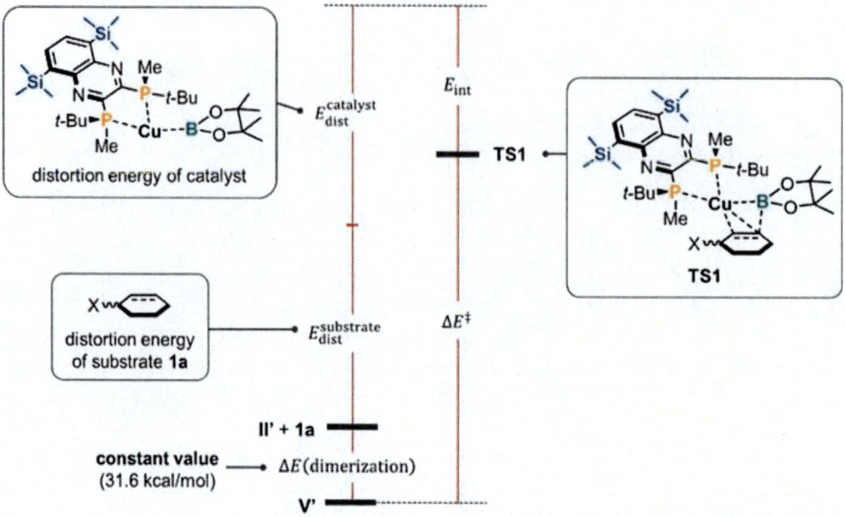

Fig. 2.94 Schematic diagram of distortion–interaction analysis of the transition state TS1

which is in good agreement with the experimental kinetic resolution. As well as the TSs of the cyclic substrates, the leaving group, and the surroundings are positioned in the 3rd quadrant of the catalyst system in **TS1**-*a*′ and **TS1**-*b*′ or the 4th quadrant in **TS1**-*c*′ and **TS1**-*d*′. Because the steric bulkiness around the leaving group is larger than that of the cyclic substrate, the steric repulsion between the catalyst and the substrate can be increased. In **TS1**-*a*′, the clash of the leaving group and the methyl group in 3rd quadrant of the ligand should destabilize the structure, albeit this TS is favored in the case of the direct enantioconvergent borylation of the cyclic substrates.

2.5.4 *Computational Analysis on Conformation of (R,R)-5,8-TMS-QuinoxP**

2.5.4.1 Analysis of Ligand Structure Distribution on Transition States of Borylcupration Based on Descriptive Dihedral Angles

The in-depth analysis of the conformations of (*R*,*R*)-5,8-TMS-QuinoxP* ligand in the calculated transition state was conducted. I defined two dihedral angles Ψ_1 and Ψ_2 (Figs. 2.16 and 2.98). These are the descriptive angles of the pseudorotation in the Cu–P_1–C_1–C_2–P_2 ring. The positions of substituents on the phosphorus atoms, methyl and *tert*-butyl groups on the chiral phosphorus centers, depended on those dihedral angles. Moreover, the combination of the two dihedral angles (Ψ_1 and Ψ_2) could describe the conformation of bisphosphine: (1) Bottom-opened conformers, with a positive Ψ_1 and negative Ψ_2 ($\Psi_1 > 0$, $\Psi_2 < 0$), (2) top-opened conformers

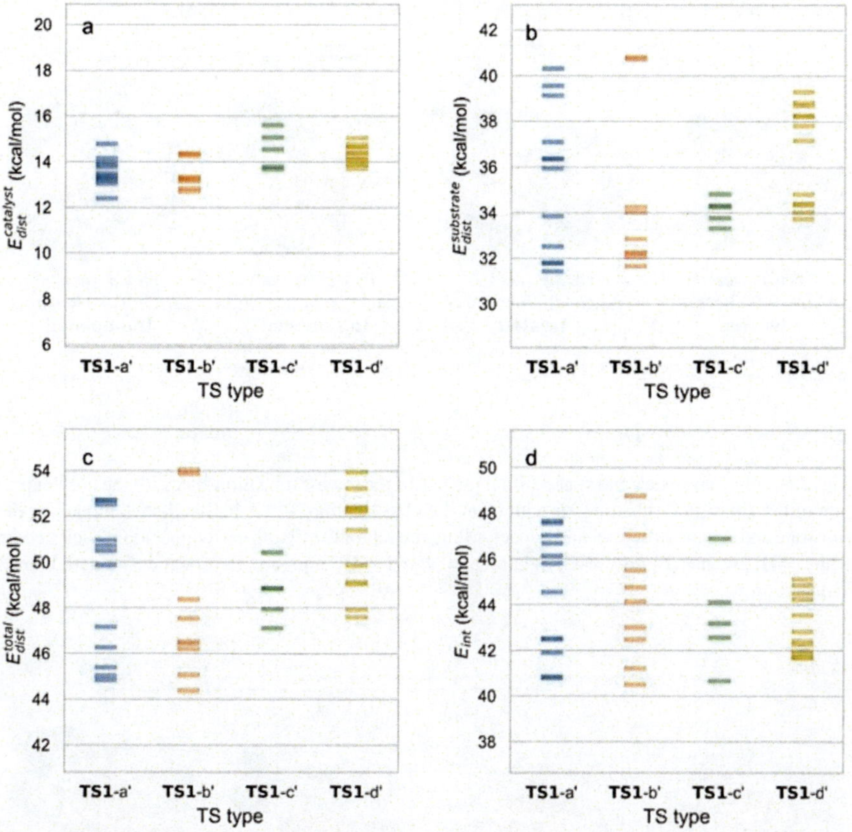

Fig. 2.95 Diagram of the distortion energy of boryl copper(I) catalyst species in the transition state **TS1**

with a negative Ψ_1 and positive Ψ_2 ($\Psi_1 < 0$, $\Psi_2 > 0$), (3) C_2-symmetric conformers, in which both dihedral angles were close to zero ($\Psi_1 \approx \Psi_2 \approx 0$), and (4) twisted conformers, in which both dihedral angles were negative ($\Psi_1 < 0$, $\Psi_2 < 0$) (Fig. 2.99). The correlation diagrams of relative free energies of the transition states against Ψ_1 and Ψ_2 were shown in Figs. 2.17, 2.100, and 2.101. In the case of (*R*,*R*)-5,8-TMS-QuinoxP*, traces of the IRCs were clearly divided into three regions which can be categorized as *bottom-opened*, *twisted,* and *top-opened* conformer groups (Fig. 2.100). The separated traces implied that the categorization of the phosphine conformations is not only effective on a point of the transition states but also during the reaction coordinates. (Fig. 2.100B). Contrary, in the case of (*R*,*R*)-QuinoxP*, the transition states and traces of the IRCs were linearly distributed (Fig. 2.101). Therefore, it is assumed that the TMS groups affect the phosphine conformations.

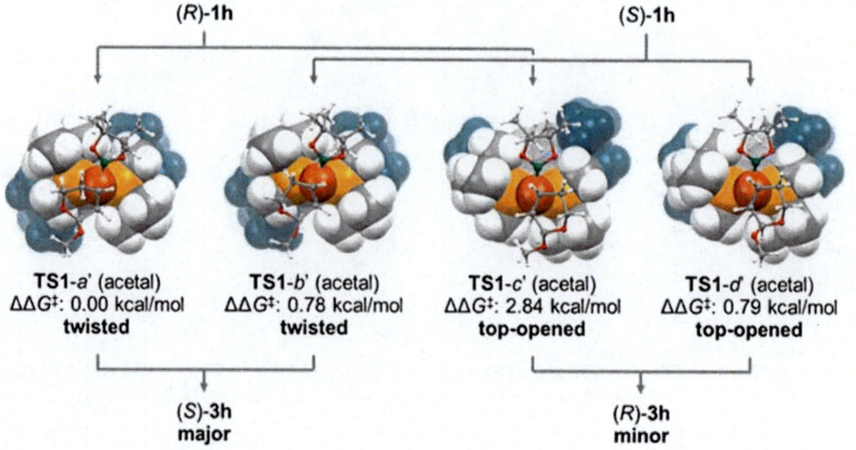

Fig. 2.96 Ball-and-stick and space-filling models of the lowest transition states for the borylcupration of the simplified substrate **1h** with (*R*,*R*)-5,8-TMS-QuinoxP*: A ball-and-stick model for the substrate and B(pin) structure, and a space-filling model for the ligand and copper atom. Color code: white = H; green = B; gray = C; light blue = N; red = O; orange = P; brown = Cu; dark blue = TMS group

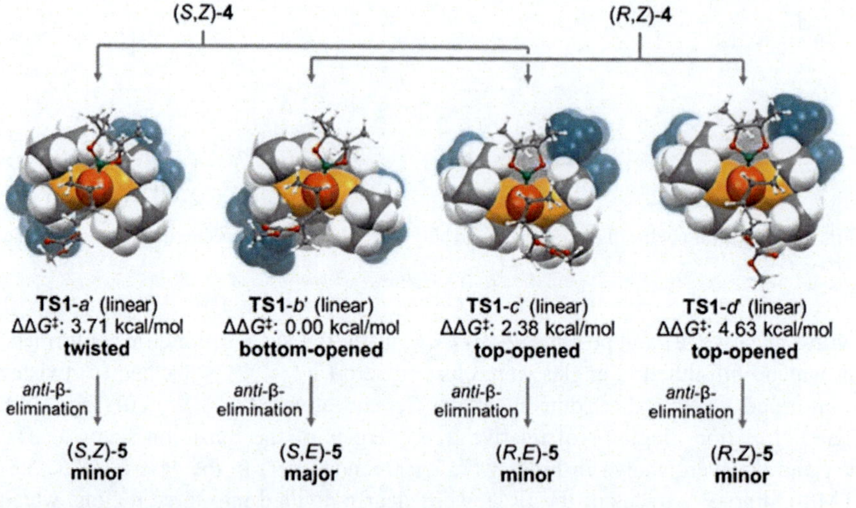

Fig. 2.97 Ball-and-stick and space-filling models of the lowest transition states for the borylcupration of the simplified linear substrate with (*R*,*R*)-5,8-TMS-QuinoxP*: A ball-and-stick model for the substrate and B(pin) structure, and a space-filling model for the ligand and copper atom. Color code: white = H; green = B; gray = C; light blue = N; red = O; orange = P; brown = Cu; dark blue = TMS group

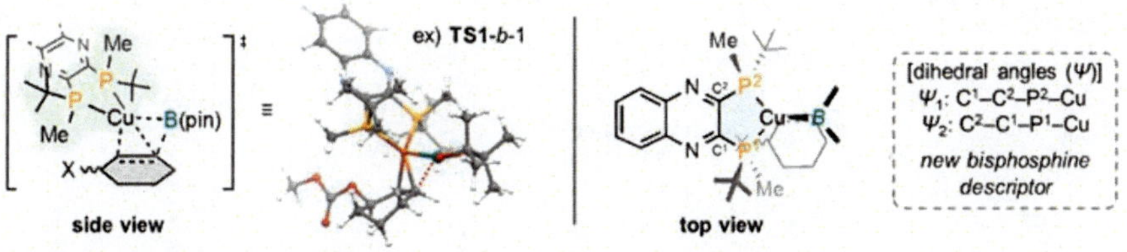

Fig. 2.98 Definition of dihedral angles for the phosphine conformation in boryl copper(I) species

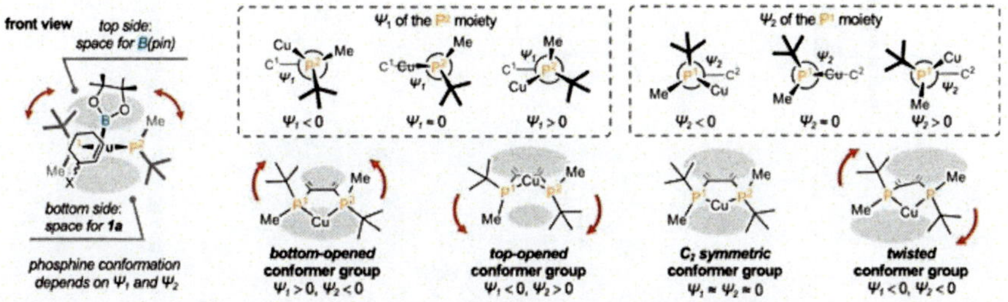

Fig. 2.99 Relationship between dihedral angles for the phosphine conformation

2.5.4.2 Analysis of Ligand Structure Distribution on Transition States of Borylcupration Based on Clustering Approach

The ligand structures in **TS1** were categorized automatically using a clustering approach. A preprocessing of the input data was conducted based on interatomic distance matrixes of the transition state structures (**TS1**) to handle all the transition state structures in a unified manner. First, two types of input structures were prepared according to the following two different procedures: A) the substrate 1a was removed from each transition state to build a distorted boryl copper(I) complex and B) the substrate 1a and the pinacolate structure were removed from each transition state with leaving the boron atom to build a pinacolate-free distorted boryl copper(I) complex (Fig. 2.102). Next, interatomic distance matrixes of those structures were prepared and arranged in lines. Then, each of the sequences was sorted according to the atomic numbers (1st and 2nd priority) and the measured distances (3rd priority). In the respective two measured atoms, the first atomic number is less than or equal to the second atomic number. For example, H–H (atomic number = 1, 1) distances are listed first in ascending order. Then, H–B (atomic number = 1, 5), H–C (atomic number = 1, 6), and H–N (atomic number = 1, 7) distances follow also in ascending order. The two P–Cu (atomic number = 15, 29) distances are the final elements of the sorted list. This sorting makes the input data invariant to translational and rotational displacements and exchanges of the atomic labels during the manual construction of the molecules. Moreover, this preprocess should retain the commutative property between the structure of the molecules and the processed data. Heatmaps of the input data for visualization were shown in Figs. 2.103A, 104A,

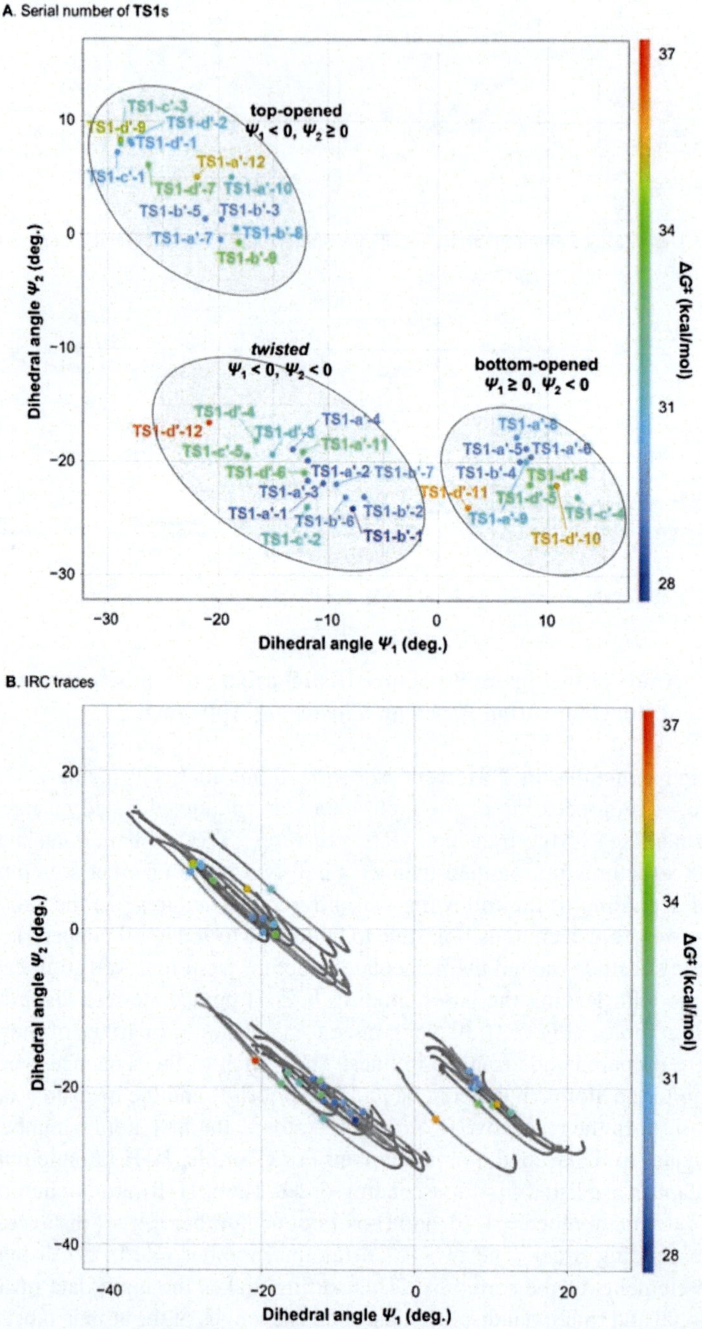

Fig. 2.100 Correlation diagram of relative free energy of the transition states against Ψ_1 and Ψ_2 with (*R*,*R*)-5,8-TMS-QuinoxP*. **a** The serial numbers of the plotted transition states. **b** Traces of the IRCs of each transition state

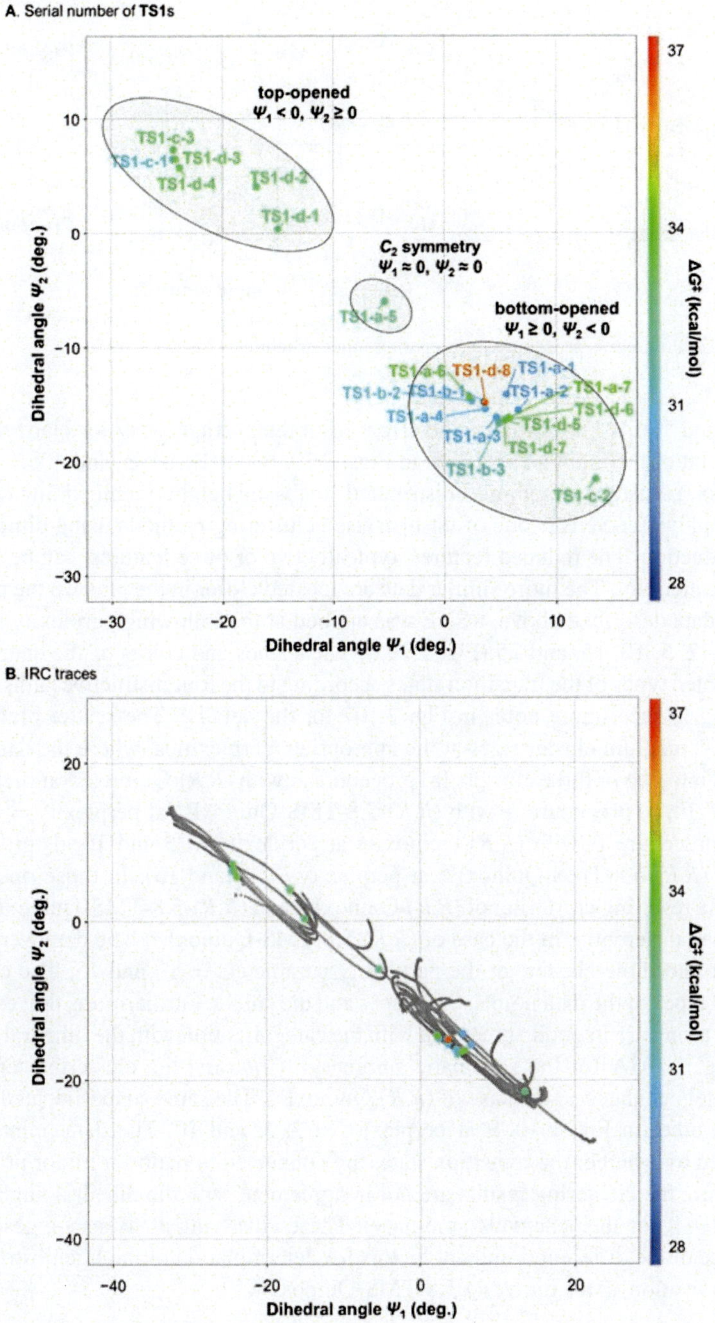

Fig. 2.101 Correlation diagram of relative free energy of the transition states against Ψ_1 and Ψ_2 with (*R*,*R*)-QuinoxP*. **a** The serial numbers of the plotted transition states. **b** Traces of the IRCs of each transition state

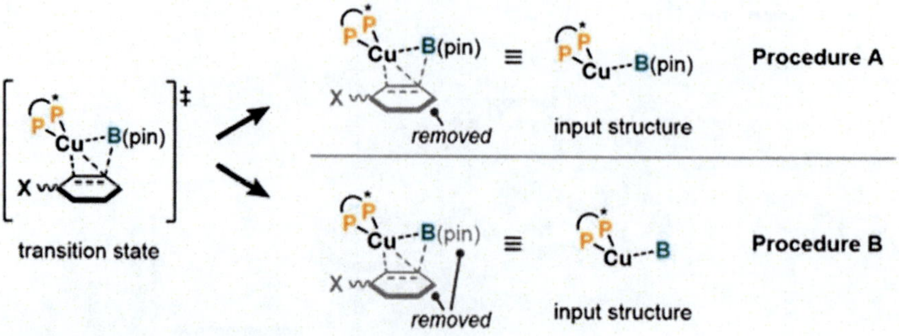

Fig. 2.102 Preprocess procedure for preparing input structures

105A, and 106A. Furthermore, the errors from the average array for clarifying the fingerprint of the input were shown in Figs. 2.103B, 104B, 105B, and 106B.

As for a clustering method, t-distributed stochastic neighbor embedding (t-SNE) was employed. t-SNE is one of unsupervised clustering methods using dimensionality reduction. The reduced features, typically two or three features, can be plotted like a scatter plot. The more similar data are located closer in the plot. To the preprocessed data described above, t-SNE was applied at the following perplexity parameters: 1, 2, 5, 10, 15, and 25 (Fig. 2.107). The shapes and colors of the data points represented types of the transition states according to the four distinctive paths shown in Fig. 2.15 (see figure notes in Fig. 2.107 for the details). The scatter plots were shown as medium cluster sizes at the appropriate perplexities, where the data were divided into two or three clusters [a) procedure A with (*R*,*R*)-QuinoxP* at perplexity = 5 and 10, b) procedure A with (*R*,*R*)-5,8-TMS-QuinoxP* at perplexity = 10 and 15, c) procedure B with (*R*,*R*)-QuinoxP* at perplexity = 5 and 10, d) procedure B with (*R*,*R*)-5,8-TMS-QuinoxP* at perplexity = 10 and 15]. In those optimized plots, the resulting clustering of (*R*,*R*)-QuinoxP* and (*R*,*R*)-5,8-TMS-QuinoxP* was expressed differently. In the case of (*R*,*R*)-5,8-TMS-QuinoxP*, The data were categorized into three clusters at the perplexity parameters = 10 and 15. The clusters have all types of the data points: the shapes and the colors. Furthermore, the results of the clustering are in good agreement with the categorization with the dihedral angles (see Fig. 2.108A for the schematic summary). Contrary, the clustering tendency completely changed in the case of (*R*,*R*)-QuinoxP*. The same shapes located closer to each other in Fig. 2.107B at perplexity = 2, 5, and 10. The data points were clustered by whether the transition states are connected to major or minor products. Therefore, the clustering results are not in agreement with the dihedral angles (see Fig. 2.108B for the schematic summary). These clustering analyses suggested the presence of the different dominant factors for determination of the ligand structures in the transition states for (*R*,*R*)-5,8-TMS-QuinoxP*.

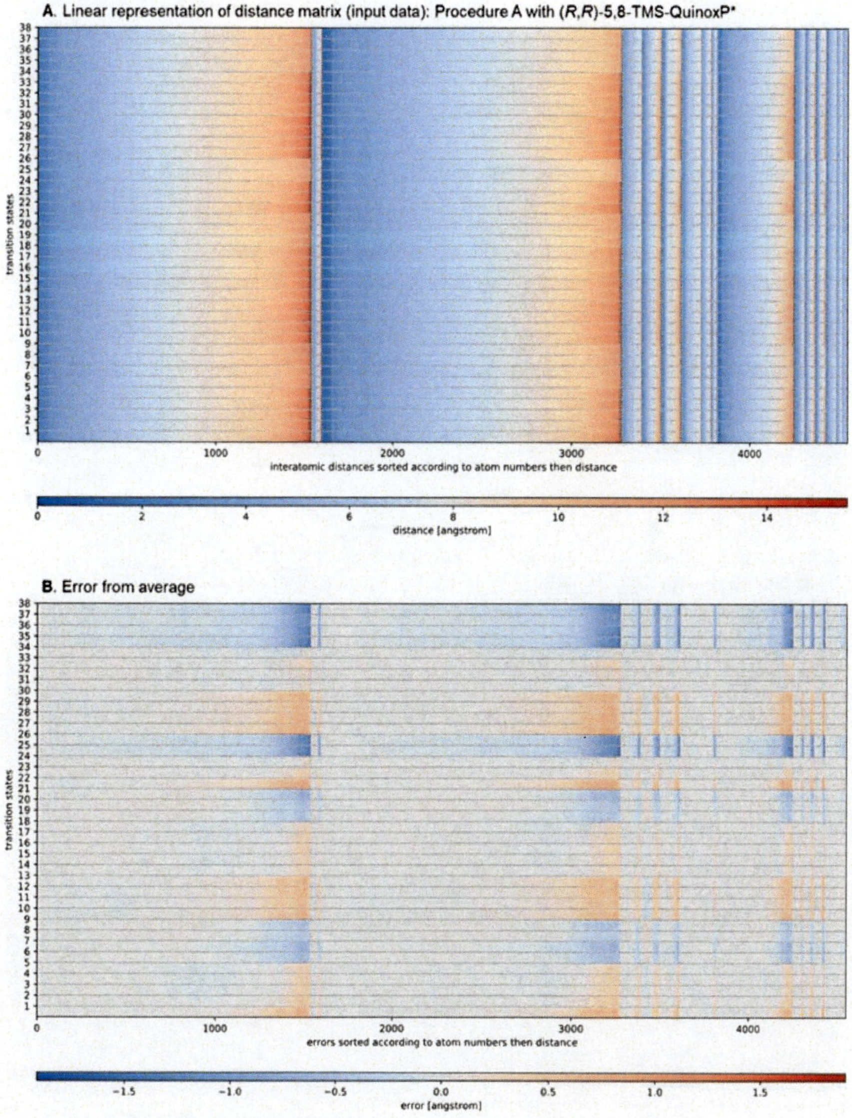

Fig. 2.103 Heatmap of input data for procedure A with (*R*,*R*)-5,8-TMS-QuinoxP*

2.5.4.3 Intramolecular Symmetry-Adapted Perturbation Theory (I-SAPT) Analysis

I investigated the effect of the silyl-modulators on the ligand backbone against the ligand conformation in the transition state **TS1** (Fig. 2.109, Table 2.9). To estimate the steric interactions between the silyl-modulators and the phosphine moieties,

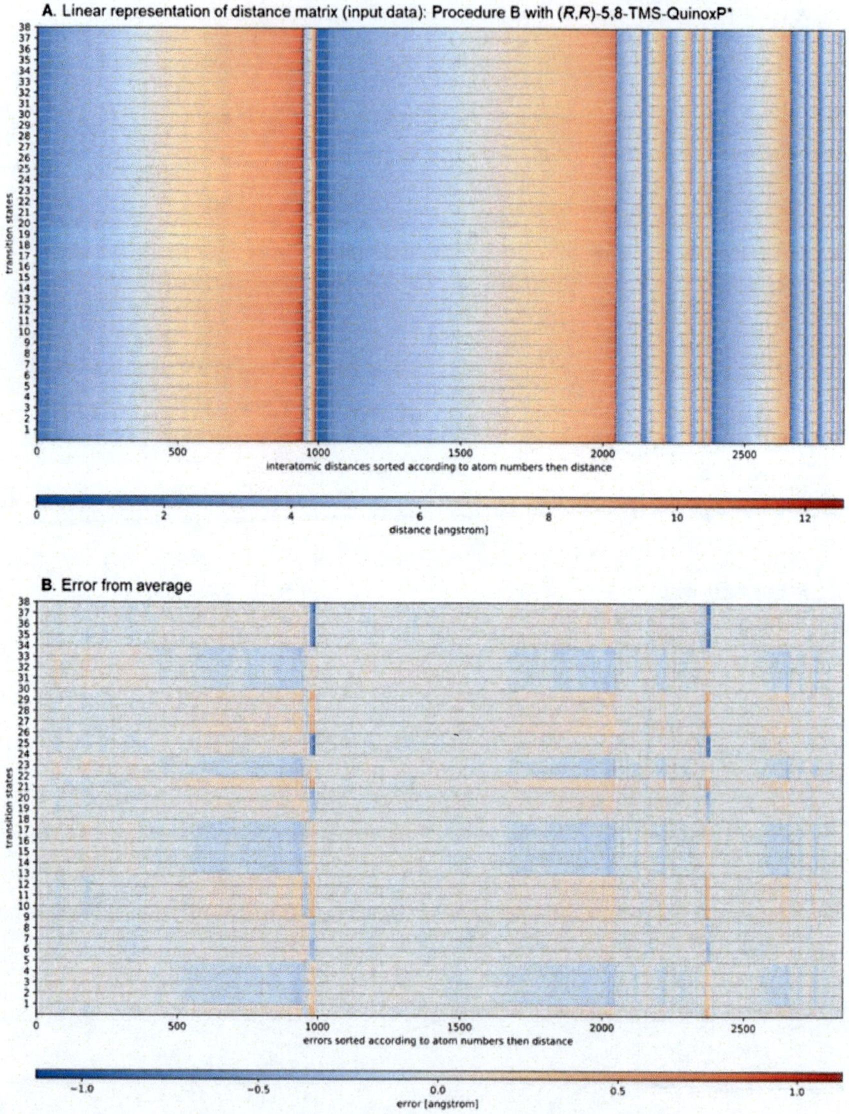

Fig. 2.104 Heatmap of input data for procedure B with (*R*,*R*)-5,8-TMS-QuinoxP*

I employed the energy decomposition analysis between the silyl-modulators and substituents of the phosphine moieties of each conformer (twisted conformation, top-opened, and bottom-opened conformer) by using intramolecular symmetry-adapted perturbation theory (I-SAPT) [104, 105]. This I-SAPT calculation can decompose the interaction energy between the two functional groups to physically

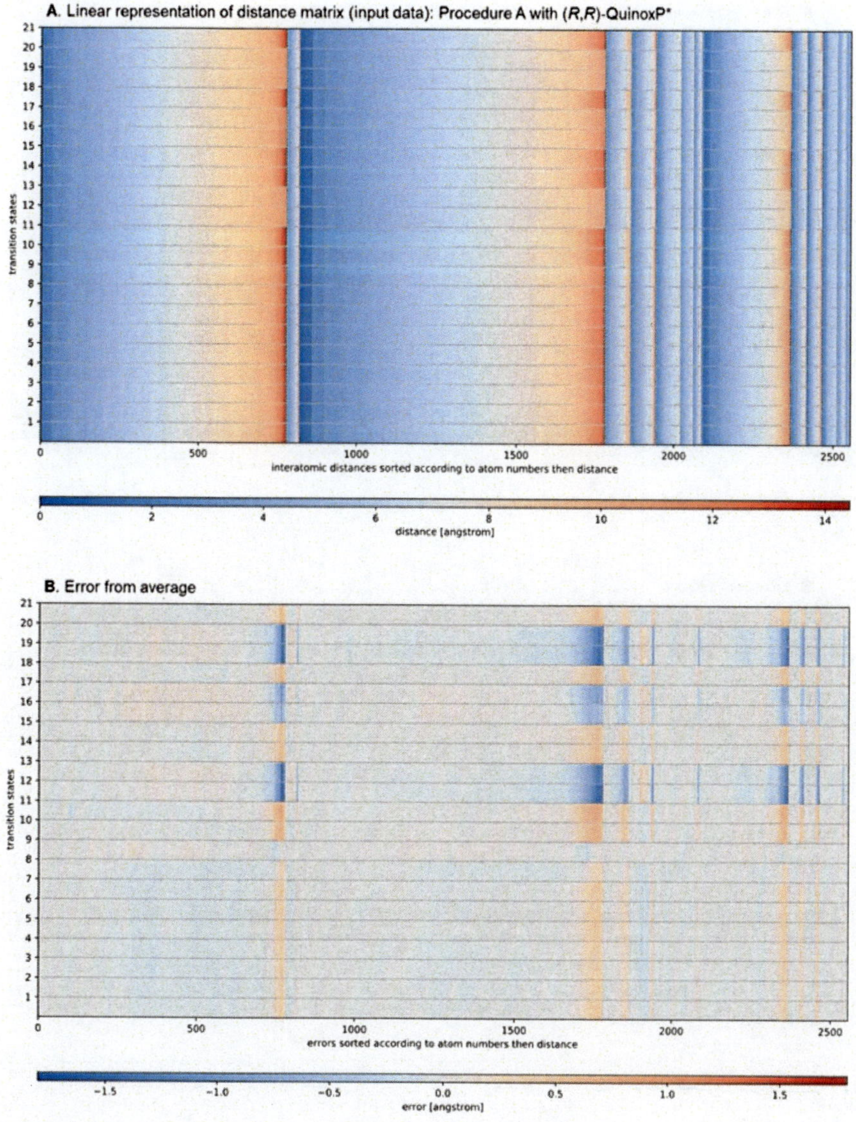

Fig. 2.105 Heatmap of input data for procedure A with (*R*,*R*)-QuinoxP*

meaningful components, including electrostatic interaction (Elst), exchange repulsion (Exch), induction effect (Ind), and dispersion interaction (Disp) terms. The I-SAPT analysis of the twisted conformation (TS1-*a*′-1) consisted by upshift (engagement A) structures indicated that the exchange interactions, namely steric repulsion, in the quadrants II and IV were higher than those of the quadrant I and III because the bulky *tert*-butyl groups strongly interact with the silyl-modulators

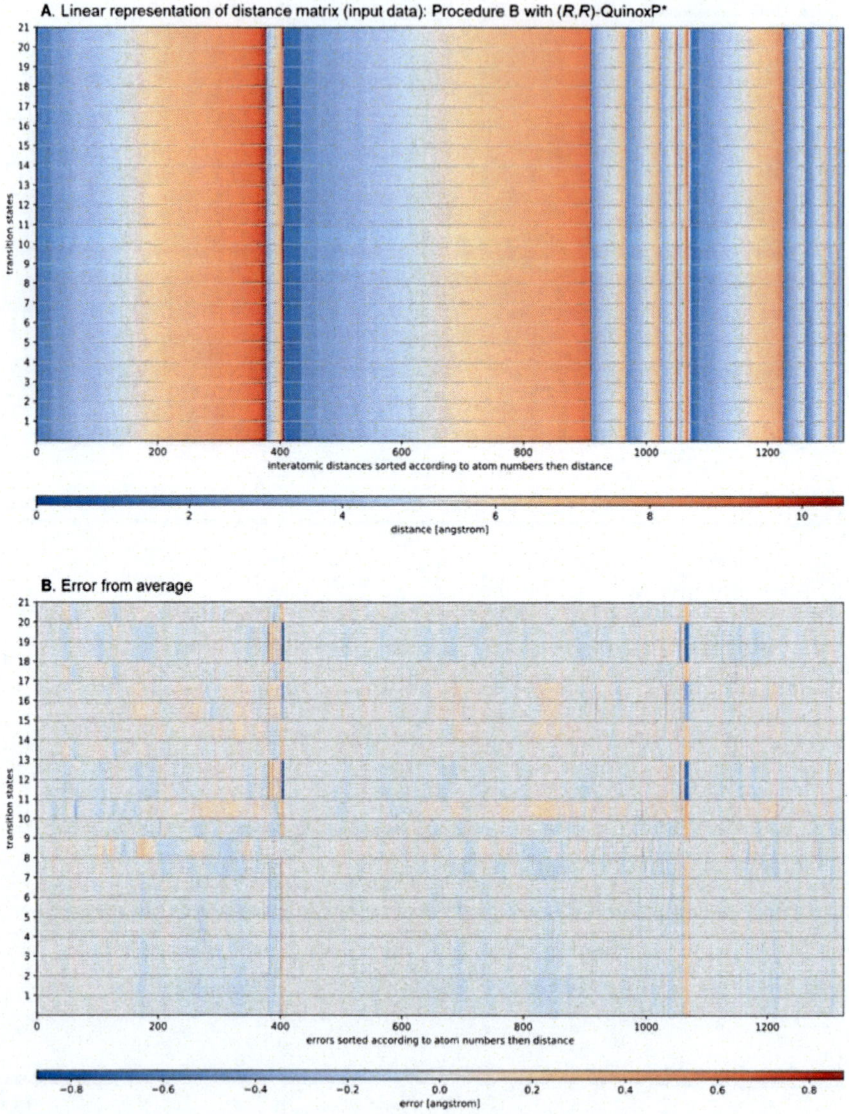

Fig. 2.106 Heatmap of input data for procedure B with (*R*,*R*)-QuinoxP*

(Fig. 2.109A, $E_{\text{exch}} = +1.8$ kcal/mol for quadrant II, $E_{\text{exch}} = +2.4$ kcal/mol for quadrant IV). However, the destabilization energy was almost canceled by the attractive dispersion interactions (Fig. 2.109A, $E_{\text{disp}} = -1.3$ kcal/mol for quadrant II, $E_{\text{disp}} = -1.4$ kcal/mol for quadrant IV; $E_{\text{total}} = +0.3$ kcal/mol for quadrant II, $E_{\text{total}} = +0.5$ kcal/mol for quadrant IV). Whereas, in the case of the top-opened conformer consisted by engagement B (downshift) for P^1 and engagement A (upshift) for P^2

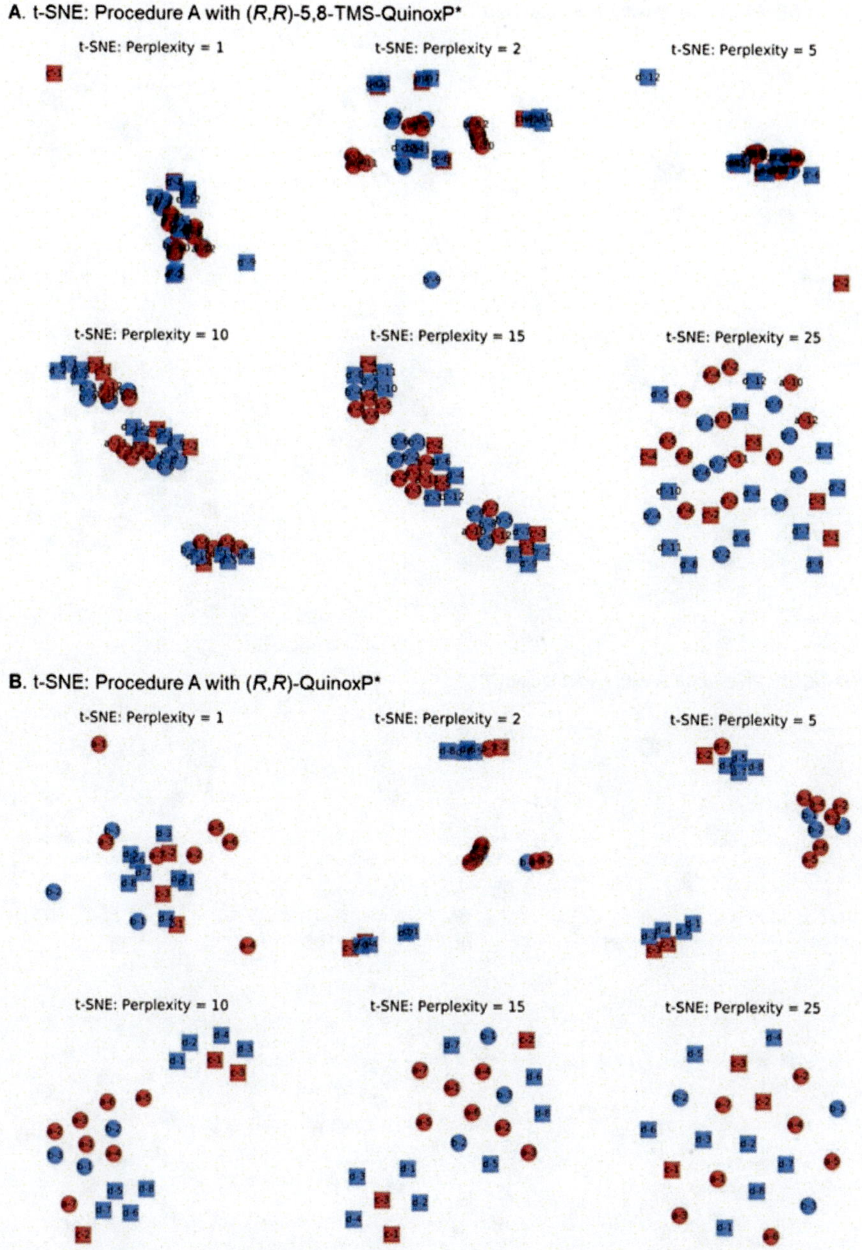

Fig. 2.107 Clustering of transition state structures using t-distributed stochastic neighbor embedding. Labels in the data points represent the indexes of the transition state ("**TS1**" is abbreviated). The shapes and colors of the data points represent as follows; red is the transition state from the (*S*)-substrate (*S*)-**1a**, blue is the transition state from the (*R*)-substrate (*R*)-**1a**, circle shape is the transition state connecting to the major product (*S*)-**3a**, and square shape is the transition state connecting to the minor product (*R*)-**3a**

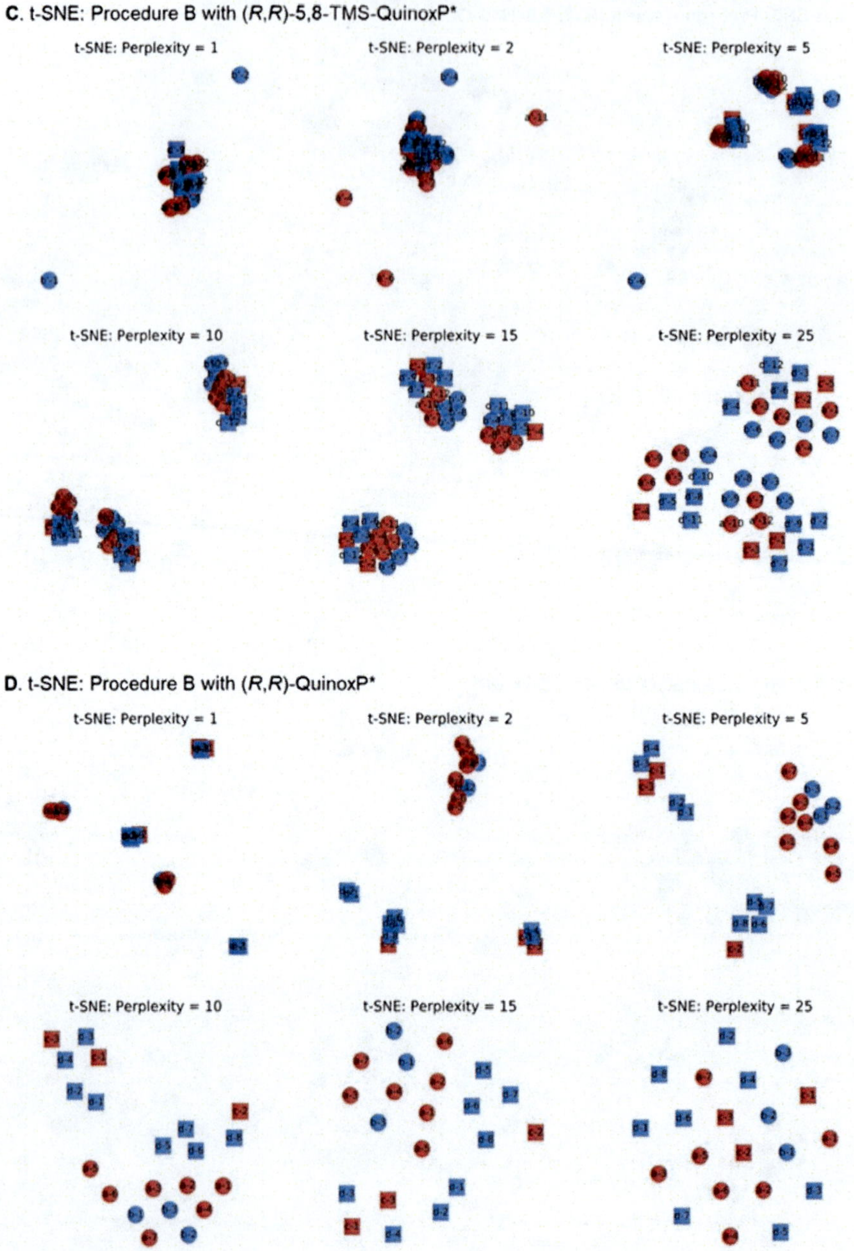

Fig. 2.107 (continued)

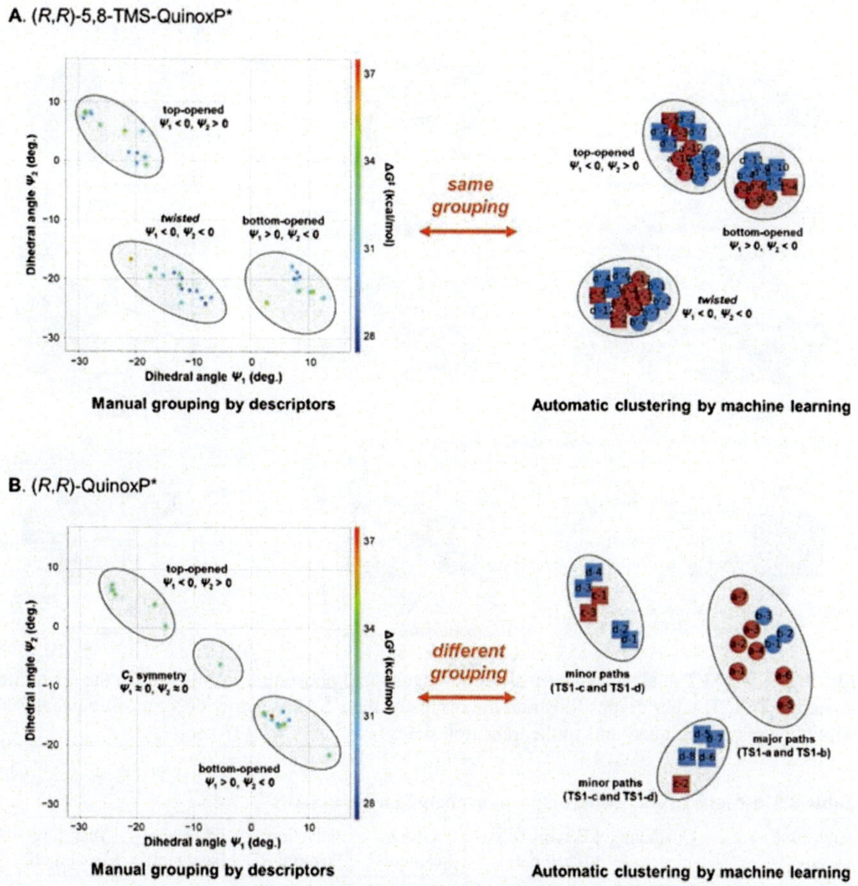

Fig. 2.108 Comparison between distribution of dihedral angles and clustering analysis

(TS1-*d*′-1), the analysis indicated that the exchange interaction in the quadrant II was significantly larger than the other interaction energy components (Fig. 2.109B, $E_{\text{exch}} = +4.5$ kcal/mol for quadrant II). Furthermore, the exchange interaction in the quadrant IV of the bottom-opened structure consisted of engagement A (upshift) for P^1 and engagement B (downshift) for P^2 (TS1-*a*′-5) was also large significantly (Fig. 2.109C, $E_{\text{exch}} = +4.7$ kcal/mol for quadrant IV). Although these large steric repulsion energies can also be canceled by the large dispersive attraction, the total interaction energy values in the quadrant II for top-opened conformation (TS1-*d*′-1) and the quadrant IV for bottom-opened conformation (TS1-*a*′-5) were still higher than those of the twisted conformation (Fig. 2.109B, $E_{\text{total}} = +0.6$ kcal/mol for quadrant II in TS1-*d*′-1; Fig. 2.109C, $E_{\text{total}} = +0.8$ kcal/mol for quadrant IV in TS1-*a*′-5). These results suggested that the engagement B (downshift), in which the silyl group meshes with one of the methyl groups in the *tert*-butyl of the ligand closely, should be unfavored compared to the engagement A (upshift).

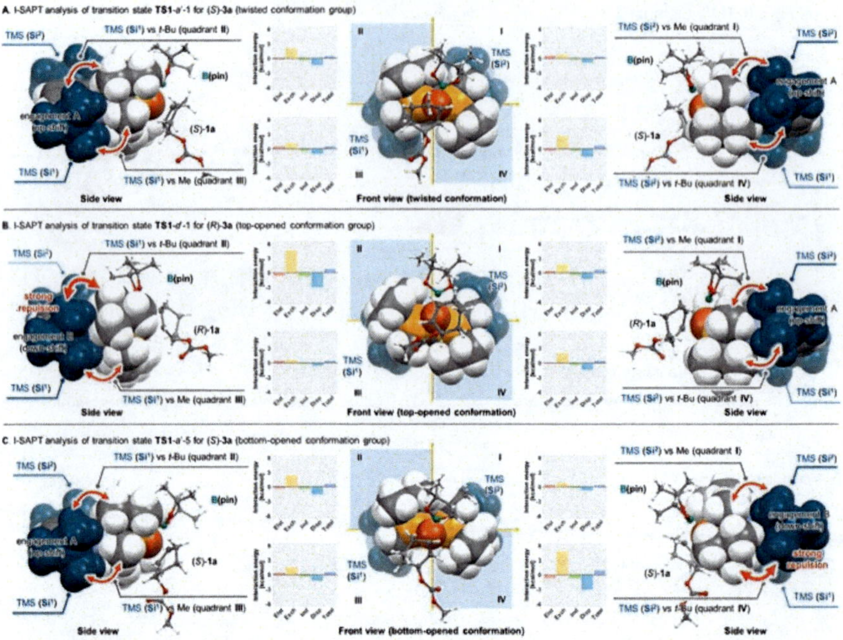

Fig. 2.109 I-SAPT analysis between silyl-modulators and phosphine substituents in the quadrant model of **TS1**. The bar graphs illustrate the results of the I-SAPT energy decomposition analysis within the respective quadrants in the transition states

Table 2.9 Results of energy decomposition analysis using I-SAPT

Transition state	Quadrant	Electrostatic [kcal/mol]	Exchange [kcal/mol]	Induction [kcal/mol]	Dispersion [kcal/mol]	Total [kcal/mol]
TS1-*a*′-1	I	0.40	0.58	−0.09	−0.68	0.20
	II	−0.10	1.84	−0.19	−1.25	0.30
	III	0.40	1.02	−0.15	−0.89	0.37
	IV	−0.22	2.43	−0.24	−1.44	0.54
TS1-*d*′-1	I	0.25	1.57	−0.20	−1.10	0.51
	II	−0.46	4.47	−0.51	−2.86	0.64
	III	0.33	0.44	−0.08	−0.45	0.25
	IV	−0.10	1.81	−0.20	−1.23	0.28
TS1-*a*′-5	I	0.28	0.57	−0.10	−0.49	0.26
	II	−0.22	2.26	−0.22	−1.37	0.46
	III	0.27	1.40	−0.1	−1.03	0.45
	IV	−0.50	4.66	−0.49	−2.87	0.79

In conclusion, the steric interactions of engagement A (upshift) between the silyl-modulators and the phosphine moiety can generate the new twisted conformation group, which cannot be generated from the catalyst bearing (*R*,*R*)-QuinoxP* without the silyl modulator.

2.5.5 Calculated Properties of All Structures (Table 2.10)

See Table 2.10.

Table 2.10 Calculated energies and thermochemical parameters of the optimized structures

Structure	*E* [Hartree]	*H* [Hartree]	*TS* [Hartree]	*G* [Hartree]
Catalyst, substrate, and product				
[(*R*,*R*)-5,8-TMS-QuinoxP*]CuOMe I′	−4068.259191	−4067.548347	0.134271	−4067.682618
[(*R*,*R*)-5,8-TMS-QuinoxP*]CuB(pin) II′	−4364.415331	−4363.558939	0.153549	−4363.712489
{[(*R*,*R*)-5,8-TMS-QuinoxP*]CuB(pin)}$_2$ V′	−8728.880994	−8727.163828	0.265018	−8727.428846
[(*R*,*R*)-5,8-TMS-QuinoxP*]CuOCO$_2$Me VI′	−4256.903225	−4256.173219	0.141818	−4256.315036
[(*R*,*R*)-QuinoxP*]CuB(pin) II	−3547.009806	−3546.368577	0.115003	−3546.483579
{[(*R*,*R*)-QuinoxP*]CuB(pin)}$_2$ V	−7094.067462	−7092.782844	0.197533	−7092.980377
Substrate 1a	−537.8027632	−537.5983632	0.05346	−537.6518222
Product 3a	−645.388873	−645.058216	0.063845	−645.12206
B_2(pin)$_2$	−822.6277948	−822.2474558	0.073005	−822.3204608
MeOBpin	−526.506886	−526.27166	0.056114	−526.327774
CO_2	−188.6068524	−188.5912704	0.026419	−188.6176894
Equilibrium structures with (*R*,*R*)-5,8-TMS-QuinoxP*				
III-a′	−4902.22478	−4901.160304	0.176127	−4901.33643
III-b′	−4902.223234	−4901.15799	0.173627	−4901.331617
III-c′	−4902.224064	−4901.159626	0.174719	−4901.334345
III-d′	−4902.225293	−4901.160566	0.17569	−4901.336256
IV-a′	−4902.271983	−4901.207659	0.174593	−4901.382252
IV-b′	−4902.268098	−4901.203711	0.174672	−4901.378383
IV-c′	−4902.266399	−4901.201197	0.173211	−4901.374408
IV-d′	−4902.26899	−4901.204448	0.174268	−4901.378716
Transition state structures of borylcupration with (*R*,*R*)-QuinoxP*				
TS1-a-1	−4084.80778	−4083.961675	0.139418	−4084.101093
TS1-a-2	−4084.808776	−4083.962104	0.138949	−4084.101052
TS1-a-3	−4084.808426	−4083.961612	0.138409	−4084.100021
TS1-a-4	−4084.807729	−4083.961286	0.138658	−4084.099944
TS1-a-5	−4084.804555	−4083.957971	0.138745	−4084.096716
TS1-a-6	−4084.803088	−4083.956409	0.138	−4084.094409
TS1-a-7	−4084.80348	−4083.956177	0.137726	−4084.093903
TS1-b-1	−4084.807813	−4083.961255	0.138858	−4084.100113
TS1-b-2	−4084.80668	−4083.96042	0.139079	−4084.099499
TS1-b-3	−4084.804912	−4083.958487	0.139491	−4084.097978
TS1-c-1	−4084.806979	−4083.960292	0.138808	−4084.0991
TS1-c-2	−4084.805774	−4083.959013	0.137324	−4084.096338
TS1-c-3	−4084.803175	−4083.956398	0.139022	−4084.09542

(continued)

Table 2.10 (continued)

Structure	*E* [Hartree]	*H* [Hartree]	*TS* [Hartree]	*G* [Hartree]
TS1-d-1	−4084.802699	−4083.956375	0.139699	−4084.096075
TS1-d-2	−4084.803418	−4083.956922	0.138739	−4084.095661
TS1-d-3	−4084.80285	−4083.956125	0.139062	−4084.095187
TS1-d-4	−4084.803472	−4083.956628	0.138517	−4084.095145
TS1-d-5	−4084.80172	−4083.955172	0.139796	−4084.094969
TS1-d-6	−4084.80172	−4083.955169	0.139756	−4084.094924
TS1-d-7	−4084.801719	−4083.955165	0.139738	−4084.094903
TS1-d-8	−4084.79633	−4083.949728	0.139364	−4084.089092
Transition state structures of borylcupration with (*R*,*R*)-5,8-TMS-QuinoxP*				
TS1-a′-1	−4902.211274	−4901.148784	0.174384	−4901.323168
TS1-a′-2	−4902.211731	−4901.149377	0.173645	−4901.323022
TS1-a′-3	−4902.212143	−4901.149053	0.17368	−4901.322732
TS1-a′-4	−4902.211339	−4901.148984	0.173742	−4901.322726
TS1-a′-5	−4902.212539	−4901.149672	0.172876	−4901.322548
TS1-a′-6	−4902.213586	−4901.149962	0.17205	−4901.322012
TS1-a′-7	−4902.210623	−4901.147634	0.173263	−4901.320897
TS1-a′-8	−4902.212139	−4901.148765	0.172013	−4901.320779
TS1-a′-9	−4902.213086	−4901.148975	0.170506	−4901.319481
TS1-a′-10	−4902.209026	−4901.145755	0.173193	−4901.318948
TS1-a′-11	−4902.207074	−4901.144117	0.172744	−4901.316861
TS1-a′-12	−4902.205385	−4901.141169	0.171375	−4901.312544
TS1-b′-1	−4902.215418	−4901.15244	0.171825	−4901.324266
TS1-b′-2	−4902.214255	−4901.151155	0.171281	−4901.322435
TS1-b′-3	−4902.210029	−4901.147464	0.174765	−4901.322229
TS1-b′-4	−4902.212991	−4901.149914	0.172292	−4901.322206
TS1-b′-5	−4902.213498	−4901.150171	0.171811	−4901.321982
TS1-b′-6	−4902.211973	−4901.149207	0.17259	−4901.321796
TS1-b′-7	−4902.211907	−4901.149322	0.172364	−4901.321685
TS1-b′-8	−4902.209923	−4901.146802	0.17293	−4901.319732
TS1-b′-9	−4902.206952	−4901.143149	0.172276	−4901.315424
TS1-c′-1	−4902.213307	−4901.149603	0.171168	−4901.320771
TS1-c′-2	−4902.209085	−4901.146003	0.172471	−4901.318474
TS1-c′-3	−4902.209525	−4901.146122	0.172251	−4901.318373
TS1-c′-4	−4902.212449	−4901.148496	0.169485	−4901.317981
TS1-c′-5	−4902.205018	−4901.142363	0.174679	−4901.317042
TS1-d′-1	−4902.209606	−4901.146651	0.173774	−4901.320426
TS1-d′-2	−4902.20932	−4901.146356	0.173209	−4901.319565
TS1-d′-3	−4902.206626	−4901.144087	0.17467	−4901.318758
TS1-d′-4	−4902.206217	−4901.143565	0.174222	−4901.317787
TS1-d′-5	−4902.206861	−4901.143814	0.173696	−4901.31751

(continued)

Table 2.10 (continued)

Structure	*E* [Hartree]	*H* [Hartree]	*TS* [Hartree]	*G* [Hartree]
TS1-d′-6	−4902.205463	−4901.14247	0.174016	−4901.316485
TS1-d′-7	−4902.2062	−4901.143011	0.173257	−4901.316268
TS1-d′-8	−4902.206526	−4901.143264	0.172913	−4901.316176
TS1-d′-9	−4902.20308	−4901.140206	0.17297	−4901.313176
TS1-d′-10	−4902.204272	−4901.140344	0.171992	−4901.312335
TS1-d′-11	−4902.202627	−4901.139356	0.172353	−4901.311709
TS1-d′-12	−4902.199734	−4901.137188	0.173277	−4901.310465
Transition state structures of β-elimination with (*R*,*R*)-5,8-TMS-QuinoxP*				
TS2-a′	−4902.263833	−4901.20229	0.175875	−4901.378165
TS2-b′	−4902.235471	−4901.173185	0.173449	−4901.346634
TS2-c′	−4902.240498	−4901.178063	0.173064	−4901.351127
TS2-d′	−4902.258302	−4901.196367	0.175232	−4901.371599
Transition state structures of borylcupration to acetal substrate				
Substrate 1 h (LG = OMe)	−385.0969904	−384.9349464	0.045281	−384.9802
TS1-a′ (acetal)	−4749.508283	−4748.488737	0.167264	−4748.6560
TS1-b′ (acetal)	−4749.506961	−4748.487983	0.166262	−4748.6542
TS1-c′ (acetal)	−4749.505413	−4748.486026	0.165457	−4748.6515
TS1-d′ (acetal)	−4749.508222	−4748.48849	0.166036	−4748.6545
Transition state structures of borylcupration to linear substrate				
Linear substrate	−499.6880583	−499.4919753	0.056	−499.5479753
TS1-a′ (linear)	−4864.10069	−4863.047314	0.174369	−4863.221683
TS1-b′ (linear)	−4864.105482	−4863.052248	0.175343	−4863.227591
TS1-c′ (linear)	−4864.102283	−4863.048559	0.175231	−4863.22379
TS1-d′ (linear)	−4864.098836	−4863.045375	0.174832	−4863.220207
SCF energies at DLPNO-CCSD(T)/Def2TZVPP/SMD(THF) level of theory				
TS1-a-1	−4079.899729	–	–	–
TS1-b-1	−4079.899977	–	–	–
TS1-c-1	−4079.899658	–	–	–
TS1-d-1	−4079.894295	–	–	–
TS1-a′-1	−4895.953133	–	–	–
TS1-b′-1	−4895.956917	–	–	–
TS1-c′-1	−4895.954895	–	–	–
TS1-d′-1	−4895.951016	–	–	–

References

1. Hall DG (ed) (2011) Boronic acids: preparation, applications in organic synthesis and medicine, vols 1–2. Wiley-VCH, Weinheim
2. Diner C, Szabó KJ (2017) Recent advances in the preparation and application of allylboron species in organic synthesis. J Am Chem Soc 139:2–14
3. Hoffmann RW (1982) diastereogenic addition of crotylmetal compounds to aldehydes. Angew Chem Int Ed 21: 555–566

4. Kennedy JWJ, Hall DG (2003) Recent advances in the activation of boron and silicon reagents for stereocontrolled allylation reactions. Angew Chem Int Ed 42:4732–4739
5. Marek I, Sklute G (2007) Creation of quaternary stereocenters in carbonyl allylation reactions. Chem Commun 1683–1691
6. Hall DG (2008) New preparative methods for allylic boronates and their application in stereoselective catalytic allylborations. Pure Appl Chem 80:913–927
7. Yus M, González-Gómez JC, Foubelo F (2013) Diastereoselective allylation of carbonyl compounds and imines: application to the synthesis of natural products. Chem Rev 113:5595–5698
8. Huo H-X, Duvall JR, Huang M-Y, Hong R (2014) Catalytic asymmetric allylation of carbonyl compounds and imines with allylic boronates. Org Chem Front 1:303–320
9. Scott HK, Aggarwal VK (2011) Highly enantioselective synthesis of tertiary boronic esters and their stereospecific conversion to other functional groups and quaternary stereocentres. Chem Eur J 17:13124–13132
10. Leonori D, Aggarwal VK (2014) Lithiation-borylation methodology and its application in synthesis. Acc Chem Res 47:3174–3183
11. Leonori D, Aggarwal VK (2015) Stereospecific couplings of secondary and tertiary boronic esters. Angew Chem Int Ed 54:1082–1096
12. Mlynarski SN, Karns AS, Morken JP (2012) Direct stereospecific amination of alkyl and aryl pinacol boronates. J Am Chem Soc 134:16449–16451
13. Alam R, Vollgraff T, Eriksson L, Szabó KJ (2015) Synthesis of adjacent quaternary stereocenters by catalytic asymmetric allylboration. J Am Chem Soc 137:11262–11265
14. Alam R, Diner C, Jonker S, Eriksson L, Szabó KJ (2016) Catalytic asymmetric allylboration of indoles and dihydroisoquinolines with allylboronic acids: stereodivergent synthesis of up to three contiguous stereocenters. Angew Chem Int Ed 55:14417–14421
15. García-Ruiz C, Chen JL-Y, Sandford C, Feeney K, Lorenzo P, Berionni G, Mayr H, Aggarwal VK (2017) Stereospecific allylic functionalization: the reactions of allylboronate complexes with electrophiles. J Am Chem Soc 139:15324–15327
16. Jonker SJT, Diner C, Schulz G, Iwamoto H, Eriksson L, Szabó KJ (2018) Catalytic asymmetric propargyl- and allylboration of hydrazonoesters: a metal-free approach to sterically encumbered chiral α-amino acid derivatives. Chem Commun 54:12852–12855
17. Kiyokawa K, Hata S, Kainuma S, Minakata S (2019) Electrophilic cyanation of allylic boranes: synthesis of β, γ-unsaturated nitriles containing allylic quaternary carbon centers. Chem Commun 55:458–461
18. Hari DP, Madhavachary R, Fasano V, Haire J, Aggarwal VK (2021) Highly diastereoselective strain-increase allylborations: rapid access to alkylidenecyclopropanes and alkylidenecyclobutanes. J Am Chem Soc 143:7462–7470
19. Börner A (ed) (2008) Phosphorus ligands in asymmetric catalysis: synthesis and application, vols 1–3. Wiley-VCH, Weinheim
20. Zhou Q-L (ed) (2011) Privileged chiral ligands and catalysts. Wiley-VCH, Weinheim
21. Phosphorus(III) Ligands in Homogeneous Catalysis Design and Synthesis; Kamer, P. C. J., van Leeuwen, P. W. N. M., Eds.; Wiley-VCH: Weinheim, 2012.
22. Stradiotto M, Lundgren RJ (eds) (2016) Ligand design in metal chemistry: reactivity and catalysis. Wiley-VCH, Weinheim
23. Van Leeuwen PWNM, Kamer PCJ, Reek JNH, Dierkes P (2000) Ligand bite angle effects in metal-catalyzed C-C bond formation. Chem Rev 100:2741–2769
24. O'Neill MJ, Riesebeck T, Cornella J (2018) Thorpe–ingold effect in branch-selective alkylation of unactivated aryl fluorides. Angew Chem Int Ed 57:9103–9107
25. Noyori R (2002) Asymmetric catalysis: science and opportunities (Nobel lecture). Angew Chem Int Ed 41:2008–2022
26. Berthod M, Mignani G, Woodward G, Lemaire M (2005) Modified BINAP: the how and the why. Chem Rev 105:1801–1836
27. Shimizu H, Nagasaki I, Matsumura K, Sayo N, Saito T (2007) Developments in asymmetric hydrogenation from an industrial perspective. Acc Chem Res 40:1385–1393

28. Burk MJ (1991) C_2-symmetric bis(phospholanes) and their use in highly enantioselective hydrogenation reactions. J Am Chem Soc 113:8518–8519
29. Burk MJ, Feaster JE, Nugent WA, Harlow RL (1993) Preparation and use of C_2-symmetric bis(phospholanes): production of α-amino acid derivatives via highly enantioselective hydrogenation reactions. J Am Chem Soc 115:10125–10138
30. Burk MJ (2000) Modular phospholane ligands in asymmetric catalysis. Acc Chem Res 33:363–372
31. Wada R, Shibuguchi T, Makino S, Oisaki K, Kanai M, Shibasaki M (2006) Catalytic enantioselective allylation of ketoimines. J Am Chem Soc 128:7687–7691
32. Lu G, Liu RY, Yang Y, Fang C, Lambrecht DS, Buchwald SL, Liu P (2017) Ligand-substrate dispersion facilitates the copper-catalyzed hydroamination of unactivated olefins. J Am Chem Soc 139:16548–16555
33. Imamoto T (2016) Searching for practically useful P-chirogenic phosphine ligands. Chem Rec 16:2659–2673
34. Dutartre M, Bayardon J, Jugé S (2016) Applications and stereoselective syntheses of P-chirogenic phosphorus compounds. Chem Soc Rev 45:5771–5794
35. Imamoto T, Sugita K, Yoshida K (2005) An air-stable P-chiral phosphine ligand for highly enantioselective transition-metal-catalyzed reactions. J Am Chem Soc 127:11934–11935
36. Iwamoto H, Imamoto T, Ito H (2018) Computational design of high-performance ligand for enantioselective markovnikov hydroboration of aliphatic terminal alkenes. Nat Commun 9:2290
37. In 2019, I and co-workers in the Ito group achieved a copper(I)-catalyzed enantioconvergent boryl substitution of racemic benzyl chloride using a modified QuinoxP*-type ligand. See, Ref. [38]
38. Iwamoto H, Endo K, Ozawa Y, Watanabe Y, Kubota K, Imamoto T, Ito H (2019) Copper(I)-catalyzed enantioconvergent borylation of racemic benzyl chlorides enabled by quadrant-by-quadrant structure modification of chiral bisphosphine ligands. Angew Chem Int Ed 58:11112–11117
39. Hu A, Ngo HL, Lin W (2004) Remarkable 4,4′-substituent effects on binap: highly enantioselective Ru catalysts for asymmetric hydrogenation of β-aryl ketoesters and their immobilization in room-temperature ionic liquids. Angew Chem Int Ed 43:2501–2504
40. Hu A, Ngo HL, Lin W (2004) 4,4′-Disubstituted BINAPs for highly enantioselective Ru-catalyzed asymmetric hydrogenation of ketones. Org Lett 6:2937–2940
41. Ogasawara M, Ngo HL, Sakamoto T, Takahashi T, Lin W (2005) Applications of 4,4′-$(Me_3Si)_2$-BINAP in transition-metal-catalyzed asymmetric carbon−carbon bond-forming reactions. Org Lett 7:2881–2884
42. Storch G, Trapp O (2015) Temperature-controlled bidirectional enantioselectivity in a dynamic catalyst for asymmetric hydrogenation. Angew Chem Int Ed 54:3580–3586
43. Ito H, Kunii S, Sawamura M (2010) Direct enantio-convergent transformation of racemic substrates without racemization or symmetrization. Nat Chem 2:972–976
44. Nafe J, Herbert S, Auras F, Karaghiosoff K, Bein T, Knochel P (2015) Functionalization of quinoxalines by using TMP bases: preparation of tetracyclic heterocycles with high photoluminescene quantum yields. Chem Eur J 21:1102–1107
45. Zhang Z, Tamura K, Mayama D, Sugiya M, Imamoto T (2012) Three-hindered quadrant phosphine ligands with an aromatic ring backbone for the rhodium-catalyzed asymmetric hydrogenation of functionalized alkenes. J Org Chem 77:4184–4188
46. Mohr JT, Moore JT, Stoltz BM (2016) Enantioconvergent catalysis. Beilstein J Org Chem 12:2038–2045
47. Delvos LB, Oestreich M (2015) Temperature-dependent direct enantioconvergent silylation of a racemic cyclic allylic phosphate by copper(I)-catalyzed allylic substitution. Synthesis 47:924–933
48. Bohan PT, Toste FD (2017) Well-defined chiral gold(III) complex catalyzed direct enantio-convergent kinetic resolution of 1,5-enynes. J Am Chem Soc 139:11016–11019

49. Ge Y, Cui XY, Tan SM, Jiang H, Ren J, Lee N, Lee R, Tan CH (2019) Guanidine–copper complex catalyzed allylic borylation for the enantioconvergent synthesis of tertiary cyclic allylboronates. Angew Chem Int Ed 58:2382–2386
50. Ito H, Kawakami C, Sawamura M (2005) Copper-catalyzed γ-selective and stereospecific substitution reaction of allylic carbonates with diboron: efficient route to chiral allylboron compounds. J Am Chem Soc 127:16034–16035
51. The perfect *E*/*Z*-selectivity of this borylative kinetic resolution stands in sharp contrast to that of the copper(I)-catalyzed enantioselective alkylation of linear allyl electrophiles; for details, see: Langlois JB, Alexakis A (2011) Identification of a valuable kinetic process in copper-catalyzed asymmetric allylic alkylation. Angew Chem Int Ed 50:1877–1881
52. Imamoto T, Tamura K, Zhang Z, Horiuchi Y, Sugiya M, Yoshida K, Yanagisawa A, Gridnev ID (2012) Rigid P-chiral phosphine ligands with *tert*-butylmethylphosphino groups for rhodium-catalyzed asymmetric hydrogenation of functionalized alkenes. J Am Chem Soc 134:1754–1769
53. Lessard S, Peng F, Hall DG (2009) α-hydroxyalkyl heterocycles via chiral allylic boronates: Pd-catalyzed borylation leading to a formal enantioselective isomerization of allylic ether and amine. J Am Chem Soc 131:9612–9613
54. The slight enantioselectivity erosion compared to the results of (*R*)-**3c** should be caused by a small amount of remaining substrate or catalyst.
55. Goering HL, Singleton VD (1976) On the stereochemistry of conversion of allylic esters to olefins with lithium dialkylcuprates. J Am Chem Soc 98:7854–7855
56. Yamanaka M, Kato S, Nakamura E (2004) Mechanism and regioselectivity of reductive elimination of π-allylcopper (III) intermediates. J Am Chem Soc 126:6287–6293
57. Langlois JB, Emery D, Mareda J, Alexakis A (2012) Mechanistic identification and improvement of a direct enantioconvergent transformation in copper-catalyzed asymmetric allylic alkylation. Chem Sci 3:1062–1069
58. Rideau E, You H, Sidera M, Claridge TDW, Fletcher SP (2017) Mechanistic studies on a Cu-catalyzed asymmetric allylic alkylation with cyclic racemic starting materials. J Am Chem Soc 139:5614–5624
59. Park JK, Lackey HH, Ondrusek BA, McQuade DT (2011) Stereoconvergent synthesis of chiral allylboronates from an *E*/*Z* mixture of allylic aryl ethers using a 6-NHC-Cu(I) catalyst. J Am Chem Soc 133:2410–2413
60. Guzman-Martinez A, Hoveyda AH (2010) Enantioselective synthesis of allylboronates bearing a tertiary or quaternary B-substituted stereogenic carbon by NHC-Cu-catalyzed substitution reactions. J Am Chem Soc 132:10634–10637
61. Yamamoto E, Takenouchi Y, Ozaki T, Miya T, Ito H (2014) Copper(I)-catalyzed enantioselective synthesis of α-chiral linear or carbocyclic (*E*)-(γ-alkoxyallyl)boronates. J Am Chem Soc 136:16515–16521
62. Borner C, Anders L, Brandhorst K, Kleeberg C (2017) Elusive phosphine copper(I) boryl complexes: synthesis, structures, and reactivity. Organometallics 36:4687–4690
63. Tobisch S (2017) Copper-catalysed aminoboration of vinylarenes with hydroxylamine esters-A computational mechanistic study. Chem Eur J 23:17800–17809
64. The DFT calculation of the borylcupration steps of the borylative kinetic resolution was also in good agreement with the experimental results. For details, see chapter 2.5.
65. Durand DJ, Fey N (2019) Computational ligand descriptors for catalyst design. Chem Rev 119:6561–6594
66. This classification produced results similar to the clustering based on the atom coordinates of the catalyst structures in the TSs by clustering. Detailed results and discussions regarding the analysis are summarized in chapter 2.5.
67. Sheldrick GM (2015) Crystal structure refinement with SHELXL. Acta Cryst Sect C 71:3–8
68. Hatano M, Kamiya S, Ishihara K (2012) In situ generated "Lanthanum(III) Nitrate Alkoxide" as a highly active and nearly neutral transesterification catalyst. Chem Commun 48:9465–9467
69. Takahata H, Suto Y, Kato E, Yoshimura Y, Ouchi H (2007) A new preparation of homochiral n-protected 5-hydroxy-3-piperidenes, promising chiral building blocks, by palladium-catalyzed deracemization of their alkyl carbonates. Adv Synth Catal 349:685–693

70. Trost BM, Richardson J, Yong K (2006) Palladium-catalyzed chemo- and enantioselective oxidation of allylic esters and carbonates. J Am Chem Soc 128:2540–2541
71. Kinderman SS, Doodeman R, van Beijma JW, Russcher JC, Tjen KCMF, Kooistra TM, Mohaselzadeh H, van Maarseveen JH, Hiemstra H, Schoemaker HE, Rutjes FPJT (2002) Ring-closing metathesis of allylic O, O- and N, O-acetals. Adv Synth Catal 344:736–748
72. Villar L, Orlov NV, Kondratyev NS, Uria U, Vicario JL, Malkov AV (2018) Kinetic resolution of secondary allyl boronates and their application in the synthesis of homoallylic amines. Chem Eur J 24:16262–16265
73. Sasaki Y, Zhong C, Sawamura M, Ito H (2010) Copper(I)-catalyzed asymmetric monoborylation of 1,3-dienes: synthesis of enantioenriched cyclic homoallyl- and allylboronates. J Am Chem Soc 132:1226–1227
74. Ye J, Zhao J, Xu J, Mao Y, Zhang YJ (2013) Pd-catalyzed stereospecific allyl-aryl coupling of allylic alcohols with arylboronic acids. Chem Commun 49:9761–9763
75. Chen F, Zhang Y, Yu L, Zhu S (2017) Enantioselective NiH/Pmrox-catalyzed 1,2-reduction of α,β-unsaturated ketones. Angew Chem Int Ed 56:2022–2025
76. Peters R, Xin Z, Maier F (2010) Catalyst versus substrate induced selectivity: kinetic resolution by palladacycle catalyzed allylic imidate rearrangements. Chem Asian J 5:1770–1774
77. Johansson C, Lloyd-Jones GC, Norrby PO (2010) Memory and dynamics in Pd-catalyzed allylic alkylation with P,N-ligands. Tetrahedron Asymmetry 21:1585–1592
78. Ito H, Ito S, Sasaki Y, Matsuura K, Sawamura M (2007) Copper-catalyzed enantioselective substitution of allylic carbonates with diboron: an efficient route to optically active α-chiral allylboronates. J Am Chem Soc 129:14856–14857
79. Nagata K, Matsukawa S, Imamoto T (2000) Stereoselective, oxidative one-carbon degradation of alkyl(dimethyl)phosphine-boranes. Synthesis of enantiomerically enriched secondary phosphine-boranes. J Org Chem 65:4185–4188
80. Gaussian 09, Revision D.01, Frisch MJ, Trucks GW, Schlegel HB, Scuseria GE, Robb MA, Cheeseman JR, Scalmani G, Barone V, Petersson GA, Nakatsuji H, Li X, Caricato M, Marenich A, Bloino J, Janesko BG, Gomperts R, Mennucci B, Hratchian HP, Ortiz JV, Izmaylov AF, Sonnenberg JL, Williams-Young D, Ding F, Lipparini F, Egidi F, Goings J, Peng B, Petrone A, Henderson T, Ranasinghe D, Zakrzewski VG, Gao J, Rega N, Zheng G, Liang W, Hada M, Ehara M, Toyota K, Fukuda R, Hasegawa J, Ishida M, Nakajima T, Honda Y, Kitao Y, Nakai H, Vreven T, Throssell K, Montgomery JA Jr, Peralta JE, Ogliaro F, Bearpark M, Heyd JJ, Brothers E, Kudin KN, Staroverov VN, Keith T, Kobayashi R, Normand J, Raghavachari K, Rendell A, Burant JC, Iyengar SS, Tomasi J, Cossi M, Millam JM, Klene M, Adamo C, Cammi R, Ochterski JW, Martin RL, Morokuma K, Farkas O, Foresman JB, Fox DJ (2009) Gaussian, Inc., Wallingford CT
81. Gaussian 16, Revision B.01, Frisch MJ, Trucks GW, Schlegel HB, Scuseria GE, Robb MA, Cheeseman JR, Scalmani G, Barone V, Petersson GA, Nakatsuji H, Li X, Caricato M, Marenich AV, Bloino J, Janesko BG, Gomperts R, Mennucci B, Hratchian HP, Ortiz JV, Izmaylov AF, Sonnenberg JL, Williams-Young D, Ding F, Lipparini F, Egidi F, Goings J, Peng B, Petrone A, Henderson T, Ranasinghe D, Zakrzewski VG, Gao J, Rega N, Zheng G, Liang W, Hada M, Ehara M, Toyota K, Fukuda R, Hasegawa J, Ishida M, Nakajima T, Honda Y, Kitao O, Nakai H, Vreven T, Throssell K, Montgomery JA Jr, Peralta JE, Ogliaro F, Bearpark MJ, Heyd JJ, Brothers EN, Kudin KN, Staroverov VN, Keith TA, Kobayashi R, Normand J, Raghavachari K, Rendell AP, Burant JC, Iyengar SS, Tomasi J, Cossi M, Millam JM, Klene M, Adamo C, Cammi R, Ochterski JW, Martin RL, Morokuma K, Farkas O, Foresman JB, Fox DJ (2016) Gaussian, Inc., Wallingford CT
82. Chai J-D, Head-Gordon M (2008) Long-range corrected hybrid density functionals with damped atom-atom dispersion corrections. Phys Chem Chem Phys 10:6615–6620
83. Marenich AV, Cramer CJ, Truhlar DG (2009) Universal solvation model based on solute electron density and on a continuum model of the solvent defined by the bulk dielectric constant and atomic surface tensions. J Phys Chem B 113:6378–6396
84. Weigend F, Ahlrichs R (2005) Balanced basis sets of split valence, triple zeta valence and quadruple zeta valence quality for H to Rn: design and assessment of accuracy. Phys Chem Chem Phys 7:3297–3305

85. Weigend F (2006) Accurate coulomb-fitting basis sets for H to Rn. Phys Chem Chem Phys 8:1057–1065
86. Fukui K (1981) The path of chemical reactions - the IRC approach. Acc Chem Res 14:363–368
87. Funes-Ardoiz I, Paton RS (2018) GoodVibes. Version 2.0.1
88. Grimme S (2012) Supramolecular binding thermodynamics by dispersion-corrected density functional theory. Chem Eur J 18:9955–9964
89. Riplinger C, Neese F (2013) An efficient and near linear scaling pair natural orbital based local coupled cluster method. J Chem Phys 138:034106
90. Riplinger C, Sandhoefer B, Hansen A, Neese F (2013) Natural triple excitations in local coupled cluster calculations with pair natural orbitals. J Chem Phys 139:134101
91. Riplinger C, Pinski P, Becker U, Valeev EF, Neese F (2016) Sparse maps–a systematic infrastructure for reduced-scaling electronic structure methods. II. Linear scaling domain based pair natural orbital coupled cluster theory. J Chem Phys 144:024109
92. Neese F (2012) The ORCA program system. Wiley Interdiscip Rev Comput Mol Sci 2:73
93. Neese F (2017) Software update: the ORCA program system. Version 4.0. Wiley Interdiscip Rev Comput Mol Sci 8:e1327
94. Python Software Foundation. Python Language Reference, Version 3.6.5. http://www.python.org
95. Van Der Walt S, Colbert SC, Varoquaux G (2011) The NumPy array: a structure for efficient numerical computation. Comput Sci Eng 13:22–30
96. Bengio Y, Courville A, Vincent P (2013) Representation learning: a review and new perspectives. IEEE Trans Pattern Anal Mach Intell 35:1798–1828
97. Hagberg AA, Schult DA, Swart PJ (2008) Proceedings of the 7th Python in science conference (SciPy 2008), Varoquaux G, Vaught T, Millman J (eds),11
98. Hunter JD (2007) Matplotlib: a 2D graphics environment. Comput Sci Eng 9:90–95
99. Maaten LVD, Hinton G (2008) Visualizing data using t-SNE. J Mach Learn Res 9:2579
100. Johnson ER, Keinan S, Mori-Sánchez P, Contreras-García J, Cohen AJ, Yang W (2010) Revealing noncovalent interactions. J Am Chem Soc 132:6498–6506
101. Lu T, Chen F (2012) Multiwfn: a multifunctional wavefunction analyzer. J Comput Chem 33:580–592
102. Contreras-García J, Johnson ER, Keinan S, Chaudret R, Piquemal JP, Beratan DN, Yang W (2011) NCIPLOT: a program for plotting noncovalent interaction regions. J Chem Theory Comput 7:625–632
103. Humphrey W, Dalke A, Schulten K (1996) VMD: visual molecular dynamics. J Mol Graph 14:33–38
104. Parrish RM, Gonthier JF, Corminbœuf C, Sherrill CD (2015) Communication: practical intramolecular symmetry adapted perturbation theory via Hartree-Fock embedding. J Chem Phys 143:051103
105. Parrish RM, Burns LA, Smith DGA, Simmonett AC, DePrince AE, Hohenstein EG, Bozkaya U, Sokolov AY, Di Remigio R, Richard RM, Gonthier JF, James AM, McAlexander HR, Kumar A, Saitow M, Wang X, Pritchard BP, Verma P, Schaefer III HF, Patkowski K, King RA, Valeev EF, Evangelista FA, Turney JM, Crawford TD, Sherrill CD (2017) Psi4 1.1: an open-source electronic structure program emphasizing automation, advanced libraries, and interoperability. J Chem Theory Comput 13:3185–3197

Chapter 3
Allylcopper(I) Isomerization-Enabled Copper(I)-Catalyzed Intramolecular Alkylboration of Terminal Allenes

3.1 Introduction

Alkenyl boronates are widely used organometallic compounds as transmetallation reagents for transition metal-catalyzed reactions [1, 2], e.g., Suzuki–Miyaura cross-coupling reaction (Fig. 3.1) [3–5]. The reaction is commonly used for the synthesis of styrene and diene derivatives, including multi-substituted ones. Thus, expansion of the skeletal diversity of the alkenyl boronates benefits the accessible structure of the resulting coupling products.

A copper(I)-catalyzed carboboration of carbon–carbon multiple bonds is a three-component coupling reaction between the multiple bonds, including alkenes, alkynes, allenes, and dienes, carbon electrophiles, and a boron source (Fig. 3.2) [6–8]. The reaction proceeds via the construction of carbon–carbon and carbon–boron bonds and is able to afford multi-functional organoboron compounds. In particular, the reaction of alkynes and allenes can produce the multi-substituted alkenyl boronates in one step (Fig. 3.2).

An intramolecular carboboration, called a borylative cyclization, furnishes the organoboron compounds bearing cyclic functionality, which are useful building blocks for the preparation of complex skeletons [9]. In 2013, the Ito group developed the first intramolecular alkylboration reaction of terminal alkenes and synthesized the alkyl boronates bearing cyclic moiety including small rings (Fig. 3.3A) [10–12]. The reaction using copper(I) catalyst and diboron reagent as the boron source proceeds through the borylcupration of the alkene moiety to generate the alkyl copper(I) intermediate, and subsequent intramolecular α-alkylation with the alkyl halide moiety. In 2015, the Ito group also reported the borylative cyclization of alkynylsilanes tethered to alkyl halides (Fig. 3.3B) [13]. In 2017, I and co-workers in the Ito group then developed the intramolecular alkylboration of propargyl ethers and amines, which proceeded via alkenyl copper(I) intermediate and furnished the heterocyclic compounds having exo-alkenyl boronates (Fig. 3.3C) [14].

Hence, I planned to develop the intramolecular alkylboration reaction of allenes to synthesize the exo-cyclization product (Fig. 3.4A). Since the Hoveyda group reported

Y. Ozawa, *Copper(I)-Catalyzed Stereoselective Borylation Reactions*, Springer Theses,
https://doi.org/10.1007/978-981-99-1098-4_3

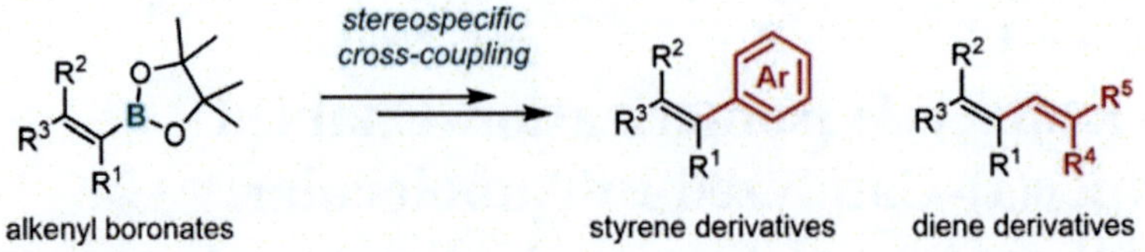

Fig. 3.1 Alkenyl boronates as coupling partner in Suzuki–Miyaura cross-coupling reaction

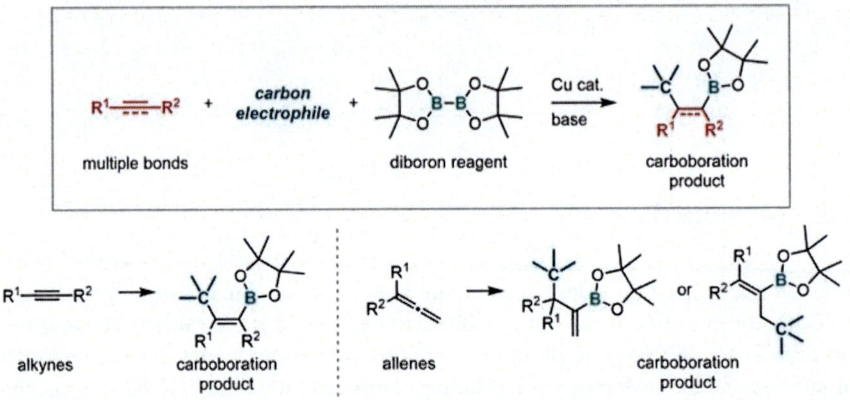

Fig. 3.2 Carboboration reaction of carbon–carbon multiple bonds, including alkenes, alkynes, allenes, and dienes

the first carboboration reaction of allenes using carbonyl compounds as the electrophile in 2013 [15], some types of electrophiles that have an electrophilic C=X bond were found to be applicable to the carboboration reaction of internal double bond of the allene substrates (Fig. 3.4B) [16–33]. These reactions proceed through the γ-addition of the electrophile to the allylcopper(I) intermediate via a chair-like six-membered transition state [7]. However, the use of alkyl halide as the electrophile has not been developed and expected to be challenging because the alkyl halide does not have γ-addition reactivity. Thus, I anticipated that the 1,3-allylic isomerization of the allylcopper(I) intermediate can form the internal allylcopper(I) $\mathbf{I_A}$ from more stable terminal isomer $\mathbf{I_B}$, and the following intramolecular α-alkylation affords the desired exo-cyclization product (Fig. 3.4A) [7, 16, 34–38].

3.2 Results and Discussion

An aliphatic terminal allene **1a** was chosen as the model allene substrate for optimizing the reaction conditions (Table 3.1). The reaction optimization was started from the condition used in the borylative cyclization of alkynes which I and co-workers reported in 2017. The reaction with (o-tol)$_3$P ligand for the copper(I) catalyst in 0.5 M of THF resulted in full conversion of substrate **1a** but a low yielding of the desired

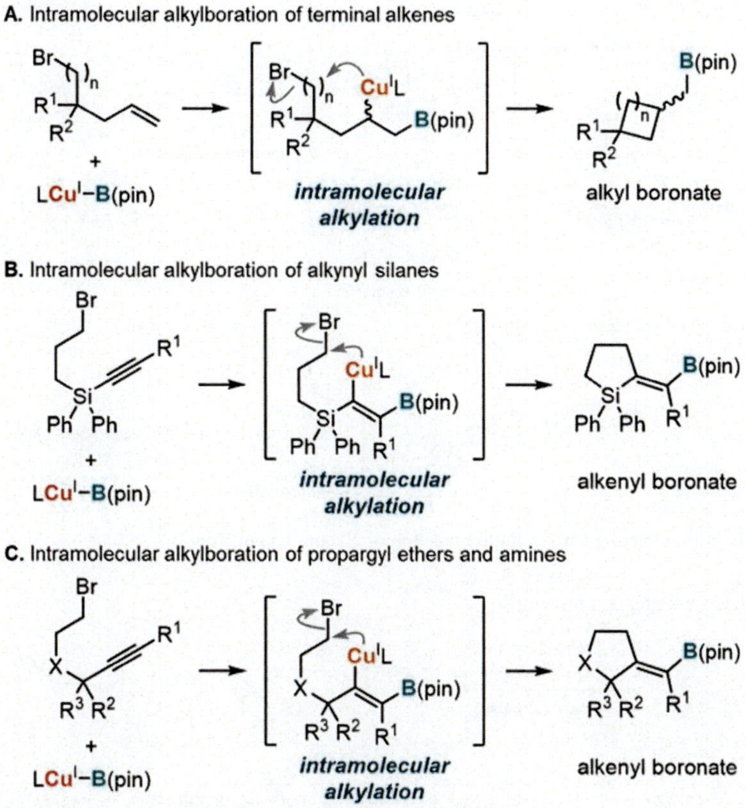

Fig. 3.3 Intramolecular carboboration reaction of alkenes and alkynes

four-membered cyclization product **3a** (entry 1: 32%). The lower concentration of the reaction was found to furnish the product in higher yield with high regioselectivity of the intermolecular alkylation (entry 2: 70%, **3a**/**4a** = 98:2; entry 3: 75%, **3a**/**4a** = 98:2). This implies that unidentified intermolecular side reactions were suppressed at low concentrations. The reaction with the monodentate phosphine ligands (PPh_3 and PCy_3), whose cone angles are smaller than that of $(o\text{-tol})_3P$ [39], furnished the product in moderate to good yield (entry 4: 46%, **3a**/**4a** = 96:4, entry 5: 62%, **3a**/**4a** = 99:1), suggesting that the steric hindrance of the ligand plays important role in production of the product. Bulkier ligands $(mes)_3P$ [40] and XPhos [41, 42] were also effective in this reaction to afford the product in high yields (entry 6: 74%, **3a**/**4a** = 97:3, entry 7: 70%, **3a**/**4a** = 98:2). Although the bidentate phosphine ligand dppp gave the product in moderate yield (entry 8: 44%, **3a**/**4a** = 97:3), Xantphos was found to be not suitable for this reaction (entry 9: 8%). Then, I investigated the effect of the copper(I) salt for the catalyst precursor. The use of CuI instead of CuCl improved the yield, whereas cationic copper(I) salt [$(MeCN)_4CuBF_4$] was less effective (entry 10: 81%, **3a**/**4a** = 98:2, entry 11: 63%, **3a**/**4a** = 98:2). Thus, I selected the conditions

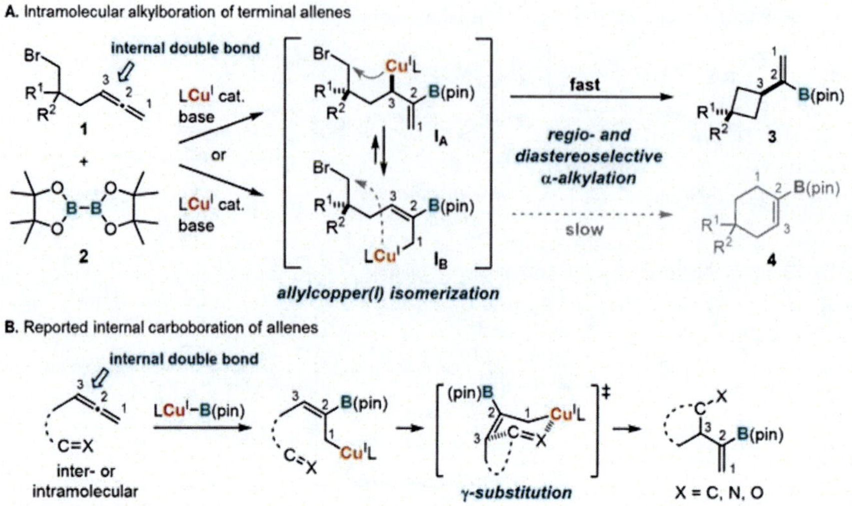

Fig. 3.4 Intramolecular carboboration reaction of alkenes and alkynes

Table 3.1 Reaction optimization[a]

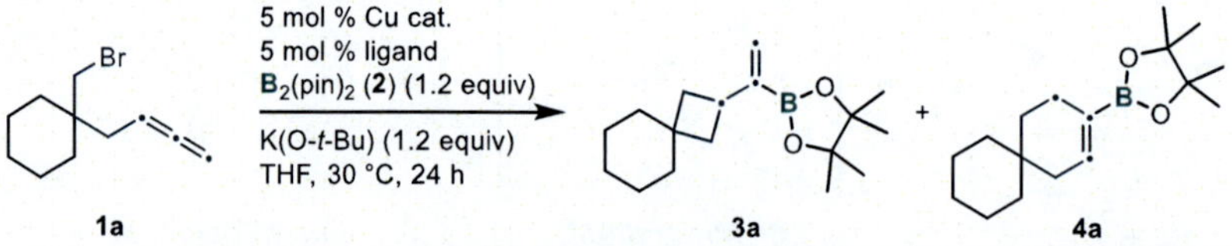

Entry	Cu cat	Ligand	Conc. [M]	Yield [%][b]	**3a/4a**[%][b]
1	CuCl	(*o*-tol)$_3$P	0.50	32[c]	nd[d]
2[e]	CuCl	(*o*-tol)$_3$P	0.25	70	98:2
3	CuCl	(*o*-tol)$_3$P	0.10	75	98:2
4	CuCl	PPh$_3$	0.10	46	96:4
5	CuCl	PCy$_3$	0.10	62	99:1
6	CuCl	(mes)$_3$P	0.10	74	97:3
7	CuCl	XPhos	0.10	70	98:2
8	CuCl	Dppp	0.10	44	97:3
9	CuCl	Xantphos	0.10	8	nd[d]
10[f]	CuI	(*o*-tol)$_3$P	0.10	81	98:2
11	[Cu(MeCN)$_4$]BF$_4$	(*o*-tol)$_3$P	0.10	63	98:2

[a] Conditions: CuI (0.0125 mmol), ligand (0.0125 mmol), **1a** (0.25 mmol), $B_2(pin)_2$ (**2**) (0.30 mmol), and K(O-*t*-Bu) (0.30 mmol) in THF (2.5 mL) at 50 °C for 24 h. [b]Determined by GC analysis of the reaction mixture using an internal standard. [c]Determined by 1H NMR analysis of the crude material using an internal standard. [d]nd = not determined. [e]020 mmol of 1a was used. [f]0.50 mmol of 1a was used

in entry 10 as the optimal conditions for the intramolecular borylative cyclization of terminal allenes.

I then investigated the scope of the substrate in this borylative cyclization reaction with a variety of allene substrates bearing bromine as the leaving group (Table 3.2). The model allene substrate **1a** could successfully be converted to the corresponding alkenylboronate **3a**, which possesses a spiro[3.5]nonane structure (**3a**: 68%, **3a**/**4a** = 98:2). The borylation products from the substrate bearing heterocycles were also obtained with high regioselectivity (**3b**: 60%, **3b**/**4b** = 98:2, **3c**: 67%, **3c**/**4c** = 98:2). The substrate with non-cyclic substituents R^1 and R^2 was also applicable to this reaction, affording the product in lower yield with high regioselectivity (**3d**: 61%, **3d**/**4d** = 95:5). However, in the absence of those substituents, the cyclobutene product was obtained merely in low yield with moderate regioselectivity (**3e**: 30%, **3e**/**4e** = 70:30). Next, I investigated the diastereoselective cyclization of allenes for the synthesis of 1,3-disubstituted cyclobutanes (**3f–3h**). Using a mono-phenyl-substituted allene substrate **1f**, a *trans*-disubstituted cyclobutane (*trans*-**3f**) was obtained with good regio- and diastereoselectivity (**3f**: 64%, **3f**/**4f** = 91:9, dr = 86:14). In contrast, the reactions with mono-alkyl-substituted substrates proceeded with higher diastereoselectivity (**3g**: 71%, **3g**/**4g** = 92:8, dr = 93:7, **3h**: 71%, **3h**/**4h** = 73:27, dr = 91:9).

Table 3.2 Substrate scope[a]

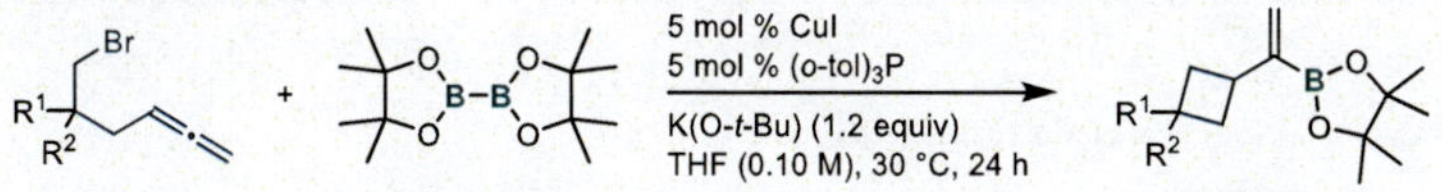

1a–h **2** (1.2 equiv) **3a–h**

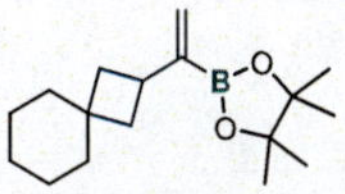

3a: 68%
3/4 = 98:2

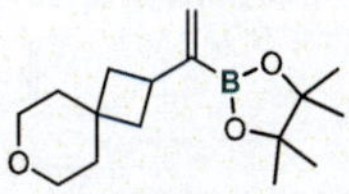

3b: 60%
3/4 = 98:2

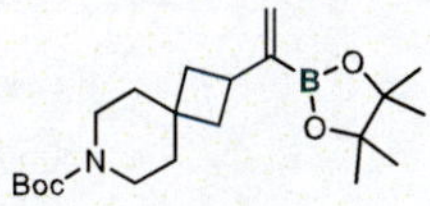

3c: 67%
3/4 = 98:2

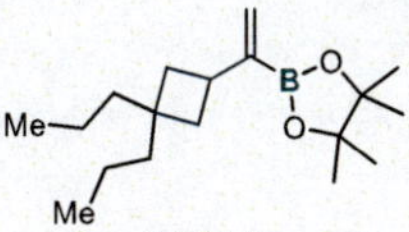

3d: 61%
3/4 = 95:5

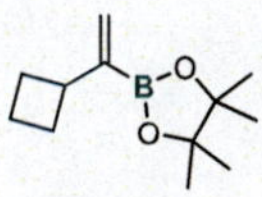

3e: 30%
3/4 = 70:30

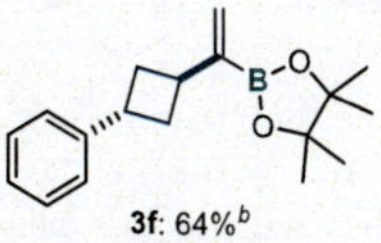

3f: 64%[b]
3/4 = 91:9
dr = 86:14

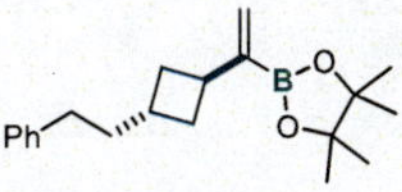

3g: 71%[b]
3/4 = 92:8
dr = 93:7

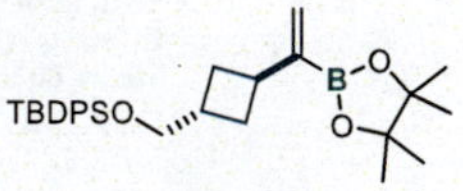

3h: 71%[b]
3/4 = 73:27
dr = 91:9

[a]Conditions: CuI (0.025 mmol), ligand (0.025 mmol), **1a** (0.50 mmol), $B_2(pin)_2$ (**2**) (0.60 mmol), and K(O-*t*-Bu) (0.60 mmol) in THF (5.0 mL) at 30 °C for 10–24 h. Isolated yield. **3/4** ratio and diastereoselectivity were determined by GC analysis of the reaction mixture. [b]The product was isolated as a mixture of the regioisomer and the diastereomer

To demonstrate the synthetic utility of the alkenyl boronates synthesized via this intermolecular alkylboration of terminal allenes, Suzuki–Miyaura cross-coupling reactions were performed with the borylation products and aryl and alkenyl halides (Fig. 3.5A). The conditions using SPhos [43] ligand were found to be suitable for this coupling reactions to produce the corresponding product in moderate to good yields (**5**: 87% yield, **6**: 83% yield). Then, the synthesis of the acylboron compound was conducted by applying the conditions reported from the Ito group (Fig. 3.5B) [44]. Acylboron compound **8** was obtained in high yield via ozonolysis of alkenyl MIDA boronate **7** (83%).

For the mechanistic investigation, I conducted protonolysis of the allylic copper(I) intermediate generated in situ (Fig. 3.6). The protoboration product, alkenylboron compound **9**, was obtained in high yield retaining the alkyl halide moiety (**9**: 79%). Thus, it is assumed that the allene moiety is more reactive than the alkyl halide moiety under these reaction conditions. The reaction would be initiated from borylcupration of allene moiety rather than a radical generation from the C–Br bond [45–48].

The proposed catalytic cycle in this reaction is shown in Fig. 3.7. The boryl copper(I) species **V** are generated in situ from CuI and the diboron reagent in the presence of base. According to the mechanistic study shown above, the boryl copper(I) intermediate **V** reacts with the allene rather than the alkyl halide moiety of substrate **1** to form the allylic copper(I) intermediate $\mathbf{I_A}$ or $\mathbf{I_B}$ through borylcupration of the internal or terminal double bond in the allene, respectively (the subscript A denotes species for four-membered ring product **3**; the subscript B denotes species

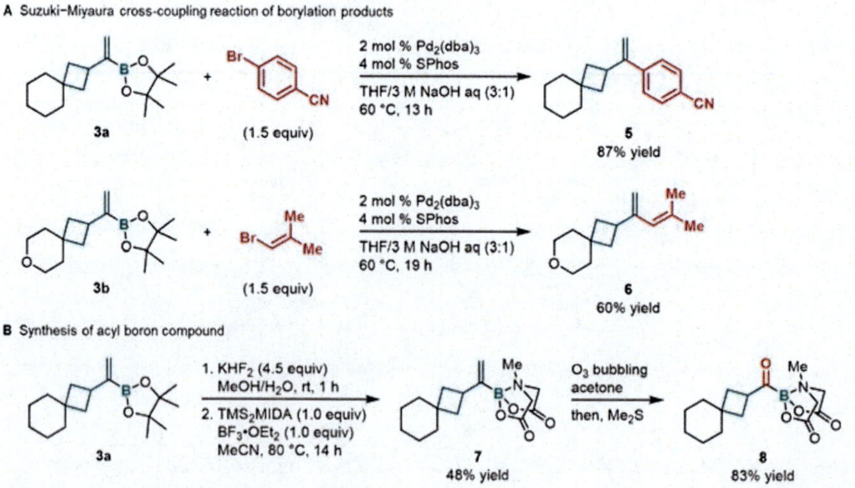

Fig. 3.5 Derivatization reactions of the alkenyl boronates synthesized via this alkylboration reaction of allenes

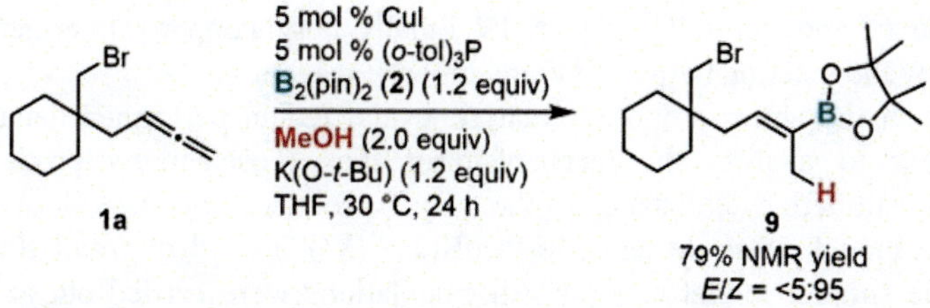

Fig. 3.6 Protonolysis of the allylcopper(I) intermediate

for six-membered ring product **4**). The resulting allylic copper(I) species $\mathbf{I_A}$ and $\mathbf{I_B}$ are easily interconverted at room temperature. Then, reversible coordination of an alkoxide base to the copper(I) center formed cuprates $\mathbf{II_A}$ and $\mathbf{II_B}$ [49, 50]. The subsequent intramolecular S_N2-type oxidative addition of the copper(I) to the C–Br bond produces cupracycles $\mathbf{III_A}$ and $\mathbf{III_B}$ [36, 51]. In this step, the formation of five-membered cupracycle $\mathbf{III_A}$ is kinetically favored over that of seven-membered cupracycle $\mathbf{III_B}$ (see the Supporting Information for the detailed discussion) [52]. As potassium bromide (KBr) is precipitated from the solution, this S_N2-type oxidative addition step would be irreversible and thus determines the regioselectivity of the C–C bond formation. Reductive elimination of $\mathbf{III_A}$ forms the four-membered

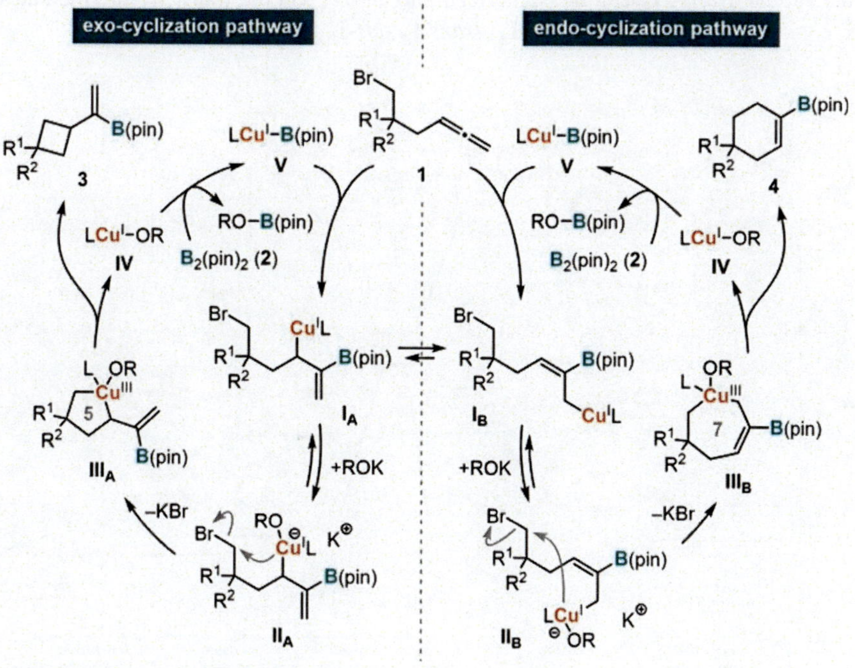

Fig. 3.7 Proposed catalytic cycle

ring product **3** and copper(I)-alkoxide **IV**. Finally, boryl copper(I) intermediate **V** is restored by the reaction between **IV** and diboron reagent **2**.

Based on the above-proposed catalytic cycle, I then performed detailed DFT calculations to analyze the mechanism of diastereoselectivity producing the *trans*-disubstituted cyclobutane as the major product (Fig. 3.8). I used allene **1x** as the model substrate and $Me_3PCuB(en)$ (**V′**) instead of (o-tol)$_3$PCuB(pin) (**V**) as the simpler model catalyst. All calculations were carried out at ωB97X-D/def2-SVP/SMD(THF) level of theory (see the Supporting Information for the computational details). The energy diagram, which the relative energies are with respect to **1x** and **V′**, is shown in Fig. 3.2. Borylcupration of alkene moieties in the allene substrate **1x** can potentially form four different isomers of allylcopper(I) species: *trans*-$\mathbf{I_A}$ (–31.4 kcal/mol) and *cis*-$\mathbf{I_A}$ (–31.7 kcal/mol) as the internal allylcopper(I) species and *trans*-$\mathbf{I_B}$ (–36.7 kcal/mol) and *cis*-$\mathbf{I_B}$ (–33.9 kcal/mol) as the terminal allylcopper(I) species. The *trans*-$\mathbf{I_A}$ can directly isomerize to *trans*-$\mathbf{I_B}$ and *cis*-$\mathbf{I_B}$ through a low energy barrier, but not to *cis*-$\mathbf{I_A}$ (see the Supporting Information for a detailed discussion). In the same manner, *cis*-$\mathbf{I_A}$ can directly isomerize to *trans*-$\mathbf{I_B}$ and *cis*-$\mathbf{I_B}$, but not to *trans*-$\mathbf{I_A}$. Coordination of an alkoxide to the copper(I) center in intermediates **I** forms the metastable intermediates **II**, which is an exergonic reaction by 14.2–15.0 kcal/mol. The copper(I) center in **II** has a trigonal planar geometry. Also, the potassium cation is chelated by two oxygen atoms in the alkoxide base and B(pin) group. According to the energy profiles described in this paragraph, the cuprate *trans*-$\mathbf{II_A}$ can be isomerized into *cis*-$\mathbf{II_A}$ via the transient intermediates **I**, e.g., in the sequence as *trans*-$\mathbf{II_A}$, *trans*-$\mathbf{I_A}$, *cis*-$\mathbf{I_B}$, *cis*-$\mathbf{I_A}$, and *cis*-$\mathbf{II_A}$.

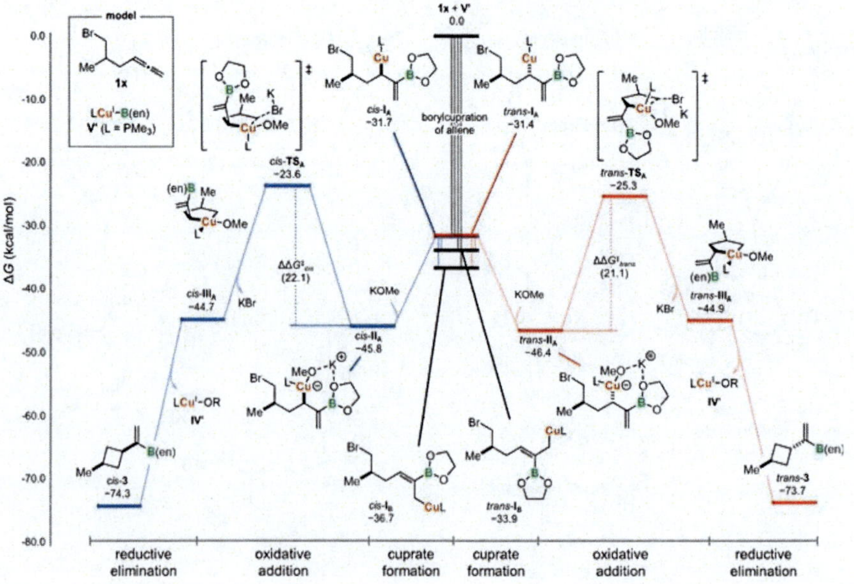

Fig. 3.8 Free energy diagram of the intramolecular alkylboration of terminal allenes

I then elucidated the mechanism of diastereoselectivity in the intramolecular cyclization step. The oxidative additions from *trans*-$\mathbf{II_A}$ and *cis*-$\mathbf{II_A}$ generate more unstable cupracycle *trans*-$\mathbf{III_A}$ and *cis*-$\mathbf{III_A}$, respectively, with the release of KBr. However, KBr should be precipitated from the solution, thus those steps would be irreversible. Along this step, copper(I) is oxidized to copper(III), which has a square planar geometry. Focused on the activation energy difference between these two steps, the value ($\Delta\Delta G^{\ddagger}{}_{cis}$–$\Delta\Delta G^{\ddagger}{}_{trans}$) is 1.0 kcal/mol, which is in good agreement with the value calculated from the experimental results (experimentally: 1.1–1.5 kcal/mol, *cis*/*trans* = 86:14–93:7). It is considered that the stability of these transition states is affected by the configuration of the substituent on cupracycle in the transition state. The *cis*-oriented substituents in *cis*-$\mathbf{TS_A}$ would cause the steric repulsion to destabilize the structure. Moreover, I could not find an energetically reasonable isomerization path from *cis*-$\mathbf{III_A}$ to *trans*-$\mathbf{III_A}$, which implies allylic isomerization of allylcopper(III) would not occur [36, 51]. Thus, it is assumed that the diastereoselectivity is determined at the S_N2-type oxidative addition step under kinetic control.

3.3 Conclusions

I have developed an intramolecular borylative cyclization of terminal allenes for the synthesis of alkenylboronates having a four-membered ring structure. The products were obtained with high regio- and diastereoselectivity, which was enabled by facile allylic isomerization of allylcopper(I) species. The DFT studies could explain these kinetic control selectivities.

3.4 Experimental Details

3.4.1 Instrumentation and Chemicals

Materials were obtained from commercial suppliers and purified by standard procedures unless otherwise noted. Solvents were also purchased from commercial suppliers, degassed via three freeze–pump–thaw cycles, and further dried over molecular sieves (MS 4A). ^{1}H NMR spectra were recorded on JEOL JNM-ECX400P (400 MHz) and JNM-ECS400 (400 MHz) spectrometer, and spectra were referenced to tetramethylsilane (0.00 ppm). ^{13}C NMR spectra were recorded on JEOL JNM-ECX400P (100 MHz) and JNM-ECS400 (100 MHz) spectrometer, and spectra were referenced to the solvent ($CDCl_3$: 77.0 ppm; acetone-d_6: 29.92 ppm). CuCl (ReagentPlus® grade, 224332-25G, ≥99%) and K(O-*t*-Bu)/THF (1.0 M, 328650-50ML) were purchased from Sigma-Aldrich Co. and used as received. GC analyses were conducted with a Shimadzu GC-2014 or GC-2025 equipped with a ULBON

HR-1 glass capillary column (Shinwa Chemical Industries) and an FID detector. High-resolution mass spectra were recorded at the Global Facility Center, Hokkaido University.

3.4.2 General Experimental Procedure

Procedure for the Copper(I)-Catalyzed Intramolecular Alkylboration of 1a

Copper iodide (4.8 mg, 0.025 mmol), bis(pinacolato)diboron (**2**) (152.4 mg, 0.60 mmol), and tri(*o*-tolyl)phosphine (7.6 mg, 0.025 mmol) were placed in an oven-dried reaction vial. After the vial was sealed with a screw cap containing a Teflon-coated rubber septum, the vial was connected to a vacuum/nitrogen manifold through a needle. It was evacuated and then backfilled with nitrogen. This cycle was repeated three times. Dry THF (4.4 mL) and K(O-*t*-Bu)/THF (1.0 M, 0.6 mL, 0.6 mmol) were added in the vial through the rubber septum using a syringe. After stirring for 30 min, **1a** (114.8 mg, 0.50 mmol) was added to the mixture at 30 °C. After the reaction was completed, the reaction mixture was passed through a short silica gel column (Φ: 10 mm, height of the silica gel column: ca. 30 mm) eluting with Et_2O. The crude material was purified by flash column chromatography (SiO_2, CH_2Cl_2 /hexane, typically 0:100–8:92) to give the corresponding borylation product **3a** as a colorless oil.

3.4.3 Substrate Preparation

Preparation of 1-(bromomethyl)-1-(buta-2,3-dien-1-yl)cyclohexane (1a, Fig. 3.9)

The bromide **1a** was synthesized through the reaction of the corresponding alcohol **S1** according to the literature procedure [53]. In a vacuum-dried 50 mL two-neck round-bottomed flask, [1-(buta-2,3-dien-1-yl)cyclohexyl]methanol (**S1**) (0.856 g, 5.15 mmol) and carbon tetrabromide (1.88 g, 5.67 mmol) were dissolved in dry THF (10 mL), and the mixture was cooled to 0 °C under a nitrogen atmosphere. Triphenylphosphine (1.49 g, 5.67 mmol) was added to the mixture, and the mixture was warmed to room temperature. After the mixture was stirred for 4 h, the solvents

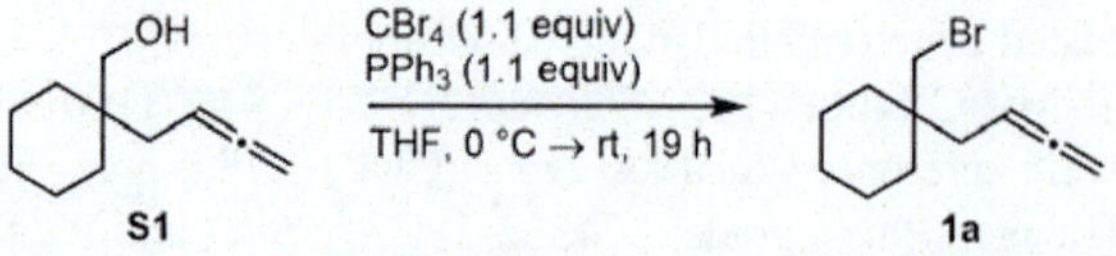

Fig. 3.9 Scheme of preparation procedure of **1a**

were removed by evaporation. The crude product was purified by flash column chromatography and bulb-to-bulb distillation to obtain the corresponding bromide **1a** (0.778 g, 3.40 mmol, 66%) as a colorless oil.

^{1}H NMR (400 MHz, $CDCl_3$, δ): 1.35–1.53 (m, 10H), 2.13 (td, J = 5.4, 2.5 Hz, 2H), 3.40 (s, 2H), 4.65 (td, J = 4.5, 2.1 Hz, 2H), 5.00 (tt, J = 9.9, 3.4 Hz, 1H). ^{13}C NMR (100 MHz, $CDCl_3$, δ): 21.5 (CH_2), 26.0 (CH_2), 33.8 (CH_2), 35.3 (CH_2), 37.2 (C), 43.5 (CH_2), 73.7 (CH_2), 84.5 (CH), 209.9 (C). HRMS-EI (m/z): [M]+ calcd for $C_{11}H_{17}Br$, 228.0514; found, 228.0508.

4-(Bromomethyl)-4-(buta-2,3-dien-1-yl)tetrahydro-2H-pyran (1b, Fig. 3.10)

1b was prepared as a colorless oil from the corresponding alcohol according to the procedure described above.

^{1}H NMR (400 MHz, $CDCl_3$, δ): 1.52–1.65 (m, 4H), 2.25 (dt, J = 8.1, 2.2 Hz, 2H), 3.44 (s, 2H), 3.67 (t, J = 5.6 Hz, 4H), 4.68 (dt, J = 6.7, 2.2 Hz, 2H), 5.00 (tt, J = 9.9, 3.2 Hz, 1H). ^{13}C NMR (100 MHz, $CDCl_3$, δ): 33.8 (CH_2), 34.2 (CH_2), 35.2 (C), 42.0 (CH_2), 63.3 (CH_2), 74.1 (CH_2), 83.7 (CH), 210.0 (C). HRMS-EI (m/z): [M]+ calcd for $C_{10}H_{15}BrO$, 230.0306; found, 230.0306.

***tert*-Butyl 4-(bromomethyl)-4-(buta-2,3-dien-1-yl)piperidine-1-carboxylate (1c, Fig. 3.11)**

1c was prepared as a colorless oil from the corresponding alcohol according to the procedure described above.

^{1}H NMR (400 MHz, $CDCl_3$, δ): 1.42–1.62 (m, 13H), 2.19 (dt, J = 8.2, 2.3 Hz, 2H), 3.33 (ddd, J = 18.0, 8.8, 4.4 Hz, 2H), 3.40 (s, 2H), 3.41–3.51 (m, 2H), 4.68 (dt, J = 6.7, 2.3 Hz, 2H), 4.99 (tt, J = 8.0, 7.2 Hz, 1H). ^{13}C NMR (100 MHz, $CDCl_3$, δ): 28.3 (CH_3), 33.0 (CH_2), 34.0 (CH_2), 36.1 (C), 38.7 (CH_2), 39.7 (CH_2), 41.7 (CH_2), 74.2 (CH_2), 79.5 (C), 83.8 (CH), 154.7 (C), 210.0 (C). HRMS-ESI (m/z): [M + Na]+ calcd for $C_{15}H_{24}BrNO_2Na$, 352.0883; found, 352.0884.

Fig. 3.10 Scheme of **1b**

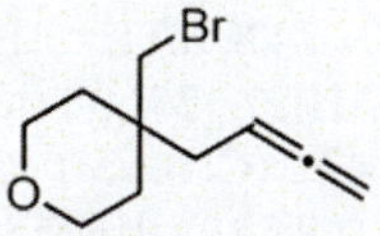

1b

Fig. 3.11 Scheme of **1c**

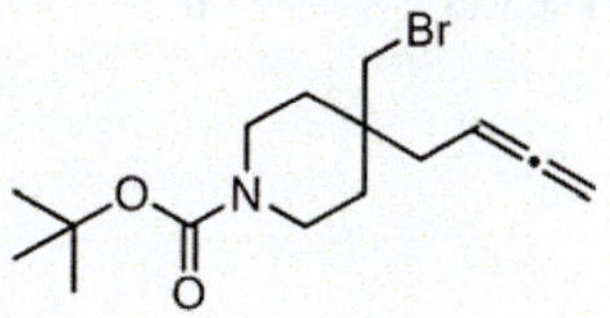

1c

Fig. 3.12 Scheme of **1d**

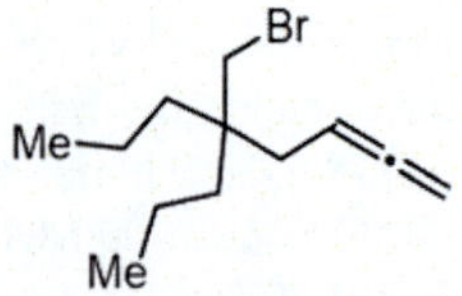

3d

5-(Bromomethyl)-5-propylocta-1,2-diene (1d, Fig. 3.12)

1d was prepared as a colorless oil from the corresponding alcohol according to the procedure described above.

^{1}H NMR (400 MHz, $CDCl_3$, δ): 0.92 (t, J = 7.0 Hz, 6H), 1.16–1.33 (m, 8H), 2.03 (dt, J = 8.1, 2.2 Hz, 2H), 3.30 (s, 2H), 4.65 (dt, J = 6.3, 2.2 Hz, 2H), 4.97 (tt, J = 8.0, 7.2 Hz, 1H). ^{13}C NMR (100 MHz, $CDCl_3$, δ): 14.7 (CH_3), 16.4 (CH_2), 34.4 (CH_2), 37.0 (CH_2), 40.1 (C), 42.1 (CH_2), 73.8 (CH_2), 84.8 (CH), 210.0 (C). HRMS-EI (m/z): [M]+ calcd for $C_{12}H_{21}Br$, 244.0827; found, 244.0830.

6-Bromohexa-1,2-diene (1e, Fig. 3.13)

1e was prepared as a colorless oil from the corresponding alcohol according to the procedure described above.

^{1}H NMR (400 MHz, $CDCl_3$, δ): 1.99 (quint, J = 7.0 Hz, 2H), 2.16 (nonet, J = 3.4 Hz, 2H), 3.46 (t, J = 6.5 Hz, 2H), 4.71 (dt, J = 6.7, 3.3 Hz, 2H), 5.10 (quint, J = 6.6 Hz, 1H). ^{13}C NMR (100 MHz, $CDCl_3$, δ): 26.5 (CH_2), 31.8 (CH_2), 33.0 (CH_2), 75.4 (CH_2), 88.3 (CH), 208.6 (C). HRMS-EI (m/z): [M]+ calcd for C_6H_9Br, 159.9888; found, 159.9885.

(1-Bromohexa-4,5-dien-2-yl)benzene (1f, Fig. 3.14)

1f was prepared as a colorless oil from the corresponding alcohol according to the procedure described above.

^{1}H NMR (400 MHz, $CDCl_3$, δ): 2.36–2.47 (m, 1H), 2.55–2.65 (m, 1H), 3.09 (dt, J = 14.8, 6.7 Hz, 1H), 3.57–3.68 (m, 2H), 4.59–4.68 (m, 2H), 4.96 (quint, J = 6.9 Hz, 1H), 7.17–7.38 (m, 5H). ^{13}C NMR (100 MHz, $CDCl_3$, δ): 33.1 (CH_2), 37.8 (CH_2), 47.5 (CH), 75.1 (CH_2), 87.0 (CH), 127.1 (CH), 127.7 (CH), 128.4 (CH), 141.6 (C), 209.1 (C). HRMS-EI (m/z): [M]+ calcd for $C_{12}H_{13}Br$, 236.0201; found, 236.0205.

Fig. 3.13 Scheme of **1e**

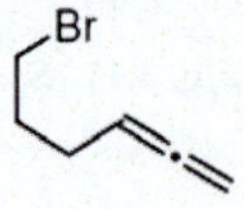

1e

Fig. 3.14 Scheme of **1f**

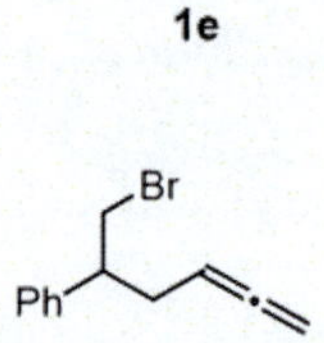

1f

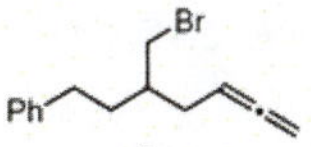

1g

Fig. 3.15 Scheme of **1g**

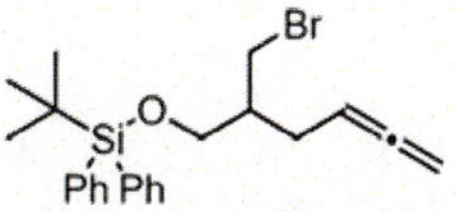

1h

Fig. 3.16 Scheme of **1h**

[3-(Bromomethyl)hepta-5,6-dien-1-yl]benzene (1g, Fig. 3.15)

1g was prepared as a colorless oil from the corresponding alcohol according to the procedure described above.

^{1}H NMR (400 MHz, $CDCl_3$, δ): 1.68–1.86 (m, 3H), 2.13–2.22 (m, 2H), 2.56–2.72 (m, 2H), 3.53 (d, J = 3.6 Hz, 2H), 4.69 (dt, J = 6.6, 2.8 Hz, 2H), 5.04 (quint, J = 7.0 Hz, 1H), 7.15–7.33 (m, 5H). ^{13}C NMR (100 MHz, $CDCl_3$, δ): 31.3 (CH_2), 32.8 (CH_2), 34.0 (CH_2), 38.3 (CH_2), 39.0 (CH), 74.9 (CH_2), 86.8 (CH), 125.9 (CH), 128.3 (CH), 128.4 (CH), 141.8 (C), 209.2 (C). HRMS-EI (m/z): [M]+ calcd for $C_{14}H_{17}Br$, 264.0514; found, 264.0517.

{[2-(Bromomethyl)hexa-4,5-dien-1-yl]oxy}(tert-butyl)diphenylsilane (1h, Fig. 3.16)

1h was prepared as a colorless oil from the corresponding alcohol according to the procedure described above.

^{1}H NMR (400 MHz, $CDCl_3$, δ): 1.06 (s, 9H), 1.93–2.04 (m, 1H), 2.09–2.17 (m, 2H), 3.59–3.75 (m, 4H), 4.64 (dt, J = 6.4, 3.1 Hz, 2H), 5.00 (quint, J = 7.0 Hz, 1H), 7.36–7.47 (m, 6H), 7.64–7.70 (m, 4H). ^{13}C NMR (100 MHz, $CDCl_3$, δ): 19.3 (C), 26.8 (CH_3), 28.4 (CH_2), 35.6 (CH_2), 42.6 (CH), 63.7 (CH_2), 75.2 (CH_2), 87.1 (CH), 127.7 (CH), 129.7 (CH), 133.4 (C), 133.4 (C), 135.6 (CH), 135.6 (CH), 209.0 (C). HRMS-ESI (m/z): [M + Na]+ calcd for $C_{23}H_{29}BrOSiNa$, 451.1063; found, 451.1071.

3.4.4 Borylation Product Characterization

4,4,5,5-Tetramethyl-2-[1-(spiro[3.5]nonan-2-yl)vinyl]-1,3,2-dioxaborolane (3a, Fig. 3.17)

The reaction was conducted with 114.8 mg (0.50 mmol) of **1a**. The product **3a** was obtained as a colorless oil in 68% yield (94.0 mg) with **3/4** = 98:2.

^{1}H NMR (400 MHz, $CDCl_3$, δ): 1.26 (s, 12H), 1.33 (s, 6H), 1.39–1.47 (m, 2H), 1.50–1.63 (m, 4H), 1.95 (td, J = 9.1, 2.5 Hz, 2H), 2.98 (quint of triplet, J = 9.0, 1.7 Hz, 1H), 5.51–5.56 (m, 1H), 5.73 (dd, J = 3.1, 1.6 Hz, 1H). ^{13}C NMR (100 MHz, $CDCl_3$, δ): 22.8 (CH_2), 23.1 (CH_2), 24.7 (CH_3), 26.2 (CH_2), 33.2 (CH), 35.3 (C), 36.9

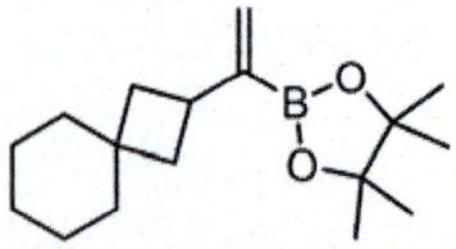

3a

Fig. 3.17 Scheme of **3a**

(CH_2), 38.2 (CH_2), 40.6 (CH_2), 83.1 (*C*), 125.4 (CH_2), 146.3 (br, B–*C*). HRMS-EI (*m*/*z*): [M]+ calcd for $C_{17}H_{29}{}^{11}BO_2$, 276.2264; found, 276.2260.

2-[1-(7-Oxaspiro[3.5]nonan-2-yl)vinyl]-4,4,5,5-tetramethyl-1,3,2-dioxaborolane (3b, Fig. 3.18)

The reaction was conducted with 115.7 mg (0.50 mmol) of **1b**. The product **3b** was obtained as a colorless oil in 60% yield (82.9 mg) with **3/4** = 98:2.

^{1}H NMR (400 MHz, $CDCl_3$, δ): 1.27 (s, 12H), 1.50 (t, *J* = 5.2 Hz, 2H), 1.66–1.74 (m, 4H), 2.06 (td, *J* = 9.1, 2.5 Hz, 2H), 2.99–3.10 (m, 1H), 3.54 (t, *J* = 5.2 Hz, 2H), 3.65 (t, *J* = 5.4 Hz, 2H), 5.52–5.57 (m, 1H), 5.76 (dd, *J* = 2.7, 1.6 Hz, 1H). ^{13}C NMR (100 MHz, $CDCl_3$, δ): 24.7 (CH_3), 32.7 (*C*), 33.2 (*C*H), 37.0 (CH_2), 37.8 (CH_2), 40.1 (CH_2), 64.6 (CH_2), 64.9 (CH_2), 83.1 (*C*), 125.8 (CH_2), 145.6 (br, B–*C*). HRMS-EI (*m*/*z*): [M]+ calcd for $C_{16}H_{27}{}^{11}BO_3$, 278.2056; found, 278.2055.

***tert*-Butyl 2-[1-(4,4,5,5-tetramethyl-1,3,2-dioxaborolan-2-yl)vinyl]-7-azaspiro[3.5]nonane-7-carboxylate (3c, Fig. 3.19)**

The reaction was conducted with 165.0 mg (0.50 mmol) of **1c**. The product **3c** was obtained as a colorless oil in 67% yield (126.7 mg) with **3/4** = 98:2.

^{1}H NMR (400 MHz, $CDCl_3$, δ): 1.26 (s, 12H), 1.38–1.48 (m, 4H), 1.45 (s, 9H), 1.59–1.70 (m, 4H), 2.02 (td, *J* = 9.1, 2.5 Hz, 2H), 3.04 (quint, *J* = 8.9 Hz, 1H), 3.26 (t, *J* = 5.6 Hz, 2H), 3.36 (t, *J* = 5.4 Hz, 2H), 5.52–5.56 (m, 1H), 5.76 (dd, *J* = 2.9, 1.8 Hz, 1H). ^{13}C NMR (100 MHz, $CDCl_3$, δ): 24.6 (CH_3), 28.3 (CH_3), 33.1 (*C*H), 33.5 (*C*), 35.8 (CH_2), 37.1 (CH_2), 39.1 (CH_2), 40.4 (br, CH_2), 40.9 (br, CH_2), 78.9

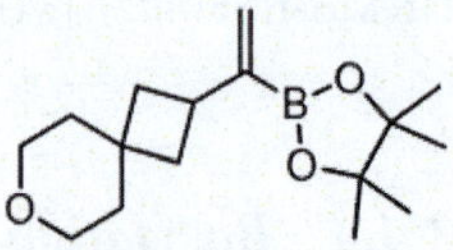

3b

Fig. 3.18 Scheme of **3b**

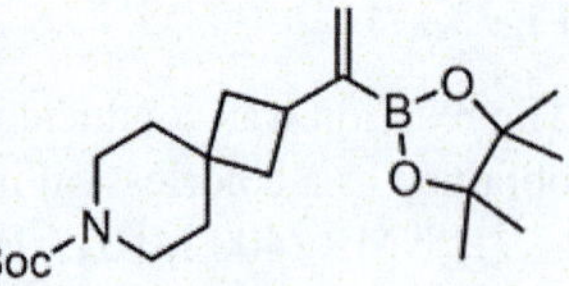

3c

Fig. 3.19 Scheme of **3c**

Fig. 3.20 Scheme of **3d**

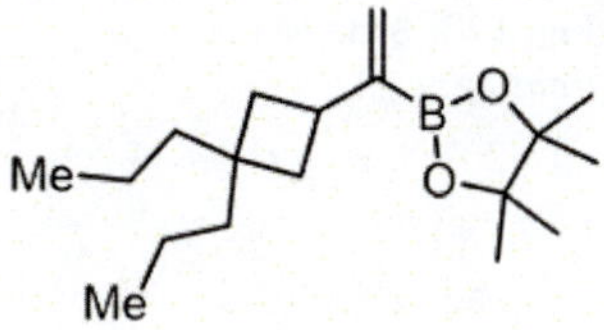

3d

(*C*), 83.1 (*C*), 125.8 (CH_2), 145.3 (br, B–*C*), 154.8 (*C*). HRMS-ESI (*m/z*): [M + Na]+ calcd for $C_{21}H_{36}{}^{11}BNO_4Na$, 400.2634; found, 400.2635.

2-[1-(3,3-Dipropylcyclobutyl)vinyl]-4,4,5,5-tetramethyl-1,3,2-dioxaborolane (3d, Fig. 3.20)

The reaction was conducted with 121.0 mg (0.50 mmol) of **1d**. The product **3d** was obtained as a colorless oil in 61% yield (87.7 mg) with **3/4** = 95:5.

^{1}H NMR (400 MHz, $CDCl_3$, δ): 0.85 (t, *J* = 7.2 Hz, 3H), 0.92 (t, *J* = 7.2 Hz, 3H), 1.08–1.31 (m, 6H), 1.26 (s, 12H), 1.43–1.50 (m, 2H), 1.61 (td, *J* = 9.5, 2.4 Hz, 2H), 1.90 (td, *J* = 9.1, 2.6 Hz, 2H), 2.92–3.04 (m, 1H), 5.50–5.54 (m, 1H), 5.72 (dd, *J* = 3.3, 1.8 Hz, 1H). ^{13}C NMR (100 MHz, $CDCl_3$, δ): 14.8 (CH_3), 14.9 (CH_3), 17.0 (CH_2), 17.4 (CH_2), 24.7 (CH_3), 33.3 (*C*H), 37.2 (*C*), 37.8 (CH_2), 39.7 (CH_2), 42.7 (CH_2), 83.1 (*C*), 125.4 (*C*H), 146.2 (br, B–*C*). HRMS-ESI (*m/z*): [M + H]+ calcd for $C_{18}H_{34}{}^{11}BO_2$, 293.2650; found, 293.2652.

2-(1-Cyclobutylvinyl)-4,4,5,5-tetramethyl-1,3,2-dioxaborolane (3e, Fig. 3.21)

The reaction was conducted with 80.5 mg (0.50 mmol) of **1e**. The product **3e** was obtained as a colorless oil in 30% yield (31.6 mg) with **3/4** = 70:30.

^{1}H NMR (400 MHz, $CDCl_3$, δ): 1.26 (s, 12H), 1.66–1.75 (m, 1H), 1.82–1.95 (m, 3H), 2.02–2.14 (m, 2H), 3.03–3.14 (m, 1H), 5.52–5.57 (m, 1H), 5.75 (dd, *J* = 3.5, 1.6 Hz, 1H). ^{13}C NMR (100 MHz, $CDCl_3$, δ): 18.1 (CH_2), 24.7 (CH_3), 28.2 (CH_2), 40.2 (*C*H), 83.1 (*C*), 125.7 (CH_2). HRMS-EI (*m/z*): [M]+ calcd for $C_{12}H_{21}{}^{11}BO_2$, 208.1637; found, 208.1643.

***trans*-4,4,5,5-Tetramethyl-2-[1-(3-phenylcyclobutyl)vinyl]-1,3,2-dioxaborolane (*trans*-3f, Fig. 3.22)**

Fig. 3.21 Scheme of **3e**

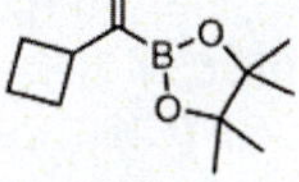

3e

Fig. 3.22 Scheme of *trans*-**3f**

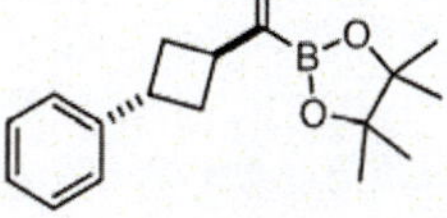

trans-**3f**

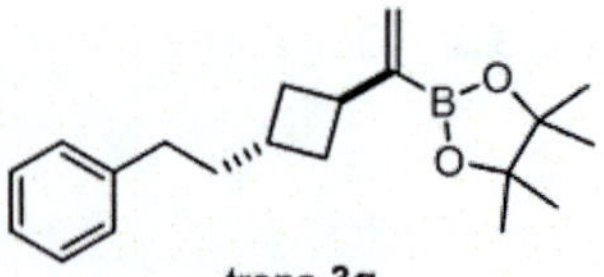

Fig. 3.23 Scheme of *trans*-**3g**

The reaction was conducted with 118.8 mg (0.50 mmol) of **1f**. The product *trans*-**3f** was obtained as a colorless oil in 64% yield (90.6 mg) with **3/4** = 91:9 and dr = 86:14. The ^{1}H and ^{13}C NMR charts include *cis*-**3f** and **4f**.

^{1}H NMR (400 MHz, $CDCl_3$, δ): 1.28 (s, 12H), 2.31–2.45 (m, 4H), 3.14–3.24 (m, 1H), 3.50 (quint, $J = 7.9$ Hz, 1H), 5.72–5.76 (m, 1H), 5.88 (dd, $J = 2.9$, 1.7 Hz, 1H), 7.14–7.36 (m, 5H). ^{13}C NMR (100 MHz, $CDCl_3$, δ): 24.7 (CH_3), 33.5 (CH_2), 35.3 (*C*H), 35.7 (*C*H), 83.2 (*C*), 125.5 (*C*H), 125.5 (CH_2), 126.5 (*C*H), 128.2 (*C*H), 146.4 (*C*). HRMS-EI (*m/z*): [M]+ calcd for $C_{18}H_{25}{}^{11}BO_2$, 284.1951; found, 284.1949.

trans-4,4,5,5-Tetramethyl-2-[1-(3-phenethylcyclobutyl)vinyl]-1,3,2-dioxaborolane (_trans_-3 g, Fig. 3.23)

The reaction was conducted with 132.6 mg (0.50 mmol) of **1g**. The product *trans*-**3g** was obtained as a colorless oil in 71% yield (111.5 mg) with **3/4** = 92:8 and dr = 97:3. The ^{1}H and ^{13}C NMR charts include *cis*-**3g** and **4g**.

^{1}H NMR (400 MHz, $CDCl_3$, δ): 1.26 (s, 12H), 1.78–1.91 (m, 4H), 2.00–2.10 (m, 2H), 2.10–2.19 (m, 1H), 2.55 (t, $J = 7.9$ Hz, 2H), 3.14 (quint, $J = 7.8$ Hz, 1H), 5.56–5.62 (m, 1H), 5.75–5.80 (m, 1H), 7.13–7.20 (m, 3H), 7.23–7.31 (m, 2H). ^{13}C NMR (100 MHz, $CDCl_3$, δ): 24.7 (CH_3), 30.4 (*C*), 32.0 (CH_2), 33.9 (CH_2), 35.9 (*C*H), 37.9 (CH_2), 83.1 (*C*), 125.3 (CH_2), 125.5 (*C*H), 128.1 (*C*H), 128.3 (*C*H), 142.7 (*C*), 145.4 (br, B–*C*). HRMS-EI (*m/z*): [M]+ calcd for $C_{20}H_{29}{}^{11}BO_2$, 312.2264; found, 312.2260.

trans-tert-Butyldiphenyl{{3-[1-(4,4,5,5-tetramethyl-1,3,2-dioxaborolan-2-yl)vinyl]cyclobutyl}methoxy}silane (_trans_-3h, Fig. 3.24)

The reaction was conducted with 214.7 mg (0.50 mmol) of **1h**. The product *trans*-**3h** was obtained as a colorless oil in 71% yield (168.3 mg) with **3/4** = 73:27 and dr = 91:9. The ^{1}H and ^{13}C NMR charts include *cis*-**3h** and **4h**.

^{1}H NMR (400 MHz, $CDCl_3$, δ): 1.06 (s, 9H), 1.26 (s, 12H), 1.99 (dd, $J = 8.3$, 6.6 Hz, 4H), 2.37 (septet, $J = 6.4$ Hz 1H), 3.09 (quint-d, $J = 8.3$, 1.4 Hz, 1H), 3.73 (d, $J = 7.0$ Hz, 2H), 5.56–5.61 (m, 1H), 5.77 (dd, $J = 3.1$, 1.6 Hz, 1H), 7.34–7.45 (m, 6H), 7.62–7.71 (m, 4H). ^{13}C NMR (100 MHz, $CDCl_3$, δ): 19.3 (*C*), 24.7 (CH_3), 26.9 (CH_3), 29.0 (CH_2), 32.7 (*C*H), 36.2 (*C*H), 67.5 (CH_2), 83.1 (*C*), 125.4 (CH_2),

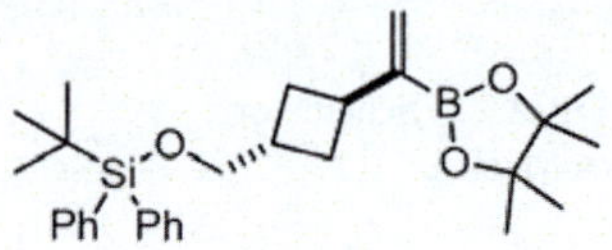

Fig. 3.24 Scheme of *trans*-**3h**

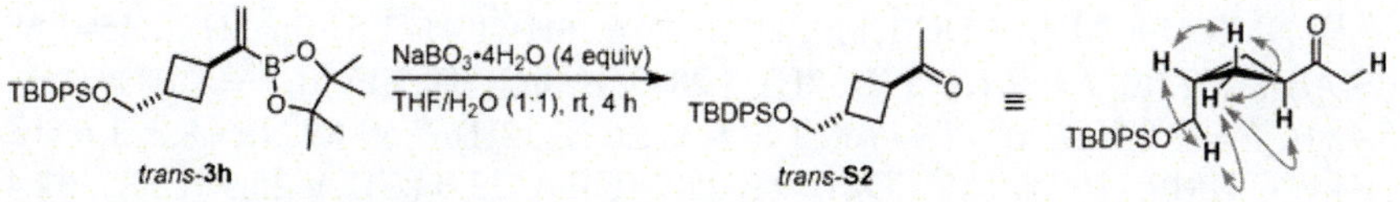

Fig. 3.25 ^{1}H NMR NOESY experiment of *trans*-**S2**

127.5 (*C*H), 129.4 (*C*H), 134.1 (*C*), 135.6 (*C*H), 145.5 (br, B–*C*). HRMS-ESI (*m*/*z*): [M + Na]+ calcd for $C_{29}H_{41}{}^{11}BO_3SiNa$, 499.2816; found, 499.2818.

The configuration of *trans*-**3h** was determined by ^{1}H NMR NOESY experiment of *trans*-**S2** after the derivatization of the boryl group in *trans*-**3h** (Fig. 3.25). The results were summarized as following schemes. A solid curved gray arrow indicates the all observed nOe except the inside of the TBDPS group.

^{1}H NMR (400 MHz, $CDCl_3$, δ): 1.06 (s, 9H), 2.00–2.12 (m, 2H), 2.09 (s, 3H), 2.22–2.32 (m, 2H), 2.35–2.49 (m, 1H), 3.19 (tt, *J* = 9.3, 6.9 Hz, 1H), 3.67 (d, *J* = 5.9 Hz, 2H), 7.35–7.48 (m, 6H), 7.61–7.72 (m, 4H). ^{13}C NMR (100 MHz, $CDCl_3$, δ): 19.3 (*C*), 25.7 (*C*H_2), 26.8 (*C*H_3), 27.2 (*C*H_3), 32.7 (*C*H), 42.9 (*C*H), 66.8 (*C*H_2), 127.6 (*C*H), 129.6 (*C*H), 133.8 (*C*), 135.6 (*C*H), 210.2 (*C*).

3.4.5 Functionalization Procedures of Borylation Products

4-[1-(Spiro[3.5]nonan-2-yl)vinyl]benzonitrile (5, Fig. 3.26)

The palladium catalyzed cross-coupling reaction with aryl halide was performed according to the literature procedure [54]. 4-Bromobenzonitrile (68.7 mg, 0.375 mmol), $Pd_2(dba)_3$ (4.6 mg, 0.005 mmol), and SPhos (4.1 mg, 0.010 mmol) were placed in an oven-dried reaction vial under a nitrogen atmosphere. A solution of **3a** (69.5 mg, 0.25 mmol) in dry THF (2.25 mL) was added to the vial followed by addition of 3 M NaOH aq (0.75 mL). Then, the mixture was warmed to 60 °C and stirred for 13 h. After the reaction mixture was cooled to room temperature, the reaction mixture was quenched by H_2O and extracted with Et_2O three times. The combined organic layer was then dried over $MgSO_4$. After filtration, the solvents were removed by evaporation. The crude mixture was purified by flash column chromatography (SiO_2, Et_2O/hexane, 0:100–14:86) to give the corresponding coupling product **5** (55.2 mg, 0.220 mmol, 87%) as a colorless oil.

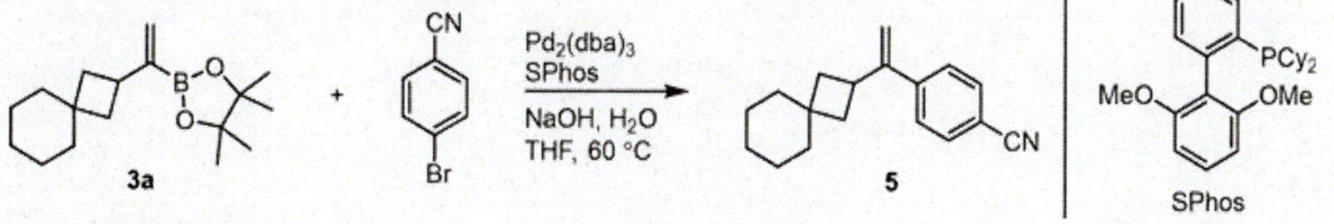

Fig. 3.26 Scheme of preparation procedure of **5**

^{1}H NMR (400 MHz, $CDCl_3$, δ): 1.29–1.40 (m, 6H), 1.41–1.51 (m, 2H), 1.55–1.68 (m, 4H), 2.07 (td, $J = 9.1, 2.5$ Hz, 2H), 3.28–3.39 (m, 1H), 5.17 (d, $J = 2.0$ Hz, 1H), 5.42 (t, $J = 0.8$ Hz, 1H), 7.44 (dt, $J = 8.3, 1.6$ Hz, 2H), 7.59 (dt, $J = 8.2, 1.6$ Hz, 2H). ^{13}C NMR (100 MHz, $CDCl_3$, δ): 22.7 (CH_2), 23.0 (CH_2), 25.9 (CH_2), 32.4 (CH), 35.1 (C), 36.8 (CH_2), 38.4 (CH_2), 40.5 (CH_2), 110.6 (C), 112.5 (CH_2), 118.9 (C), 126.7 (CH), 131.9 (CH), 145.5 (C), 151.2 (C). HRMS-EI (*m*/*z*): [M+]+ calcd for $C_{18}H_{21}N$, 251.1674; found, 251.1670.

2-(4-Methylpenta-1,3-dien-2-yl)-7-oxaspiro[3.5]nonane (6, Fig. 3.27)

The palladium catalyzed cross-coupling reaction with an aryl halide was performed according to the literature procedure [54]. 1-Bromo-2-methylprop-1-ene (68.7 mg, 0.375 mmol), $Pd_2(dba)_3$ (4.6 mg, 0.005 mmol), and SPhos (4.1 mg, 0.010 mmol) were placed in an oven-dried reaction vial under a nitrogen atmosphere. A solution of **3b** (61.8 mg, 0.22 mmol) in dry THF (2.25 mL) was added to the vial followed by addition of 3 M NaOH aq (0.75 mL). Then, the mixture was warmed to 60 °C with stirring for 19 h. After the reaction mixture was cooled to room temperature, the reaction mixture was quenched by H_2O and extracted with Et_2O three times. The combined organic layer was then dried over $MgSO_4$. After filtration, the solvents were removed by evaporation. The crude mixture was purified by flash column chromatography (SiO_2, Et_2O/hexane, 0:100–4:96) to give the corresponding coupling product **6** (27.5 mg, 0.133 mmol, 60%) as a colorless oil.

^{1}H NMR (400 MHz, $CDCl_3$, δ): 1.50 (t, $J = 5.3$ Hz, 2H), 1.64–1.72 (m, 4H), 1.75 (d, $J = 1.2$ Hz, 3H), 1.78 (d, $J = 1.2$ Hz, 3H), 2.02 (td, $J = 9.1, 2.7$ Hz, 2H), 2.89 (quint, $J = 8.8$ Hz, 1H), 3.55 (t, $J = 5.3$ Hz, 2H), 3.64 (t, $J = 5.3$ Hz, 2H), 4.77 (d, $J = 1.6$ Hz, 1H), 4.93 (t, $J = 1.9$ Hz, 1H), 5.57 (d, $J = 0.8$ Hz, 1H). ^{13}C NMR (100 MHz, $CDCl_3$, δ): 19.5 (CH_3), 26.5 (CH_3), 32.5 (C), 34.7 (CH), 37.1 (CH_2), 37.3 (CH_2), 40.2 (CH_2), 64.6 (CH_2), 65.0 (CH_2), 109.9 (CH_2), 124.1 (CH), 135.2 (C), 149.6 (C). HRMS-EI (*m*/*z*): [M]+ calcd for $C_{14}H_{22}O$, 206.1671; found, 206.1671.

6-Methyl-2-(spiro[3.5]nonane-2-carbonyl)-1,3,6,2-dioxazaborocane-4,8-dione (8, Fig. 3.28)

The acylboron compound **8** was synthesized through the literature procedure [44]. **3a** (124.4 mg, 0.45 mmol) and KHF_2 (158.2 mg, 2.03 mmol) were placed in a flask equipped with a stir bar and a rubber septum. MeOH (5 mL) and H_2O (2.5 mL) were then added into the flask using syringes through the septum, and then the resultant solution was stirred at room temperature for 1 h. After the reaction mixture was

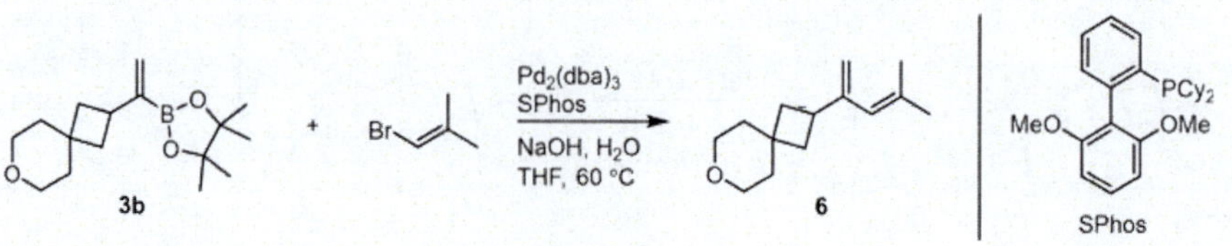

Fig. 3.27 Scheme of preparation procedure of **6**

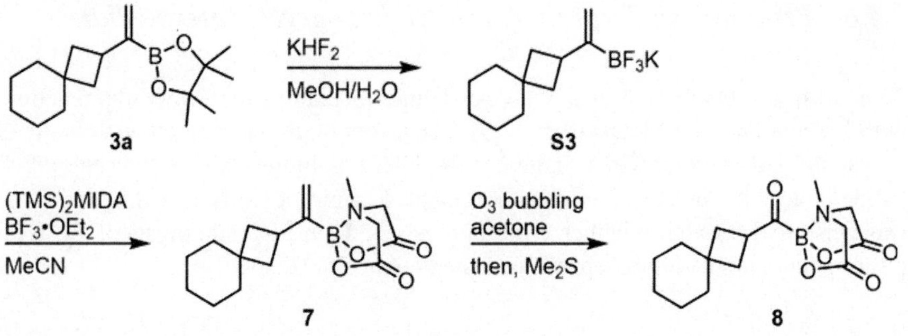

Fig. 3.28 Scheme of preparation procedure of **8**

concentrated in vacuo, the resulting solid was suspended in acetone and filtered off to remove KHF_2. The extract was concentrated in vacuo, and the resulting solid was washed with Et_2O to remove pinacol. The filtrate was concentrated in vacuo, and the product **S3** was obtained in 88% yield (101.2 mg, 0.395 mmol). The product **S3** was used in the next reaction without further purification.

S3 (89.4 mg, 0.35 mmol) was placed in an oven-dried reaction vial. After the vial was sealed with a screw cap containing a Teflon-coated rubber septum, the vial was connected to a vacuum/nitrogen manifold through a needle. It was evacuated and then backfilled with nitrogen. This cycle was repeated three times. MeCN (3.5 mL), $(TMS)_2$MIDA (102.0 mg, 0.35 mmol), and $BF_3{\cdot}Et_2O$ (44.0 μL, 0.35 mmol) were added to the vial, and then the resultant solution was stirred at 80°C for 14 h. After the reaction mixture was concentrated in vacuo, the crude mixture was purified by flash column chromatography (SiO_2, AcOEt/hexane, 20:80–100:0) to give the corresponding product **7** (59.1 mg, 0.194 mmol, 55%) as a white solid.

Alkenyl MIDA boronate **7** (45.6 mg, 0.15 mmol) and a small amount of Sudan III were placed in an oven-dried reaction vial. The vial was filled with nitrogen gas and then sealed with a screw cap containing a Teflon-coated rubber septum. Dry acetone (1.00 mL) was added in the vial through the rubber septum using a syringe, then the solution turned pink. The solution was cooled to –78 °C, at which point a stream of O_3/O_2 gas was introduced through a needle until which time the reaction mixture started to turn blue. The addition of O_3/O_2 gas was then stopped, and O_2 gas was bubbled through the solution for 5 min to remove the remained O_3 gas. The vial was charged with Me_2S (0.15 mL) and stirred at –78 °C for 5 min. The reaction mixture was then warmed to room temperature and stirred for 5 min. The solvent was then removed in vacuo. The resulting solid was washed with hexane and Et_2O to give the corresponding product **8** (38.3 mg, 0.125 mmol, 83%) as a white solid.

^{1}H NMR (400 MHz, acetone-d$_6$, δ): 1.26–1.38 (m, 6H), 1.40–1.48 (m, 2H), 1.51–1.59 (m, 2H), 1.79–1.94 (m, 4H), 3.04 (s, 3H), 3.48 (tt, $J = 9.2$, 8.2 Hz, 1H), 4.11 (d, $J = 16.9$ Hz, 2H), 4.33 (d, $J = 16.9$ Hz, 2H). ^{13}C NMR (100 MHz, acetone-d$_6$, δ): 23.2 (CH_2), 23.6 (CH_2), 26.6 (CH_2), 33.8 (CH_2), 36.0 (*C*), 38.5 (CH_2), 40.0 (CH_2), 42.9 (*CH*), 47.2 (CH_3), 63.0 (CH_2), 168.9 (*C*), 206.3 (*C*). HRMS-ESI (*m/z*): [M + Na]+ calcd for $C_{15}H_{22}{}^{11}BO_5NNa$, 330.1486; found, 330.1487.

3.4.6 Protonolysis Experiment of Allylcooper(I) Intermediate

To confirm the generation of allylcopper(I) intermediate, I performed the reaction with 2.0 equivalent of MeOH (Fig. 3.29). The protoboration product **9** was obtained in a high yield (79%), which indicates the allene is more reactive than the alkyl halide moiety under this conditions. Although I cannot identify that the product **9** was generated through whether γ-protonolysis of $\mathbf{I_A}$ or α-protonolysis of $\mathbf{I_B}$, it is clear to generate the allylcopper(I) intermediate in situ (Fig. 3.30).

3.5 Computational Details

All calculations were performed with the Gaussian 09 (revision D.01) suite of programs [55]. Geometry optimizations and transition state (TS) calculations were carried out at ωB97X-D [56] level of theory in THF (SMD solvation model) [57] with Def2-SVP [58] basis set. Frequency calculations were conducted at the same level of theory on the optimized geometries to check the all the stationary points as either minima or transition states and to obtain zero-point energy (ZPE), thermal energy,

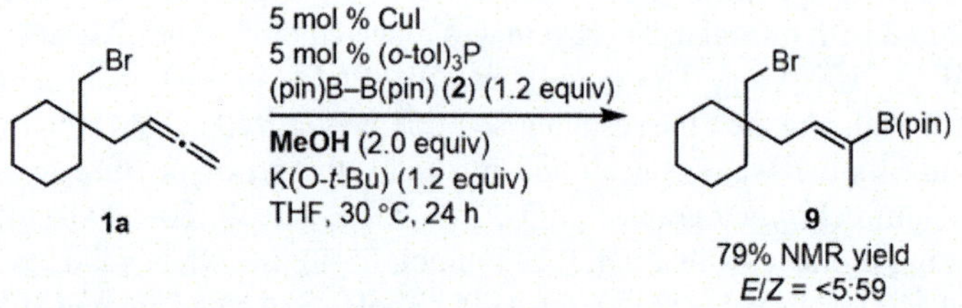

Fig. 3.29 Protoboration of substrate **1a**

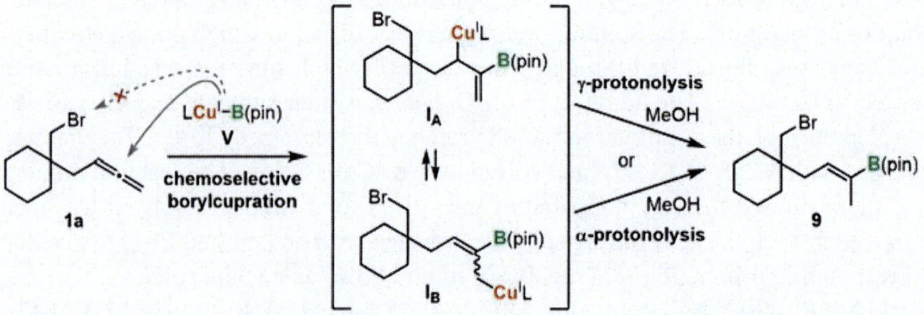

Fig. 3.30 Generation of allylcopper(I) intermediate through the chemoselective borylcupration of allene **1a**

and Gibbs free energy corrections at 298.15 K (1.0 atm). Intrinsic reaction coordinate (IRC) calculations were carried out to confirm the transition state connecting the correct reactant and product on the potential energy surface.

I choose 6-bromo-5-methylhexa-1,2-diene (**1x**) as a model substrate to describe the overall reaction mechanism and the regio- and diastereoselectivity. In the calculations, PMe_3 and B(en) group was used as a model catalyst system instead of (*o*-tol)$_3$P and B(pin) group, respectively.

3.5.1 Isomerization of Allylcopper(I) Intermediates

I conducted the DFT calculations on isomerization step of allylcopper(I) intermediates (Figs. 3.31 and 3.32). I used electron energy (E) for the energy comparison because I encountered the transition state with low frequency (25i cm^{-1}) on which the energy correction is not correct due to their high anharmonicity. The relative energies based on *cis*-$\mathbf{I_B}$, which is the most stable intermediate, were illustrated in Figs. 3.31 and 3.32, and the activation energies of the isomerization were illustrated in Fig. 3.31. The highest energy barrier [13.2 kcal/mol (*cis*-$\mathbf{I_B}$ to *cis*-$\mathbf{I_A}$ via **TS3**)] was relatively high for the isomerization at room temperature. But the isomerization could also occur via **TS4** with the second highest activation energy [11.6 kcal/mol (*cis*-$\mathbf{I_B}$ to *trans*-$\mathbf{I_A}$ via **TS4**)]. The barrier value is low enough to pass over at room temperature.

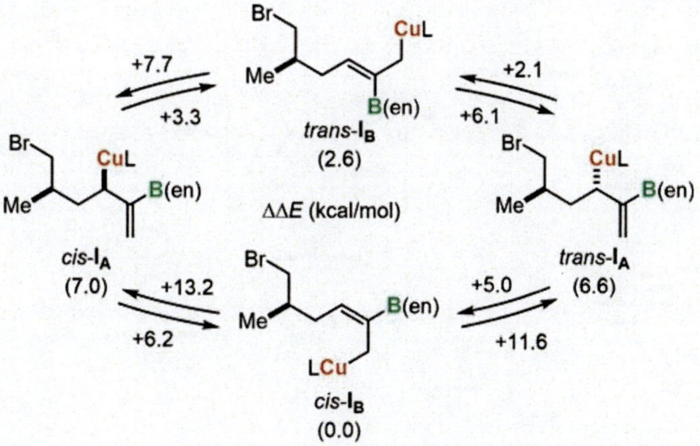

Fig. 3.31 Isomerization of allylcopper(I) intermediates and the activation energies

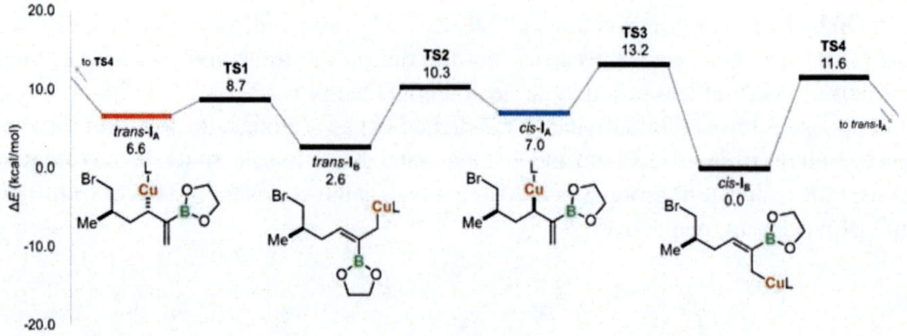

Fig. 3.32 Energy diagram of isomerization steps of allylcopper(I) intermediates

3.5.2 Energy Diagram of Six-Membered Ring Formation Pathway

I conducted the DFT calculations of six-membered ring formation pathway for **4** (Fig. 3.33). *cis*-$\mathbf{I_B}$ is formed through borylcupration of **1x** and isomerization of the allylcopper(I) intermediates **I**. After the formation of cuprate *cis*-$\mathbf{II_B}$, oxidative addition is occurred to generate cupracycle *cis*-$\mathbf{III_B}$. The activation energy of this step is 21.9 kcal/mol, which is larger than that of the pathway to produce *trans*-**3** (21.1 kcal/mol) and smaller than that of the pathway to produce *cis*-**3** (22.1 kcal/mol). This tendency is in good agreement with alkyl group-substituted substrates (**3g**: **3**/**4** =

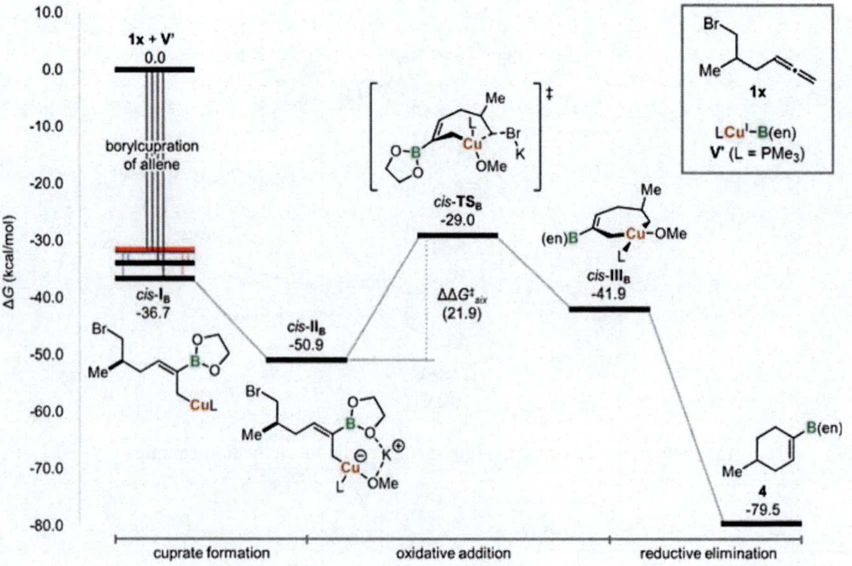

Fig. 3.33 Energy diagram of six-membered ring formation pathway

92:8, dr = 93:7, **3h**: **3/4** = 73:27, dr = 91:9). Furthermore, this result implies that the seven-membered ring formation is kinetically more disfavored than five-membered ring formation.

References

1. Hall DG (ed) (2011) Boronic acids: preparation, applications in organic synthesis and medicine, vols 1–2. Wiley-VCH, Weinheim
2. Carreras J, Caballero A, Pérez PJ (2019) Alkenyl boronates: synthesis and applications. Chem Asian J 14:329–343
3. Miyaura N, Suzuki A (1995) Palladium-catalyzed cross-coupling reactions of organoboron compounds. Chem Rev 95:2457–2483
4. Lennox AJJ, Lloyd-Jones GC (2014) Selection of boron reagents for Suzuki-Miyaura coupling. Chem Soc Rev 43:412–443
5. Xu L, Zhang S, Li P (2015) Boron-selective reactions as powerful tools for modular synthesis of diverse complex molecules. Chem Soc Rev 44:8848–8858
6. Shimizu Y, Kanai M (2014) Recent progress in copper-catalyzed difunctionalization of unactivated carboncarbon multiple bonds. Tetrahedron Lett 55:3727–3737
7. Pulis AP, Yeung K, Procter DJ (2017) Enantioselective copper catalysed, direct functionalisation of allenes: via allyl copper intermediates. Chem Sci 8:5240–5247
8. Whyte A, Torelli A, Mirabi B, Zhang A, Lautens M (2020) Copper-catalyzed borylative difunctionalization of π-systems. ACS Catal 10:11578–11622
9. Buñuel E, Cárdenas DJ (2016) Borylative cyclization reactions. Eur J Org Chem 2016:5446–5464
10. Kubota K, Yamamoto E, Ito H (2013) Copper(I)-catalyzed borylative exo-cyclization of alkenyl halides containing unactivated double bond. J Am Chem Soc 135:2635–2640
11. Recently, Fernández and co-workers reported the borylative cyclization reaction of alkenes and a computational study for the mechanistic investigation in detail. See, Ref. [12]
12. Royes J, Ni S, Farré A, La Cascia E, Carbó JJ, Cuenca AB, Maseras F, Fernández E (2018) Copper-catalyzed borylative ring closing C–C coupling toward spiro- and dispiroheterocycles. ACS Catal 8:2833–2838
13. Kubota K, Iwamoto H, Yamamoto E, Ito H (2015) Silicon-tethered strategy for copper(I)-catalyzed stereo- and regioselective alkylboration of alkynes. Org Lett 17:620–623
14. Iwamoto H, Ozawa Y, Kubota K, Ito H (2017) Copper(I)-catalyzed regio- and stereoselective intramolecular alkylboration of propargyl ethers and amines. J Org Chem 82:10563–10573
15. Meng F, Jang H, Jung B, Hoveyda AH (2013) Cu-catalyzed chemoselective preparation of 2-(pinacolato)boron-substituted allylcopper complexes and their in situ site-, diastereo-, and enantioselective additions to aldehydes and ketones. Angew Chem Int Ed 52:5046–5051
16. Rae J, Yeung K, McDouall JJW, Procter DJ (2016) Copper-catalyzed borylative cross-coupling of allenes and imines: selective three-component assembly of branched homoallyl amines. Angew Chem Int Ed 55:1102–1107
17. Yeung K, Ruscoe RE, Rae J, Pulis AP, Procter DJ (2016) Enantioselective generation of adjacent stereocenters in a copper-catalyzed three-component coupling of imines, allenes, and diboranes. Angew Chem Int Ed 55:11912–11916
18. Meng F, Li X, Torker S, Shi Y, Shen X, Hoveyda AH (2016) Catalytic enantioselective 1,6-conjugate additions of propargyl and allyl groups. Nature 537:387–393
19. Zhao Y-S, Tang X-Q, Tao J-C, Tian P, Lin G-Q (2016) Efficient access to *cis*-decalinol frameworks: copper(I)-catalyzed borylative cyclization of allene cyclohexanediones. Org Biomol Chem 14:4400–4404
20. Zhao W, Montgomery J (2016) Cascade copper-catalyzed 1,2,3-trifunctionalization of terminal allenes. J Am Chem Soc 138:9763–9766

21. Fujihara T, Sawada A, Yamaguchi T, Tani Y, Terao J, Tsuji Y (2017) Boraformylation and silaformylation of allenes. Angew Chem Int Ed 56:1539–1543
22. Boreux A, Indukuri K, Gagosz F, Riant O (2017) Acyl fluorides as efficient electrophiles for the copper-catalyzed boroacylation of allenes. ACS Catal 7:8200–8204
23. Jang H, Romiti F, Torker S, Hoveyda AH (2017) Catalytic diastereo- and enantioselective additions of versatile allyl groups to N-H ketimines. Nat Chem 9:1269–1275
24. Sawada A, Fujihara T, Tsuji Y (2018) Copper-catalyzed bora-acylation and bora-alkoxyoxalylation of allenes. Adv Synth Catal 360:2621–2625
25. After this study on intramolecular carboboration of allenes using alkyl electrophiles was reported, some groups reported the reactions using the other electrophiles. See, Refs. [26–33]
26. Yeung K, Talbot FJT, Howell GP, Pulis AP, Procter DJ (2019) Copper-catalyzed borylative multicomponent synthesis of quaternary α-amino esters. ACS Catal 9:1655–1661
27. Han J, Zhou W, Zhang P-C, Wang H, Zhang R, Wu H-H, Zhang J (2019) Design and synthesis of WJ-Phos, and application in Cu-catalyzed enantioselective boroacylation of 1,1-disubstituted allenes. ACS Catal 9:6890–6895
28. Zhang S, del Pozo J, Romiti F, Mu Y, Torker S, Hoveyda AH (2019) Delayed catalyst function enables direct enantioselective conversion of nitriles to NH_2-amines. Science 364:45–51
29. Shen B-X, Min X-T, Hu Y-C, Qian L-L, Yang S-N, Wan B, Chen Q-A (2020) Copper-catalyzed boroacylation of allenes to access tetrasubstituted vinylboronates. Org Biomol Chem 18:9253–9260
30. Li Z, Zhang L, Nishiura M, Luo G, Luo Y, Hou Z (2020) Enantioselective cyanoborylation of allenes by N-heterocyclic carbene-copper catalysts. ACS Catal 10:11685–11692
31. Zhao C-Y, Zheng H, Ji D-W, Min X-T, Hu Y-C, Chen Q-A (2020) Copper-catalyzed asymmetric carboboronation of allenes to access α-quaternary amino esters with adjacent stereocenters. Cell Rep Phys Sci 1:100067
32. del Pozo J, Zhang S, Romiti F, Xu S, Conger RP, Hoveyda AH (2020) Streamlined catalytic enantioselective synthesis of α-substituted β, γ-unsaturated ketones and either of the corresponding tertiary homoallylic alcohol diastereomers. J Am Chem Soc 142:18200–18212
33. Deng H, Dong Y, Shangguan Y, Yang F, Han S, Wu J, Liang B, Guo H, Zhang C (2021) Copper-catalyzed three-component carboboronation of allenes using highly strained cyclic ketimines as electrophiles. Org Lett 23:4431–4435
34. Yuan W, Zhang X, Yu Y, Ma S (2013) Amide-controlled highly selective catalytic borylcupration of allenes. Chem Eur J 19:7193–7202
35. Semba K, Shinomiya M, Fujihara T, Terao J, Tsuji Y (2013) Highly selective copper-catalyzed hydroboration of allenes and 1,3-dienes. Chem Eur J 19:7125–7132
36. Sheng W, Wang M, Lein M, Jiang L, Wei W, Wang J (2013) Mechanism of copper(I)-catalyzed allylic alkylation of phosphorothioate esters: influence of the leaving group on α regioselectivity. Chem Eur J 19:14126–14142
37. Semba K, Fujihara T, Terao J, Tsuji Y (2013) Copper-catalyzed borylation of α-alkoxy allenes with bis(pinacolato)diboron: efficient synthesis of 2-boryl 1,3-butadienes. Angew Chem Int Ed 52:12400–12403
38. Xiong M, Xie X, Liu Y (2017) Copper-catalyzed borylative cyclization of in situ generated o-allenylaryl nitriles with bis(pinacolato)diboron. Org Lett 19:3398–3401
39. Tolman CA (1977) Steric effects of phosphorus ligands in organometallic chemistry and homogeneous catalysis. Chem Rev 77:313–348
40. Suresh CH (2006) Molecular electrostatic potential approach to determining the steric effect of phosphine ligands in organometallic chemistry. Inorg Chem 45:4982–4986
41. Billingsley K, Buchwald SL (2007) Highly efficient monophosphine-based catalyst for the palladium-catalyzed Suzuki-Miyaura reaction of heteroaryl halides and heteroaryl boronic acids and esters. J Am Chem Soc 129:3358–3366
42. Niemeyer ZL, Milo A, Hickey DP, Sigman MS (2016) Parameterization of phosphine ligands reveals mechanistic pathways and predicts reaction outcomes. Nat Chem 8:610–617
43. Barder TE, Walker SD, Martinelli JR, Buchwald SL (2005) Catalysts for Suzuki-Miyaura coupling processes: scope and studies of the effect of ligand structure. J Am Chem Soc 127:4685–4696

44. Taguchi J, Ikeda T, Takahashi R, Sasaki I, Ogasawara Y, Dairi T, Kato N, Yamamoto Y, Bode JW, Ito H (2017) Synthesis of acylborons by ozonolysis of alkenylboronates: preparation of an enantioenriched amino acid acylboronate. Angew Chem Int Ed 56:13847–13851
45. Yang CT, Zhang ZQ, Tajuddin H, Wu CC, Liang J, Liu JH, Fu Y, Czyzewska M, Steel PG, Marder TB, Liu L (2012) Alkylboronic esters from copper-catalyzed borylation of primary and secondary alkyl halides and pseudohalides. Angew Chem Int Ed 51:528–532
46. Ito H, Kubota K (2012) Copper(I)-catalyzed boryl substitution of unactivated alkyl halides. Org Lett 14:890–893
47. Iwamoto H, Kubota K, Yamamoto E, Ito H (2015) Copper(I)-catalyzed carbon-halogen bond-selective boryl substitution of alkyl halides bearing terminal alkene moieties. Chem Commun 51:9655–9658
48. Iwamoto H, Akiyama S, Hayama K, Ito H (2017) Copper(I)-catalyzed stereo- and chemoselective borylative radical cyclization of alkyl halides bearing an alkene moiety. Org Lett 19:2614–2617
49. Yoshikai N, Nakamura E (2012) Mechanisms of nucleophilic organocopper(I) reactions. Chem Rev 112:2339–2372
50. Nakamura E, Mori S (2000) Wherefore art thou copper? Structures and reaction mechanisms of organocuprate clusters in organic chemistry. Angew Chem Int Ed 39:3750–3771
51. Bartholomew ER, Bertz SH, Cope S, Murphy M, Ogle CA (2008) Preparation of σ- and π-allylcopper(III) intermediates in S_N2 and S_N2' reactions of organocuprate(I) reagents with allylic substrates. J Am Chem Soc 130:11244–11245
52. Casadei MA, Galli C, Mandolini L (1984) Ring-closure reactions. 22. Kinetics of cyclization of diethyl (ω-bromoalkyl)malonates in the range of 4-to 21-membered rings. Role of ring strain. J Am Chem Soc 106:1051–1056
53. Wang Z, Nicolini C, Hervieu C, Wong YF, Zanoni G, Zhang L (2017) Remote cooperative group strategy enables ligands for accelerative asymmetric gold catalysis. J Am Chem Soc 139:16064–16067
54. Meng F, McGrath KP, Hoveyda AH (2014) Multifunctional organoboron compounds for scalable natural product synthesis. Nature 513:367–374
55. Gaussian 09, Revision D.01, Frisch MJ, Trucks GW, Schlegel HB, Scuseria GE, Robb MA, Cheeseman JR, Scalmani G, Barone V, Petersson GA, Nakatsuji H, Li X, Caricato M, Marenich A, Bloino J, Janesko BG, Gomperts R, Mennucci B, Hratchian HP, Ortiz JV, Izmaylov AF, Sonnenberg JL, Williams-Young D, Ding F, Lipparini F, Egidi F, Goings J, Peng B, Petrone A, Henderson T, Ranasinghe D, Zakrzewski VG, Gao J, Rega N, Zheng G, Liang W, Hada M, Ehara M, Toyota K, Fukuda R, Hasegawa J, Ishida M, Nakajima T, Honda Y, Kitao O, Nakai H, Vreven T, Throssell K, Montgomery JA Jr, Peralta JE, Ogliaro F, Bearpark M, Heyd JJ, Brothers E, Kudin KN, Staroverov VN, Keith T, Kobayashi R, Normand J, Raghavachari K, Rendell A, Burant JC, Iyengar SS, Tomasi J, Cossi M, Millam JM, Klene M, Adamo C, Cammi R, Ochterski JW, Martin RL, Morokuma K, Farkas O, Foresman JB, Fox DJ (2009) Gaussian, Inc., Wallingford CT
56. Chai J Da, Head-Gordon M (2008) Long-range corrected hybrid density functionals with damped atom-atom dispersion corrections. Phys Chem Chem Phys 10:6615–6620
57. Marenich AV, Cramer CJ, Truhlar DG (2009) Universal solvation model based on solute electron density and on a continuum model of the solvent defined by the bulk dielectric constant and atomic surface tensions. J Phys Chem B 113:6378–6396
58. Weigend F, Ahlrichs R (2005) Balanced basis sets of split valence, triple zeta valence and quadruple zeta valence quality for H to Rn: design and assessment of accuracy. Phys Chem Chem Phys 7:3297–3305

Chapter 4
Intermolecular Alkylboration of *gem*-Disubstituted Allenes for Stereoselective Synthesis of Multi-alkylated Allylic Boronates

4.1 Introduction

Allylic boronates have been recognized as essential building blocks in organic synthesis [1, 2], as the carbon–boron bond can be stereospecifically transformed into carbon–carbon and carbon–heteroatom bonds at the γ- or α-position [3–18]. Actually, these reactions are practically used in total syntheses of natural products [19–22]. Among the possible substitution patterns of multi-substituted allylic boronates, 2,3,3-all-carbon-trisubstituted allylic boronates that possess a stereodefined tetrasubstituted alkene moiety are attractive precursors of contiguous and densely substituted sp^2- and sp^3-carbon skeletons (Fig. 4.1). For example, the homoallylic alcohols bearing a quaternary carbon atom are obtained as the γ-coupling products via allylboration reaction of aldehydes. Also, the allylic functionalized tetrasubstituted alkenes such as allylic amines and alcohols are obtained via oxidative 1,2-migration of the boryl group. However, the stereoselective synthesis of sterically congested tetrasubstituted alkene moiety in the trisubstituted allylic boronates, especially having four different substituents, is highly challenging.

In 2002, the group of Hall first developed a regio- and stereoselective synthetic approach of differentially 2,3,3-all-carbon-trisubstituted allylic boronates using a conjugate addition of organocopper species to alkynyl esters and a succeeding reaction with a boron source (Fig. 4.2A) [23]. Following this development, they developed the syntheses and applications of similar trisubstituted allylic boronates [19, 24, 25]. In spite of the excellent regio- and stereoselectivities, and functional group tolerance, the electron-withdrawing group (EWG) in the substrate and a stoichiometric amount of organocopper reagent are necessary in this reaction. Furthermore, the authors stated that this reaction should be carried out at a cryogenic temperature for high stereoselectivity because the stereochemistry of the alkenyl copper(I) species gets unstable above –30 °C [23]. The groups of RajanBabu and Disier independently reported the allylic boronates embedded in cyclopropene and -butenes (Fig. 4.2B) [26–29]. Despite these excellent pioneering studies, the structural variety

Y. Ozawa, *Copper(I)-Catalyzed Stereoselective Borylation Reactions*, Springer Theses,
https://doi.org/10.1007/978-981-99-1098-4_4

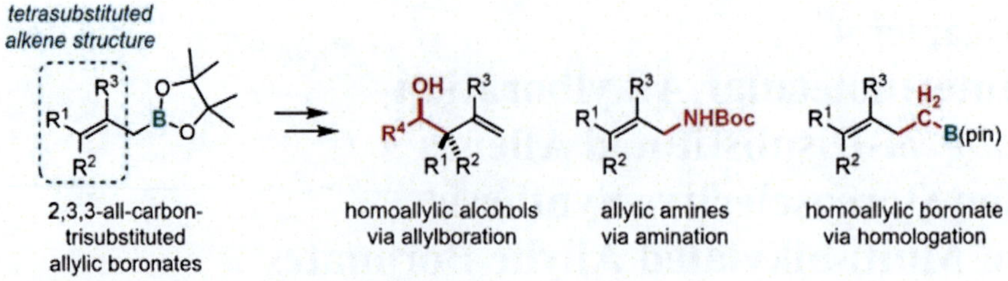

Fig. 4.1 Allylic boronates as building blocks

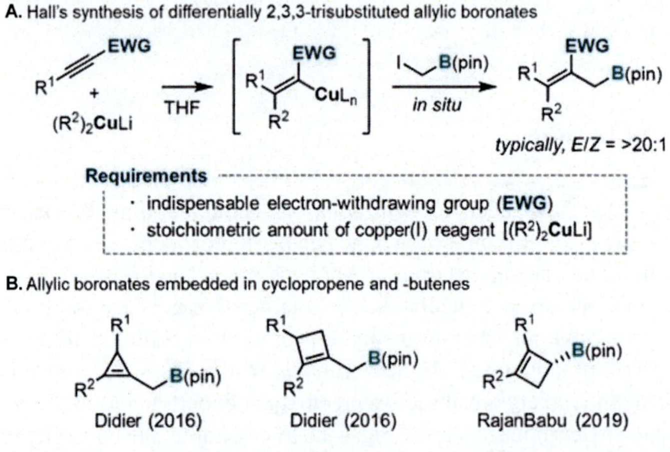

Fig. 4.2 Previous works toward synthesis of differentially 2,3,3-all-carbon-trisubstituted allylic boronates

of differentially 2,3,3-all-carbon-trisubstituted allylic boronates still remains limited [30, 31].

A copper(I)-catalyzed carboboration reaction of allenes is three-component coupling between allenes, carbon electrophiles, and a boron source to potentially produce allylic and alkenyl boronates (Fig. 4.3A) [32–35]. Thus, reaction is an attractive research target because the regio- and stereodivergence of the carbon–boron and carbon–carbon bonds formation reactions, which potentially provide various alkenyl and allylic boronates. Up to date, most research groups have focused on the 2,1-borylcupration of allenes and the synthesis of linear-type alkenyl boronates **I** and branched-type alkenyl boronates **II** (left arrow in Fig. 4.3A, B). For these reactions, as allylcopper(I) intermediate generated via the 2,1-borylcupration of the terminal double bond of allenes is the key intermediate for production of the alkenyl boronates. The Tsuji, Hoveyda, Brown, and Liu groups independently reported the inter- and intramolecular reactions for the synthesis of linear-type (*Z*)-alkenyl boronates **I** [36–39]. The other several groups including I and co-workers also reported the preparation of the branched-type alkenyl boronates **II** [40–58].

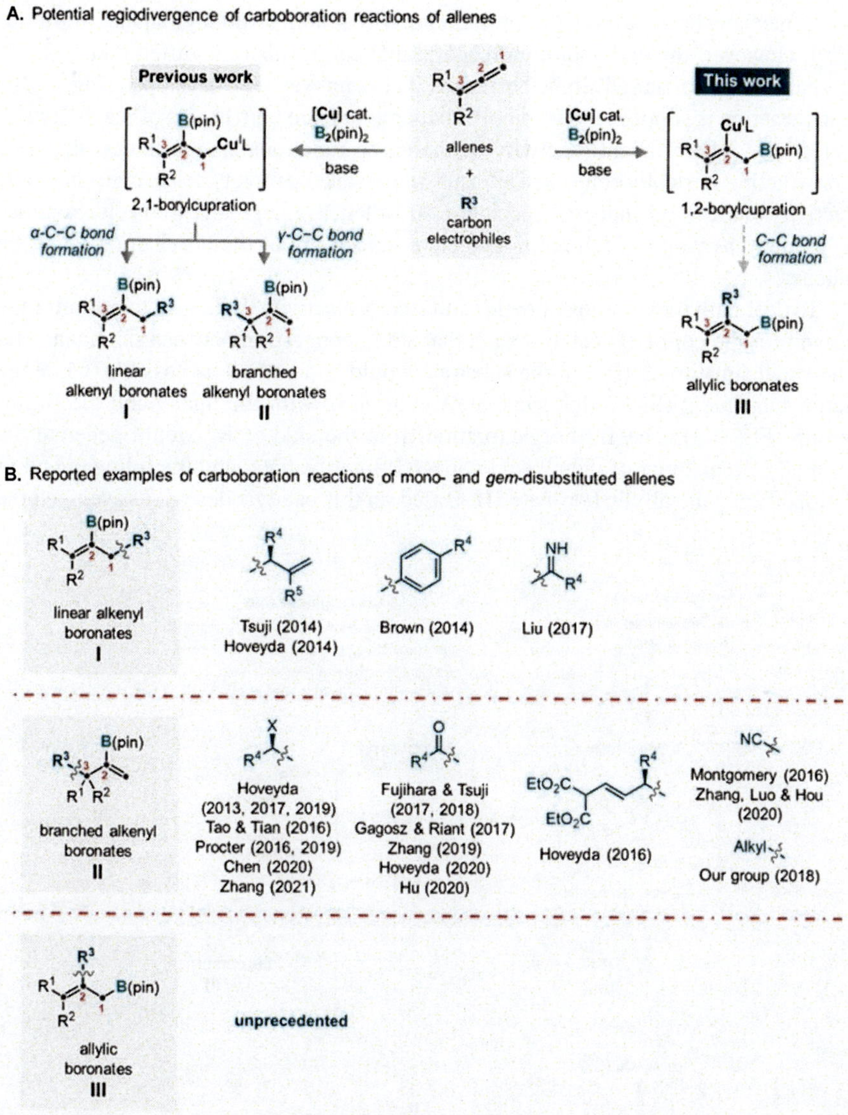

Fig. 4.3 Carboboration reactions of allenes to access alkenyl and allylic boronates

During my previous study on the intramolecular alkylboration of allenes [49], I got interested in the regiodivergence of borylcupration of the allenes. I speculated that the multi-substituted allylic boronates **III** might be synthesized via the alkenyl copper(I) intermediate generated by the 1,2-borylcupration of the allene (right arrow in Fig. 4.3A). The group of Ma has previously demonstrated the 1,2-borylcupration of *gem*-dialkylallenes and the generation of alkenylcopper(I) species as a mechanistic study on the protoboration reaction of allenyl silane substrates [59]. However, the carboboration reaction that can furnish the differentially 2,3,3-all-carbon-trisubstituted allylic boronates **III** has not yet been developed (Fig. 4.3B). This reaction is supposed to be highly challenging from both aspect of reactivity and selectivity. The efficient reactivity of the carbo-functionalization is necessary for constructing sterically congested tetrasubstituted alkene moiety. Also, high chemoselectivity of the boryl copper(I) species toward allenes rather than carbon electrophiles is required as well as high regio- and stereoselectivity of the borylcupration of the allenes.

To deal with these issues of regio- and stereoselectivity, I planned to optimize the ligand for the copper(I) catalyst such that steric congestion between the ligand and the small substituent (R^S) in the substrate would be as small as possible, while the large substituent (R^L) in the substrate would clash with the ligand and the B(pin) group (Fig. 4.4). The 1,2-borylcupration from the side of R^S would generate the desired stereoisomer of the alkenyl copper(I) intermediate, and the following alkylation affords the allylic boronate **III**. Based on this catalyst design, I developed the

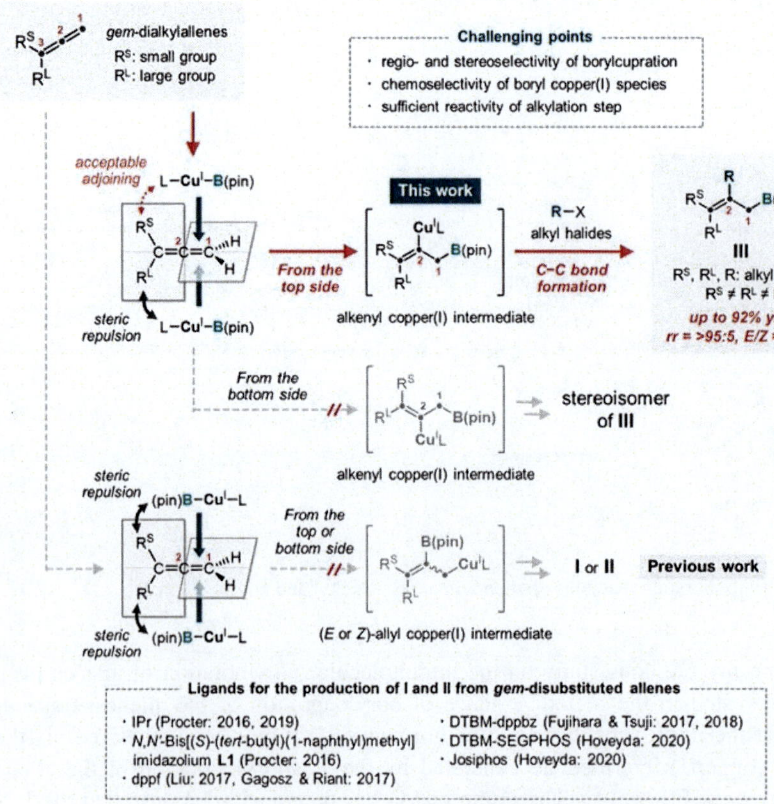

Fig. 4.4 Carboboration reactions of allenes to access alkenyl and allylic boronates

copper(I)-catalyzed the alkylboration reaction of *gem*-dialkylallenes for the synthesis of differentially 2,3,3-trialkylsubstituted allylic boronates. The SIMes and Xantphos ligands, which have pocket-like structures, realize the high regio- and stereoselectivities by recognizing the bulkiness of the two alkyl groups in the allene substrates. Those structures are not found in the previous ligands producing alkenyl boronates **I** or **II** listed at the bottom-right in Fig. 4.2C. A density functional theory (DFT)-based computational mechanistic study revealed that the transition states of the π-coordination of the allenes moiety to boryl copper(I) species are energetically comparable to that of the borylcupration step. These steps are crucial for the high regio- and stereoselectivity.

4.2 Results and Discussion

The development of the carboboration reaction was started with cyclohexyl methyl allene (**1a**) and alkyl halide (**2a**) as model substrates by applying the standard conditions described in Table 4.1. The Xantphos ligand, which is a rigid and large-bite-angle bidentate phosphine ligand, furnished allylic boronate **4aa** with perfect regioselectivity but moderate *E*/*Z*-selectivity (entry 1: 88%, *E*/*Z* = 60:40, **4**:**5** = >99:1). Using the backbone-flexible bisphosphines dppp or dppf gave the product with lower regioselectivity (entry 2: 30%, *E*/*Z* = 70:30, **4**:**5** = 44:56; entry 3: 5%, **4**:**5** = 6:94). The use of monophosphine PCy_3 resulted in production of a trace amount of the desired product (entry 4: 2%, **4**:**5** = 81:19), with generating the borylative substitution product of the alkyl halide **2a** in a significantly increased yield [60–64]. Then, *N*-heterocyclic carbene (NHC) ligands were examined with lower catalyst loadings at lower reaction temperatures because the NHC ligands showed higher reactivity than phosphine ligands. The backbone-saturated NHC SIMes afforded the product in high yield with high selectivities (entry 5: 73%, *E*/*Z* = >95:5, **4**:**5** = 92:8). Contrary, the yield was dropped when the backbone-unsaturated NHC IMes was employed, although the selectivities were high (entry 6: 55%, *E*/*Z* = >95:5, **4**:**5** = 94:6). The NHC ligand IAd, which has bulky adamantly groups, showed inverted *E*/*Z*-selectivity to generate (*Z*)-**4aa** as the major product (entry 7: 71%, *E*/*Z* = 32:68, **4**:**5** = 98:2). Thus, SIMes was chosen as the optimal ligand for the substrate **1a**. Next, the solvent was screened in order to explore the effect of solvent polarity on the reaction outcome. Although *N*,*N*-dimethylacetamide (DMA), which is also a polar aprotic solvent as well as *N*,*N*-dimethylformamide (DMF), gave the desired product in moderate yield, a significant amount of the corresponding protoboration product was generated (entry 10: 48%, *E*/*Z* = >95:5, **4**:**5** = 85:15). The acetyl group in DMA could be the proton source for the production of the protoboration product. The less-polar aprotic solvents tetrahydrofuran (THF) and acetonitrile (MeCN) gave the product in low yield (entry 8: 19%; entry 9: 13%). In contrast, toluene and *n*-hexane as the non-polar solvents furnished trace amounts of the borylation products, which is attributed to disfavor of polar S_N2-type reactions in less-polar solvents, i.e., the S_N2-type oxidative addition of alkenyl copper(I) species to alkyl halides (entry 11: <1%; entry 12: 5%).

Table 4.1 Reaction optimization[a]

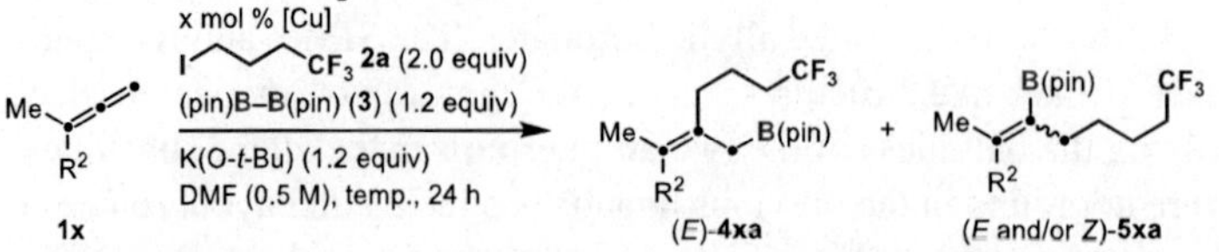

Entry	1x (R^2)	[Cu] (*x* mol %)	Temp [°C]	Variation of other conditions	Yield of 4xa (4xa + 5xa) [%][b]	*E*/*Z* of 4xa [%][c]	4xa/5xa [%][b]
1	**1a** (Cy)	CuCl/Xantphos (5)	30	None	88 (88)	60:40	>99:1
2	**1a** (Cy)	CuCl/dppp (5)	30	None	30 (68)	70:30	44:56
3	**1a** (Cy)	CuCl/dppf (5)	30	None	5 (80)	n.d	6:94
4	**1a** (Cy)	CuCl/PCy_3 (5)	30	None	2 (3)	n.d	81:19
5	**1a** (Cy)	SIMesCuCl (2)	0	None	73 (79)	>95:5	92:8
6	**1a** (Cy)	IMesCuCl (2)	0	None	55 (58)	>95:5	94:6
7	**1a** (Cy)	IAdCuCl (2)	0	None	71 (72)	32:68	98:2
8	**1a** (Cy)	SIMesCuCl (2)	0	Solvent: THF	19 (22)	n.d	87:13
9	**1a** (Cy)	SIMesCuCl (2)	0	Solvent: MeCN	13 (16)	n.d	84:16
10	**1a** (Cy)	SIMesCuCl (2)	0	Solvent: DMA	48 (56)	>95:5	85:15
11	**1a** (Cy)	SIMesCuCl (2)	0	Solvent: toluene	<1 (<1)	n.d	n.d
12	**1a** (Cy)	SIMesCuCl (2)	0	Solvent: *n*-hexane	5 (8)	n.d	60:40
13	**1a** (Cy)	SIMesCuCl (2)	0	Base: KOMe	5 (6)	n.d	79:21
14	**1a** (Cy)	SIMesCuCl (2)	−5	None	77 (83)	>95:5	93:7
15	**1a** (Cy)	SIMesCuCl (2)	−10	None	51 (54)	>95:5	94:6
16	**1a** (Cy)	SIMesCuCl (2)	0	Conc.: 0.25 M	68 (72)	>95:5	95:5
17	**1b** (*t*-Bu)	SIMesCuCl (2)	−5	None	58[d,e]	77:23 mixture of isomers[f]	
18	**1b** (*t*-Bu)	CuCl/Xantphos (5)	30	None	83[d]	>95:5	>99:1

[a] Standard conditions: Cu(I) catalyst (0.025 mmol), **1x** (0.5 mmol), **2a** (1.0 mmol), **3** (0.6 mmol), and K(O-*t*-Bu) (0.6 mmol) in DMF (1.0 mL). [b] Yield values and **4xa/5xa** selectivities were determined by GC analysis of the reaction mixture using an internal standard. [c] Determined by ^{13}C NMR analysis of the roughly purified material. [d] Isolated yield. [e] Containing small amounts of protoboration products of **1b** and a boryl substitution product of **2a**. [f] Determined by ^{1}H NMR analysis of the purified material. The structure of minor isomer could not be identified

Accordingly, I chose DMF as the appropriate solvent of this reaction. When potassium methoxide (KOMe) was tested as the base, the solution became jelly over the reaction to give a trace amount of the borylation product (entry 13: 5%). Then, the reaction temperature was screened. The yield and regioselectivity were improved when the reaction was performed at −5 °C (entry 14: 77%, *E*/*Z* = >95:5, **4**:**5** = 93:7). However, the yield was dropped at −10 °C, in spite that the regioselectivity was improved (entry 15: 51%, *E*/*Z* = >95:5, **4**:**5** = 94:6). At lower concentrations, the regioselectivity was improved, but the yield was decreased (entry 16: 68%, *E*/*Z* = >95:5, **4**:**5** = 95:5). Therefore, the conditions adopted in entry 14 were chosen as the optimal reaction conditions (condition A) for the following substrate scope investigation. Then, the reaction was applied to the bulkier model substrate **1b**, which has a *tert*-butyl group as R^2. However, Conditions **A** did not show sufficient reactivity and selectivities (entry 17: 58%, 77:23 mixture of isomers). In contrast, Xantphos gave a high yield of the desired product with perfect *E*/*Z* and regioselectivities (entry 18: 83%, *E*/*Z* = >95:5, **4**:**5** = >99:1). These high selectivities of the borylcupration agree with the mechanistic investigations reported by the Ma group [59]. Therefore, I selected the conditions applied in entry 18 as the optimized reaction conditions for bulkier substrates (Conditions **B**).

With the optimized reaction conditions in hand, the scope of allene and alkyl halide substrates was investigated (Table 4.2). For the characterization of the reaction products, some of the borylation products examined below were isolated after the boryl group was oxidized with reagents because the desired borylation products were sometimes inseparable from the byproducts, for example, the protoboration product of the allene and the borylative substitution product of the alkyl halide. The alkylboration products of allene **1a** were obtained in good yield with perfect *E*/*Z* and regioselectivity [(*E*)-**4aa**, (*E*)-**4ab**, (*E*)-**4ac**, (*E*)-**4ad**: 49–66%, *E*/*Z* = >95:5, **4**:**5** = >95:5]. As a representative borylation product produced under Conditions **A**, the stereochemistry of (*E*)-**4aa** was confirmed using a NOESY experiment (for details, see the Supporting Information). In order to investigate the effect of bulkiness of a *sec*-alkyl group R^2 in the allene, an acyclic *sec*-alkyl group was examined. Under Conditions **A**, the reaction of allene substrate **1c** resulted in complex mixture, while Conditions **B** afforded the corresponding alkylboration product in high yield with high selectivities, which were not affected by the length of the alkyl group in the halide substrate {(*E*)-**4cb**, (*E*)-**4cc**: 88–89%, *E*/*Z* = 88:12, **4**:**5** = >95:5}. Then, the substituent R^1 was investigated with *n*-propyl, *n*-octyl, and *iso*-butyl groups as the *prim*-alkyl group. Although the yields were decreased slightly in the order of *n*-propyl > *n*-octyl > *iso*-butyl group, the regio- and *E*/*Z*-selectivities were high [(*E*)-**4db**, (*E*)-**4dc**, (*E*)-**4eb**, (*E*)-**4ec**, (*E*)-**4fb**, (*E*)-**4fc**: 50–87%, *E*/*Z* = 87:13–91:9, **4**:**5** = >95:5]. The catalyst bearing SIMes ligand is able to recognize the bulkiness between α- and β-branched structure of alkyl groups R^1 and R^2, as found in (*E*)-**4fb** and (*E*)-**4fc**. Then, *prim*-alkyl *tert*-alkyl allenes were examined to afford the corresponding products with high regio- and *E*/*Z*-selectivities [(*E*)-**4ga**, (*E*)-**4ba**, (*E*)-**4ha**, (*E*)-**4hc**: 64–92%, *E*/*Z* = >95:5, **4**:**5** = >95:5]. As a representative borylation product produced under Conditions **B**, the stereochemistry of (*E*)-**4ga** was confirmed by an X-ray analysis with the single-crystal. Moreover, even more challenging methyl *prim*-alkyl

Table 4.2 Scope of allene substrate[a, b, c]

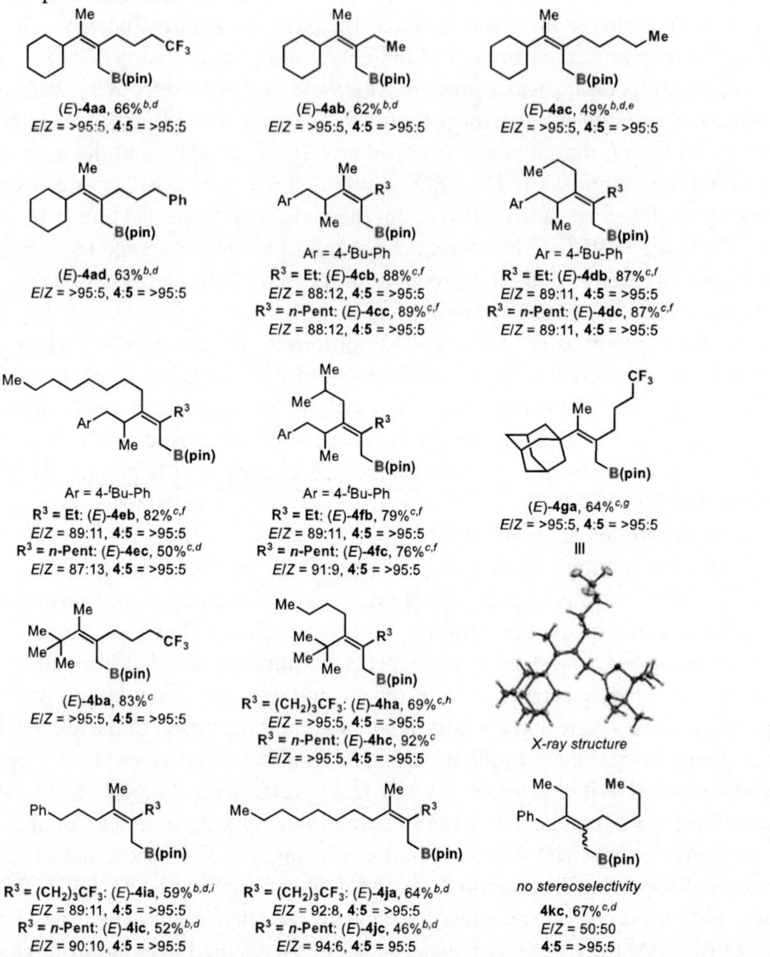

[a] Isolated yield of the borylation product unless otherwise noted. The *E*/*Z* values of **4** and the **4**/**5** selectivity were determined by ^{1}H NMR analysis after column chromatography unless otherwise noted. [b] Conditions A: SIMesCuCl (0.01 mmol), **1x** (0.5 mmol), **2y** (1.0 mmol), **3** (0.6 mmol), and K(O-*t*-Bu) (0.6 mmol) in DMF (1.0 mL) at −5 °C for 24 h. [c] Conditions B: CuCl (0.025 mmol), Xantphos (0.025 mmol), **1x** (0.5 mmol), **2y** (1.0 mmol), **3** (0.6 mmol), and K(O-*t*-Bu) (0.6 mmol) in DMF (1.0 mL) at 30 °C for 24 h. [d] Isolated yield after oxidation of the boryl group. [e] The reaction was performed at 30 °C. [f] The regio- and stereoselectivity was determined by GC analysis after silica gel column chromatography. [g] A small amount of byproduct, possibly the borylation product of the alkyl iodide, was removed by selective oxidation. [h] The reaction was performed for 48 h. [i] The reaction was performed at 0 °C

allene substrates were found to be applicable to this reaction. The corresponding borylation products were obtained in moderate to good yield with high *E*/*Z*- and regioselectivities [(*E*)-**4ia**, (*E*)-**4ic**, (*E*)-**4ja**, (*E*)-**4jc**: 46–64%, *E*/*Z* = 89:11–94:6, **4**:**5** = ≥95:5]. However, the catalysts could not distinguish two *prim*-alkyl groups, i.e., an ethyl group versus a benzyl group. The borylation product was obtained under Conditions **B** in good yield but with no *E*/*Z*-selectivity [(*E*)-**4kc**: 67%, *E*/*Z* = 50:50, **4**:**5** = >95:5], while Conditions **A** resulted in a mixture of isomers with low conversion of the substrate. It should be noted that the optimized conditions for the *gem*-dialkylallenes (Conditions **A** and **B**) are not applicable to aryl-substituted allenes (for details, see the Supporting Information).

Subsequently, I investigated the substrate scope of alkyl halides **2** in this reaction (Table 4.3). As well as ethyl and *n*-pentyl groups which has already been examined above, the *n*-decyl group as a longer alkyl chain was applicable as the R^3 substituent [(*E*)-**4bb**, (*E*)-**4bc**, (*E*)-**4be**: 75–88%, *E*/*Z* = >95:5, **4**:**5** = >95:5]. A β-branched alkyl iodide, *iso*-butyl iodide, also showed high reactivity and afford the borylation product in high yield with marginally dropped *E*/*Z*-selectivity [(*E*)-**4bf**: 81%, *E*/*Z* = 91:9, **4**:**5** = >95:5]. For a benzyl group, benzyl bromide was chosen for the carbon electrophile instead of unstable benzyl iodide. The corresponding product was obtained in good yield with high selectivities [(*E*)-**4bg**: 55%, *E*/*Z* = >95:5, **4**:**5** = >95:5]. Then, some functionalized alkyl halides were employed. Although a methoxy group could be used in this reaction [(*E*)-**4bh**: 74%, *E*/*Z* = >95:5, **4**:**5** = >95:5], the reaction of alkyl iodide containing a silyl ether group or a chloride group gave a mixture of the desired product and unidentified byproducts [(*E*)-**4bi**, (*E*)-**4bj**: 44–51%, *E*/*Z* = >95:5, **4**:**5** = >95:5]. Furthermore, iodomethane **2k** decomposed under the reaction conditions, probably via etherification of the alkoxide base or direct borylative substitution of the alkyl iodide. Also, phenethyl iodide **2l** decomposed in the reaction, probably via β-elimination because styrene was detected in the crude reaction mixture by GC–MS analysis. Besides, *sec*- and *tert*-alkyl halide substrate **2m** and **2n** were not suitable for this reaction because of the lack of reactivity to alkenyl copper(I) intermediate in the S_N2-type oxidative addition step.

To demonstrate the synthetic utility of the allylic boronates prepared via the above method, derivatization reactions of the boryl group were carried out (Fig. 4.5). First, a large-scale synthesis of (*E*)-**4hc** was conducted. Similar to the sub-mmol scale reaction, the desired product was obtained in high yield with perfect selectivities [(*E*)-**4hc**: 88% (1.86 g), *E*/*Z* = >95:5, **4**:**5** = >95:5; Fig. 4.6A]. The oxidation of the boryl group afforded the corresponding allylic alcohol in high yield [(Z)-**6hc** 89%, *E*/*Z* = >95:5; Fig. 4.6B]. Also, the amination developed by Morken group furnished the allylic amine in good yield [(Z)-**7hc**: 46%, *E*/*Z* = >95:5; Fig. 4.6B]. The Matteson homologation with CH_2ClBr as a mono-carbon source also successfully afforded the corresponding homoallylic boronate in high yield with perfect stereospecificity [(*E*)-**8hc**: 73%, *E*/*Z* = >95:5; Fig. 4.6B].

I then performed allylboration reactions constructing sterically congested quaternary carbon (Fig. 4.7 and Table 4.4). The allylboration reaction of formaldehyde with allylic boronate (*E*)-**4hc** under heating afforded the homoallylic alcohol product in high yield (**9a**: 90%; Fig. 4.7) [65]. This product has vicinal quaternary carbons,

Table 4.3 Scope of alkyl halide[a, b, c]

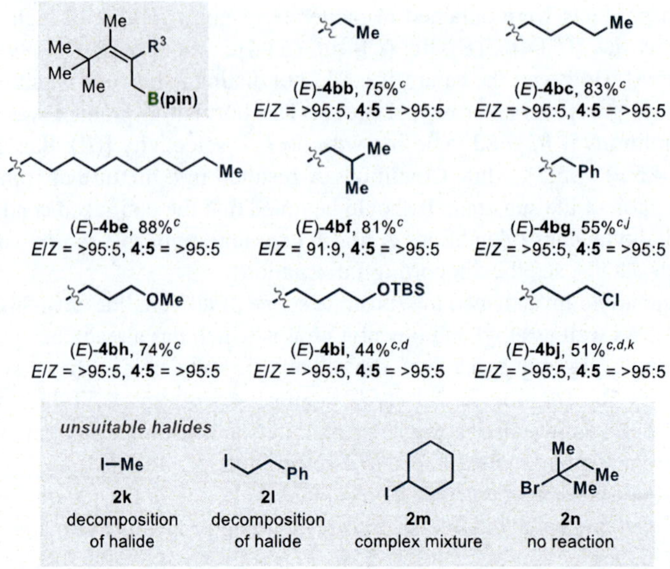

[a] Isolated yield of the borylation product unless otherwise noted. The *E*/*Z* values of **4** and the **4**/**5** selectivity were determined by ^{1}H NMR analysis after column chromatography unless otherwise noted. [b] Conditions A: SIMesCuCl (0.01 mmol), **1x** (0.5 mmol), **2y** (1.0 mmol), **3** (0.6 mmol), and K(O-*t*-Bu) (0.6 mmol) in DMF (1.0 mL) at −5 °C for 24 h. [c] Conditions B: CuCl (0.025 mmol), Xantphos (0.025 mmol), **1x** (0.5 mmol), **2y** (1.0 mmol), **3** (0.6 mmol), and K(O-*t*-Bu) (0.6 mmol) in DMF (1.0 mL) at 30 °C for 24 h. [d] Benzylbromide was used instead of benzyliodide. [e] Isolated yield after oxidation of the boryl group. [f] The selectivity of C–C bond formation at iodide and chloride was 86:14

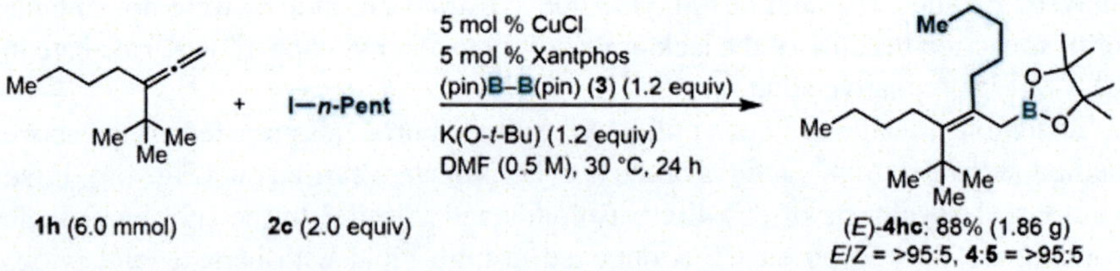

Fig. 4.5 Gram-scale synthesis of 2,3,3-trialkylsubstituted allylic boronates

which are difficult to construct in general through C–C bond formation reactions such as intermolecular aldol or Grignard reactions, owing to steric congestion [66, 67]. Then, Lewis-acid-mediated conditions for allylboration reaction allowed the reaction with aryl aldehydes (Table 4.4) [25, 68, 69]. The allylboration with benzaldehyde and other aryl aldehydes afforded the product in high yield with perfect diastereoselectivity (**9b**, **9c**, **9d**: 52–75%, dr = >95:5). I confirmed the stereochemistry of allylboration product **9d** using an X-ray analysis using the single crystal (Fig. 4.8).

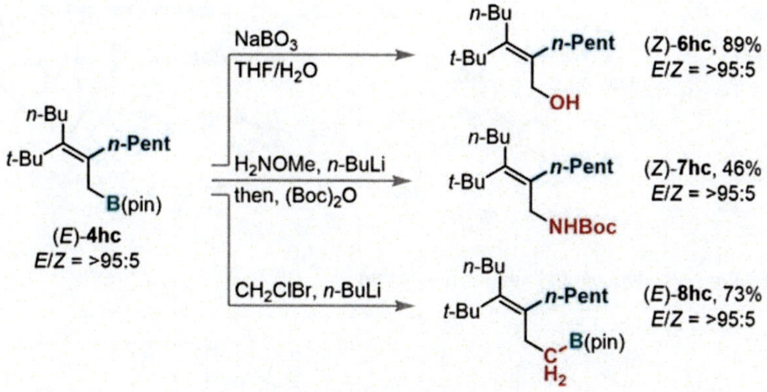

Fig. 4.6 Derivatization reactions of 2,3,3-trialkylsubstituted allylic boronates

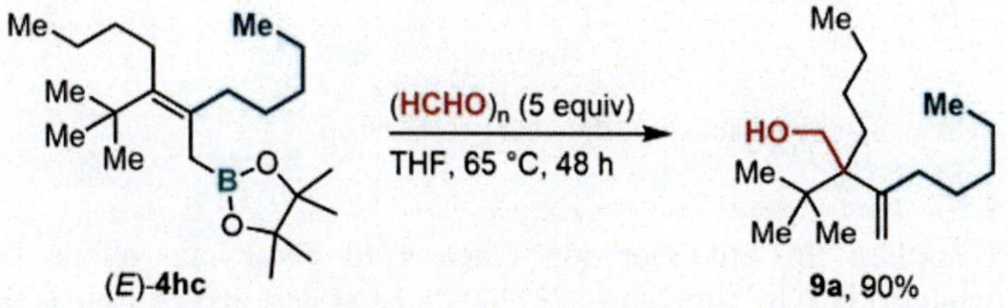

Fig. 4.7 Allylboration of formaldehyde

Table 4.4 Allylboration of aryl aldehydes

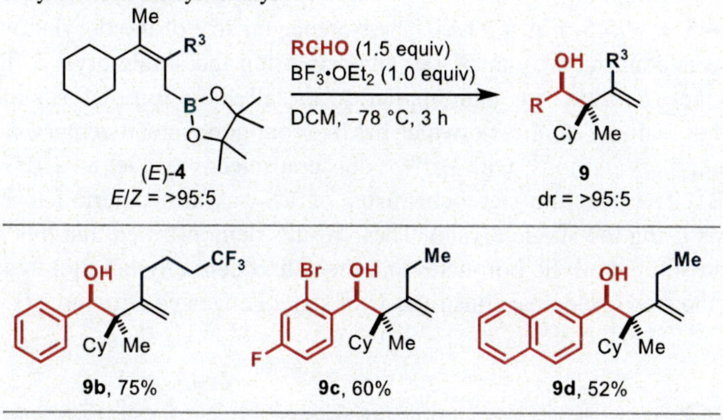

Considering the relative configuration of the vicinal stereodefined sp^3-carbons in the product, the transition state (TS) structure was estimated as shown in Fig. 4.9. Substituent R^4 in the aldehyde should be at the equatorial position in the TS for avoiding steric repulsion against R^1 and R^3 in the allylic boronate.

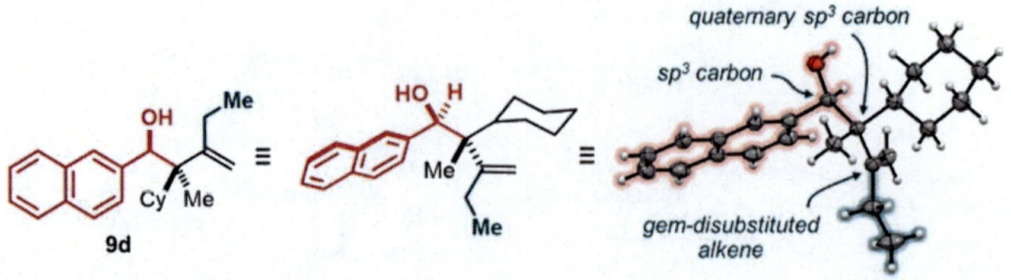

Fig. 4.8 X-ray structure of allylboration product **9d**

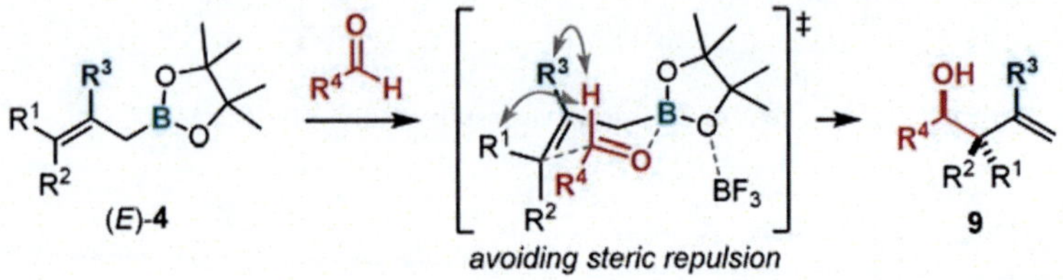

Fig. 4.9 Proposed diastereoselectivity-determining mechanism

Also, I applied this alkylboration reaction to exo-cyclic allene substrates (Figs. 4.10 and 4.11). The simple exo-cyclic allene **1l** was first chosen as the model substrate. The corresponding alkylboration product was furnished in good yield with perfect regioselectivity under Conditions **B** (**4la**: 69%, **4**:**5** = >95:5; Fig. 4.10). Thus, the reaction was employed with a steroid-type substrate to furnish the products in high yield with high regio- and stereoselectivity [(*E*)-**4ma**, (*E*)-**4mb**: 78–90%, *E*/*Z* = >95:5, **4**:**5** = >95:5; Fig. 4.11A]. The stereochemistry of the borylation product (*E*)-**4mb** was confirmed by an X-ray analysis using the single crystal. The borylation product (*E*)-**4ma** was then applied to the allylboration of formaldehyde to afford the homoallylic alcohol **9e**, which has five contiguous stereocenters, containing vicinal quaternary carbons, with perfect diastereoselectivity (**9e**: 84% dr = >95:5; Fig. 4.11B). The absolute stereochemistry of **9e** was also determined by an X-ray analysis using the single crystal. These results demonstrated that this synthesis of multi-substituted allylic boronates and the subsequent allylboration reaction are useful for the stereoselective construction of stereocongested structures (Fig. 4.11).

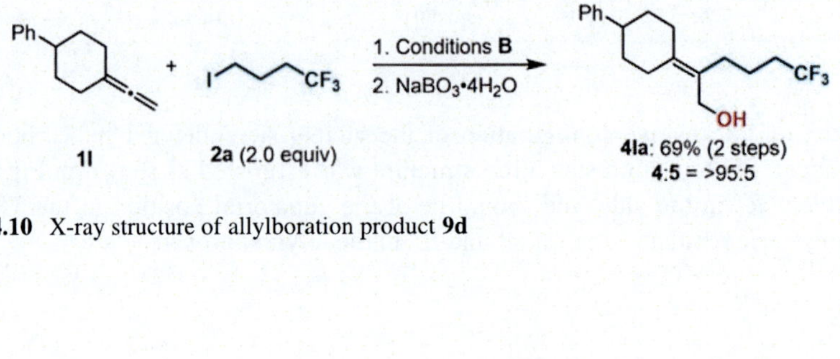

Fig. 4.10 X-ray structure of allylboration product **9d**

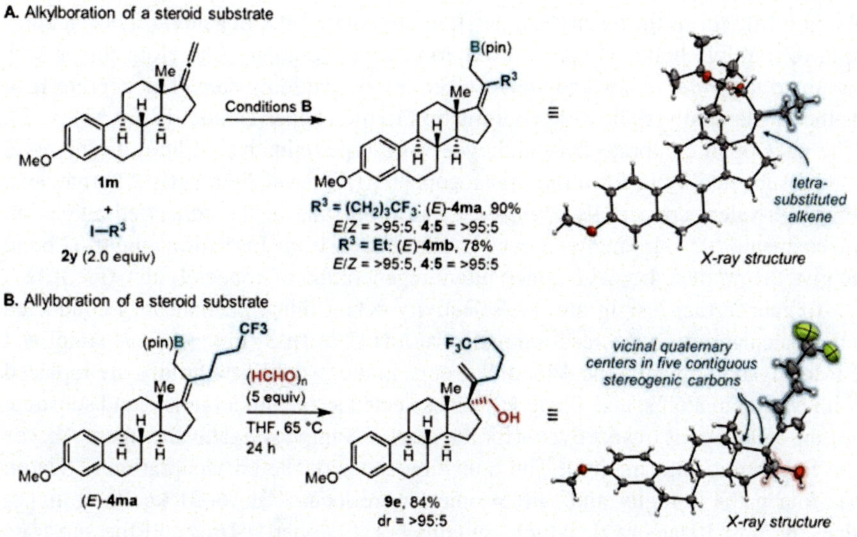

Fig. 4.11 X-ray structure of allylboration product **9d**

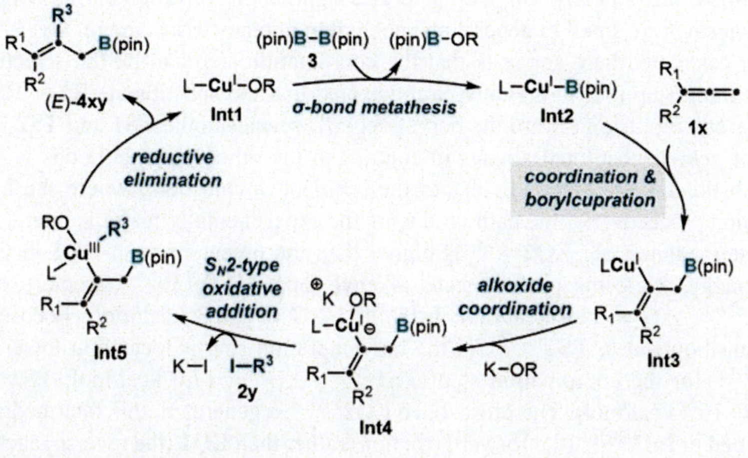

Fig. 4.12 Proposed reaction mechanism

The proposed reaction mechanism based on the past studies on copper(I)-catalyzed carboboration reactions using alkyl electrophiles is shown in Fig. 4.12 [34, 49, 70, 71]. The on-cycle species, copper(I) alkoxide **Int1**, is formed in situ through the reaction between copper(I) salt and K(O-*t*-Bu). The boryl copper(I) species **Int2**, which is the active species in copper(I) catalyst/diboron system for borylation reactions, is generated through a σ-bond metathesis between **Int1** and the diboron reagent **3**. Then, the coordination and subsequent borylcupration of allene

1x with **Int2** forms the alkenyl copper(I) intermediate **Int3**. In general, a borylcupration of multiple bonds is highly exothermic and exergonic. Therefore this step is assumed to be the regio- and stereoselectivity-determining step. The cuprate **Int4** is then generated via alkoxide coordination to the copper(I) atom in **Int3** [72–83]. The nucleophilic cuprate **Int4** undergoes S_N2-type oxidative addition to the alkyl halide substrate **2y** to form the organocopper(III) intermediate **Int5**. Alternatively, this high-valent copper(III) could be a transition state of the concerted alkylation mechanism [49, 71]. Finally, reductive elimination from **Int5** forms the C–C bond to give the product **4xy** with concomitant regeneration of copper(I) alkoxide **Int1**.

To gain further insight into the selectivity-determining mechanism, I conducted DFT calculations on the reaction paths from **Int2** to **Int3** (Fig. 4.13). A simplified SIMes ligand, in which the 4-methyl groups in the two mesityl groups are replaced with hydrogen atoms, and allene **1a** were selected as the model ligand and substrate for the calculations, respectively (for the case of Xantphos ligand and allene **1b**, see the Supporting Information). The transition states of the borylcupration (**TS2**) on the four paths to regio- and stereoisomers were located at 16–21 kcal/mol higher than the initial state **EQ1**, which contains boryl copper(I) **Int2** and the substrate **1a** (**EQ1**: $\Delta G = 0.0$ kcal/mol). During the calculation of **TS2**, the transition states for the coordination of the terminal double bond of allene **1a** to **Int2** (**TS1**s) were found to be energetically comparable to **TS2**s. Although activation barriers of coordination steps are small in general and negligible in many cases, the relative energy profile calculated here suggests that the coordination step can be the selectivity-determining step [84, 85]. Hence, I anticipated that the substituents R^1 or R^2 can cause steric repulsion toward the boryl copper(I) species in the **TS1** and **TS2** in the case of *gem*-disubstituted allenes in contrast to the other multiple bonds because the substituents are perpendicular to the terminal double bond where the borylcupration proceeds. For the path producing the experimentally major isomer, (*E*)-**4**, the intermediate state **EQ2**$^{\text{allyl-}E}$ is higher than the precursor state **EQ1** in Gibbs free energy, while the product state, alkenyl copper(I) **Int3**$^{\text{allyl-}E}$, is more stable (**EQ2**$^{\text{allyl-}E}$: $\Delta G = 12.4$ kcal/mol; **Int3**$^{\text{allyl-}E}$: $\Delta G = -25.3$ kcal/mol). Focused on the transition states, **TS2**$^{\text{allyl-}E}$ for the borylcupration step is located at lower than **TS1**$^{\text{allyl-}E}$ for the coordination step (**TS1**$^{\text{allyl-}E}$: $\Delta G^‡ = 17.0$ kcal/mol; **TS2**$^{\text{allyl-}E}$: $\Delta G^‡ = 16.4$ kcal/mol). Therefore, once **EQ2**$^{\text{allyl-}E}$ is generated, this intermediate is converted to **Int3**$^{\text{allyl-}E}$ (the forward reaction) rather than **EQ1** (the reverse reaction). Therefore, the coordination step is presumed to be irreversible. However, the difference in the energies between those two TSs (**TS1**$^{\text{allyl-}E}$ and **TS2**$^{\text{allyl-}E}$) is not large; hence, the reverse reaction from **EQ2**$^{\text{allyl-}E}$ to **EQ1** could potentially proceed. Likewise, the energy profile suggested that the reaction path for (*Z*)-**4** is also irreversible, and thus, **TS1**$^{\text{allyl-}Z}$ is the selectivity-determining step in this path (**TS1**$^{\text{allyl-}Z}$: $\Delta G^‡ = 18.9$ kcal/mol; **EQ2**$^{\text{allyl-}Z}$: $\Delta G = 13.0$ kcal/mol; **TS2**$^{\text{allyl-}Z}$: $\Delta G^‡ = 16.5$ kcal/mol; **Int3**$^{\text{allyl-}Z}$: $\Delta G = -28.0$ kcal/mol). By contrast, the reaction paths for the regioisomers (*Z*)- and (*E*)-**5** have more unstable transition state of borylcupration than those for allylic boronates (*E*)- and (*Z*)-**4**. In these paths, **TS1** is located at lower than **TS2** [for (*Z*)-**5**; **TS1**$^{\text{alkenyl-}Z}$: $\Delta G^‡ = 16.0$ kcal/mol, **EQ2**$^{\text{alkenyl-}Z}$: $\Delta G = 9.4$ kcal/mol, **TS2**$^{\text{alkenyl-}Z}$: $\Delta G^‡ = 20.1$ kcal/mol, **Int3**$^{\text{alkenyl-}Z}$: $\Delta G = -26.7$ kcal/mol; for (*E*)-**5**;

Z

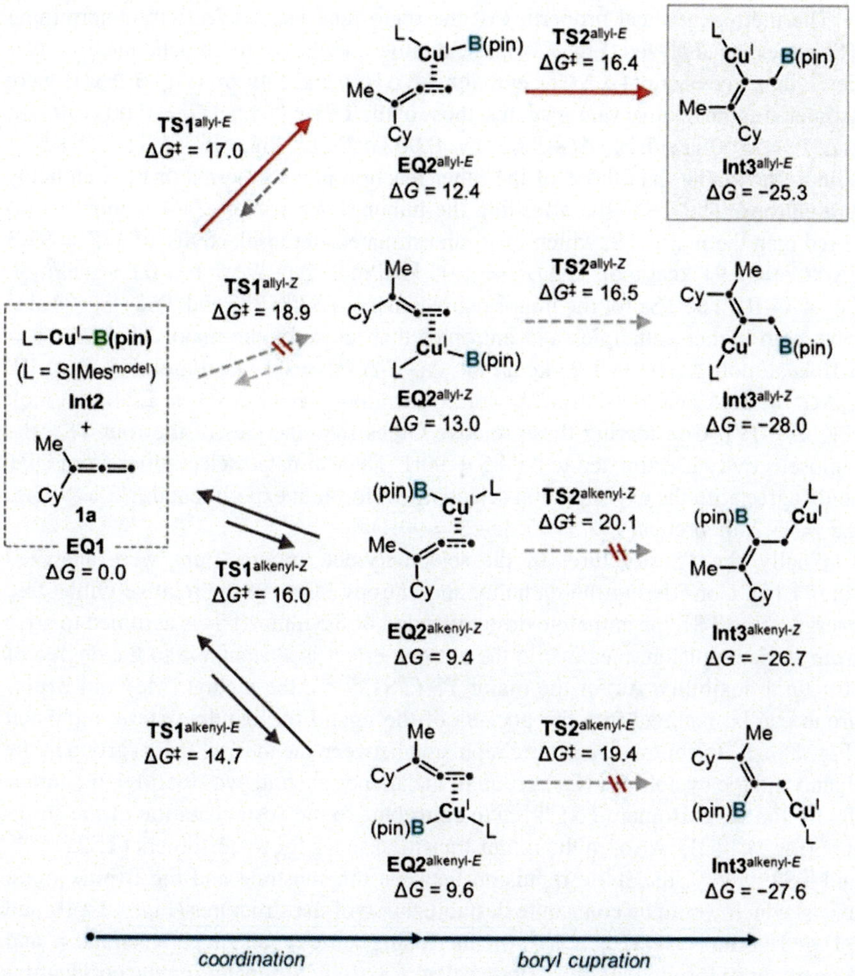

Fig. 4.13 Gibbs free energy profiles of coordination (**TS1**) and borylcupration (**TS2**) of boryl copper(I) **Int2** to allene **1a**

TS1$^{\text{alkenyl-}E}$: $\Delta G^{\ddagger} = 14.7$ kcal/mol, **EQ2**$^{\text{alkenyl-}E}$: $\Delta G = 9.6$ kcal/mol, **TS2**$^{\text{alkenyl-}E}$: $\Delta G^{\ddagger} = 19.4$ kcal/mol, **Int3**$^{\text{alkenyl-}E}$: $\Delta G = -27.6$ kcal/mol]. Because of the relatively low **TS1**^{alkenyl} and high **TS2**^{alkenyl}, the formation of intermediate **EQ2**$^{\text{alkenyl-}Z}$ and **EQ2**$^{\text{alkenyl-}E}$ is reversible, and thus, the transition states of borylcupration (**TS2**$^{\text{alkenyl-}Z}$ and **TS2**$^{\text{alkenyl-}E}$) are the selectivity-determining steps of these paths. In brief, the TSs of coordination (**TS1**$^{\text{allyl-}E}$ and **TS1**$^{\text{allyl-}Z}$) are the selectivity-determining steps for the formation of allylic boronates, whereas the TSs of borylcupration (**TS2**$^{\text{alkenyl-}Z}$ and **TS2**$^{\text{alkenyl-}E}$) are the selectivity-determining steps for the formation of alkenyl boronates. Therefore, **TS1**$^{\text{allyl-}E}$ is kinetically the most favorable for the formation of (E)-**4**, the major isomer of the product.

The thermochemical properties of the regio- and stereoselectivity-determining TSs were then analyzed (Fig. 4.14). The relative energies of thermochemical properties, Gibbs free energy ($\Delta\Delta G^{\ddagger}$), enthalpy ($\Delta\Delta H^{\ddagger}$), and entropy $\{\Delta(-T\Delta S^{\ddagger})\}$ were estimated as the relative values against those of the **TS1**$^{\text{allyl-}E}$ [$\Delta\Delta G^{\ddagger} = 0.00$ kcal/mol, $\Delta\Delta H^{\ddagger} = 0.00$ kcal/mol, $\Delta(-T\Delta S^{\ddagger}) = 0.00$ kcal/mol; Fig. 4.14-I]. The **TS1**$^{\text{allyl-}E}$ is more favorable than those of the other reaction paths in terms of both enthalpy and entropy. The **TS1**$^{\text{allyl-}Z}$ affording the minor stereoisomer (*Z*)-**4** is more disfavored than the major TS, which gives an estimated stereoselectivity of *E*/*Z* = 97:3 [$\Delta\Delta G^{\ddagger} = 1.91$ kcal/mol, $\Delta\Delta H^{\ddagger} = 1.08$ kcal/mol, $\Delta(-T\Delta S^{\ddagger}) = 0.83$ kcal/mol; Fig. 4.14-II]. The TSs for the minor regioisomers (**TS2**$^{\text{alkenyl-}Z}$ and **TS2**$^{\text{alkenyl-}E}$) also have both higher enthalpies and entropies than those of the major TS [$\Delta\Delta G^{\ddagger} = 3.10$ kcal/mol, $\Delta\Delta H^{\ddagger} = 1.93$ kcal/mol, $\Delta(-T\Delta S^{\ddagger}) = 1.17$ kcal/mol; Fig. 4.14-III, $\Delta\Delta G^{\ddagger} = 2.44$ kcal/mol, $\Delta\Delta H^{\ddagger} = 1.20$ kcal/mol, $\Delta(-T\Delta S^{\ddagger}) = 1.24$ kcal/mol; Fig. 4.14-IV]. Considering these relative Gibbs free energies of the four TSs, the regioselectivity is estimated to be **4**:**5** = 99:1. The estimated selectivities from DFT study agree with the experimental selectivities (for **4aa**: experimental *E*/*Z* = >95:5, **4**:**5** = 93:7; theoretical *E*/*Z* = 97:3, **4**:**5** = 99:1).

Finally, the TS structures of the selectivity-determining steps were analyzed (Fig. 4.14). Considering the enthalpy and entropy terms in the relative Gibbs free energy of the TSs, the enthalpic destabilization of the minor TSs is assumed to arise from steric repulsion, meanwhile the entropy effect is attributable to the degree of structural flexibility. As for the major TS (**TS1**$^{\text{allyl-}E}$), the methyl (Me) and B(pin) group can be inclined into the pockets of the ligand for avoiding steric repulsion (Fig. 4.14-I). In contrast, the steric repulsion between the mesityl (Mes) group in the ligand and the cyclohexyl (Cy) group in the allene substrate was found in the minor TS for the stereoisomer (**TS1**$^{\text{allyl-}Z}$) to contribute to the destabilization of the structure (Fig. 4.14-II). Also, in the minor transition states for regioisomers (**TS2**$^{\text{alkenyl-}Z}$ and **TS2**$^{\text{alkenyl-}E}$), the steric repulsion between the substrate and the B(pin) group exists, which would be contribute destabilization of the structures (Fig. 4.14-III and VI). Furthermore, in **TS2**$^{\text{alkenyl-}Z}$, the increasing steric repulsion between the 1- and 3-positions of forming allylic system, called 1,3-allylic strain, during the borylcupration causes additional destabilization of the structure. On the other hand, I speculated that the entropic effect would arise from the flexibility around the Cy group in the substrate. The rotation of the Cy group would not be interfered in **TS1**$^{\text{allyl-}E}$ because the Cy group is directed opposite to the catalyst (Fig. 4.14-I). In contrast, the rotation mode of the Cy group was locked in **TS1**$^{\text{allyl-}Z}$ and **TS2**$^{\text{alkenyl-}E}$ by the interactions between the Cy group and the catalyst or B(pin) group, which cause entropic destabilization of those structures (Fig. 4.14-II and IV). Also, the similar locking effect would be caused by the 1,3-allylic interaction in the minor **TS2**$^{\text{alkenyl-}Z}$, although the Cy group is positioned far from the catalyst (Fig. 4.14-III).

To understand the ligand-controlled regioselectivity, the structure of steric-hindrance environment formed by the ligands were visualized using a steric map and a newly defined "octant model" (Fig. 4.15) [86]. The "octant model" was defined as follows: All directions, as the copper(I) atom is the center, were cut into eight regions, from NNE to NNW clockwisely. Each region is colored using a five-step intensity

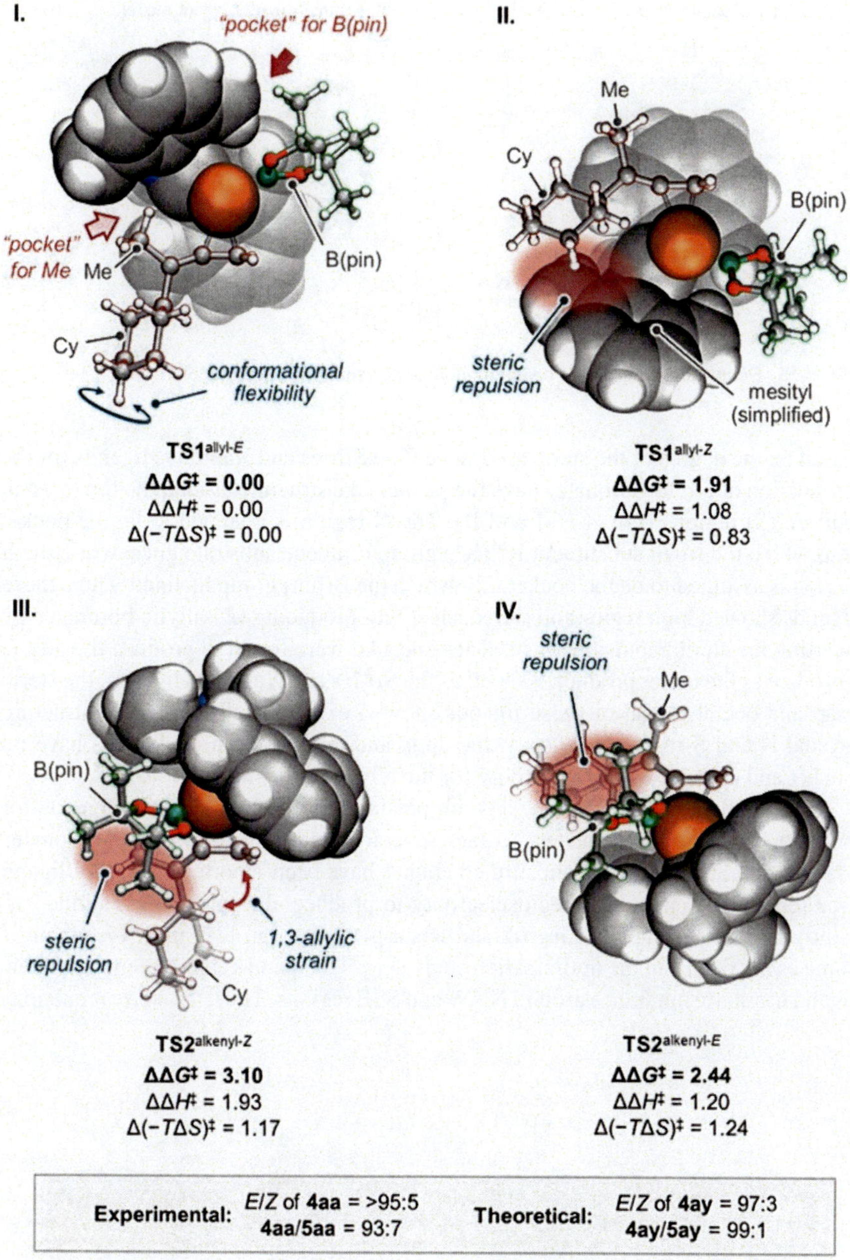

Fig. 4.14 Transition-state structures of regio- and stereodetermining steps

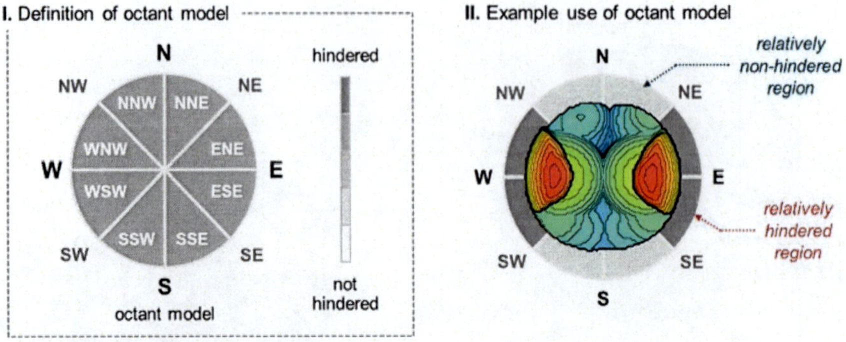

Fig. 4.15 Definition and usage of octant model for describing ligand structure

based on the degree of the steric hindrance. The SIMes and Xantphos ligands for the production of allylic boronates have the pocket-like structure around the top (N) and bottom (S) regions (Fig. 4.16-I and II). The N region is assumed to be the pocket into which the small substituent R^S (R^1) group of allene substrate goes, while the S region is assumed to be the pocket into which the B(pin) group inclines. Thus, those ligands showed high regio- and stereoselectivity producing (*E*)-allylic boronates by averting the steric repulsion. Also, IMes and IAd were found to produce the allylic boronate as the major product. As well as the SIMes and Xantphos ligands, the steric map and octant model of those ligands showed existence of pocket-like structure around N and S regions. Contrary, the dppp and dppf (vide infra) ligands have no pocket and show low regioselectivity for the allylic boronate.

On the other hand, those pockets are presumed to be lacked in the ligands for production of alkenyl boronates. To date, several ligands affording alkenyl boronates via carboboration of *gem*-disubstituted allenes have been reported. The dppf ligand was reported to show high regioselectivity to produce alkenyl boronates independently from the groups of Gagosz and Riant [47], and Liu [39] in 2017, which is same as my results in the optimization study (Fig. 4.17-I and Table 4.1, entry 3). This ligand has steric hindrance around NNW and SSE regions. Thus, the steric repulsions

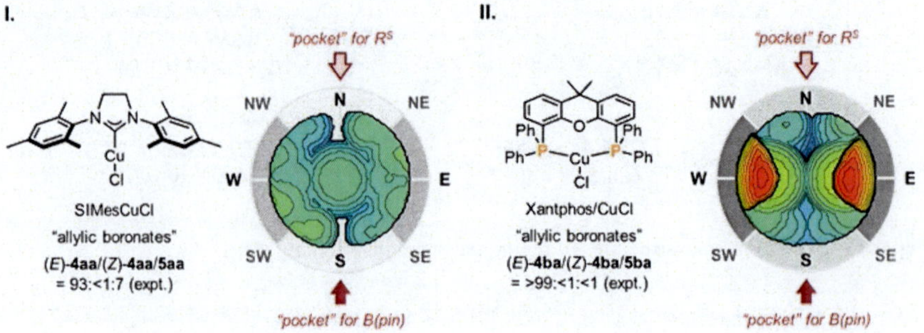

Fig. 4.16 Octant models and steric maps for SIMes and Xantphos ligand

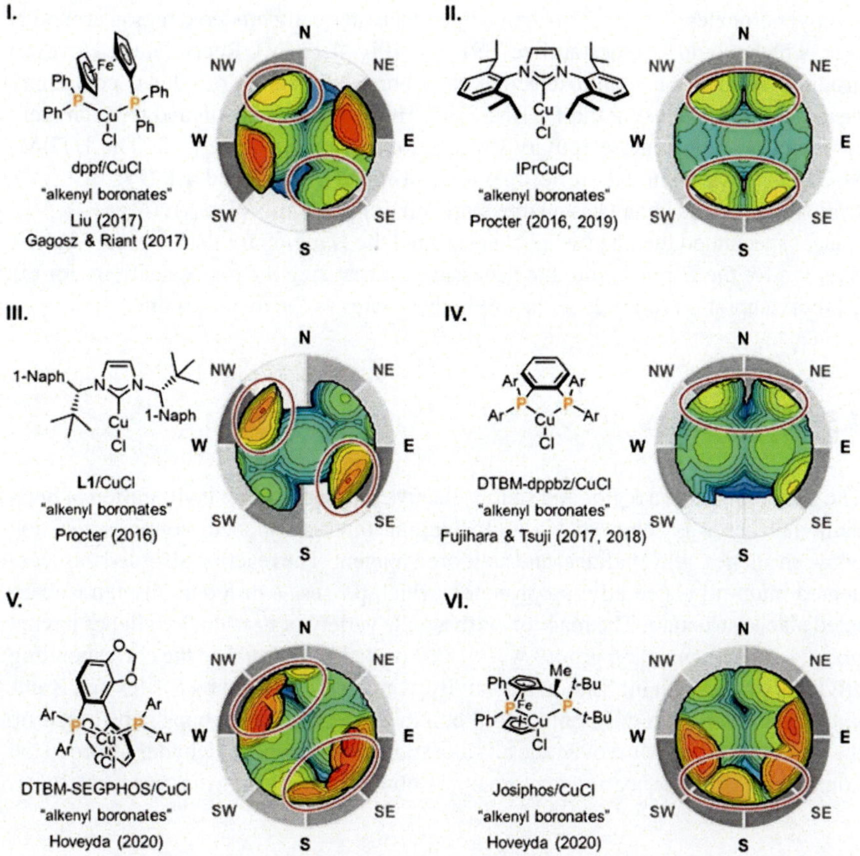

Fig. 4.17 Octant models and steric maps for ligands producing alkenyl boronate from *gem*-substituted allenes via copper(I)-catalyzed carboboration reaction

against the R^S and B(pin) group would destabilize $\mathbf{TS1}^{\text{allyl-}E}$. In 2016, IPr and **L1** were used in the racemic and enantioselective carboboration of *gem*-dialkylallenes with aldimines as the electrophile, which were reported by the Procter group [41, 42]. The IPr ligand has large sterically hindered region around N and S directions (Fig. 4.17-II); thus, the $\mathbf{TS1}^{\text{allyl-}E}$ should be destabilized. Contrary, the chiral NHC ligand (**L1**) has pockets around N and S regions. However, two protruded hindrances around NW and SE, which are the 1-naphthyl groups, might cause steric repulsion against the R^S and B(pin) groups in $\mathbf{TS1}^{\text{allyl-}E}$ (Fig. 4.17-III). The alternative assumption is that the C_2-symmetric steric environment, which is also found in the dppf ligand, stabilizes $\mathbf{TS2}^{\text{alkenyl-}Z}$ and $\mathbf{TS2}^{\text{alkenyl-}E}$. As the result, the corresponding alkenyl boronates were obtained instead of allylic boronates. The group of Fujihara and Tsuji also reported carboboration reactions of *gem*-dialkylallenes using acyl electrophiles in 2017 and 2018 [46, 50]. In their reactions, bulky DTBM-dppbz ligand was developed and found to realize high regioselectivity for the production of the corresponding alkenyl boronates. The DTBM-dppbz ligand has sterically hindered region around N side, which would also destabilize **TS1**$^{allyl\text{-}E}$ (Fig. 4.17-IV). Recently, the Hoveyda group developed the enantioselective carboboration reaction of allenes containing *gem*-dialkylallenes using the DTBM-SEGPHOS and Josiphos ligands, which were also showed high regioselectivity toward the alkenyl boronate [57]. The DTBM-SEGPHOS has a similar steric map against those of dppf and **L1** (Fig. 4.17-V), while the Josiphos ligand has a large steric hindrance from SW to NE (Fig. 4.17-VI). Thus, I speculated that the two pockets around the N region for the R^S group and the S region for the B(pin) group are necessary for realizing the regioselectivity for the 1,2-borylcupration to produce the allylic boronates as the major product.

4.3 Conclusions

The development of a regio- and stereoselective intermolecular alkylboration of *gem*-dialkylallenes was achieved via a challenging three-component coupling reaction between allenes, alkyl halides, and a diboron reagent. The reaction afforded unprecedented multi-alkylated allylic boronates, which possess a differentially tetrasubstituted alkene structure. The reaction with a wide variety of *gem*-dialkylallenes having *prim*-, *sec*-, and *tert*-alkyl groups was demonstrated and furnished the corresponding allylic boronate with high stereoselectivity, which was realized by SIMes and Xantphos ligand with can differentiate the bulkiness of the alkyl groups substituted on the allene substrate. Moreover, the allylboration reaction with aldehydes afforded the homoallylic alcohols bearing quaternary carbon atoms with high diastereoselectivity.

4.4 Experimental Details

4.4.1 Instrumentation and Chemicals

Materials were obtained from commercial suppliers and purified by standard procedures unless otherwise noted. Solvents were also purchased from commercial suppliers, degassed via three freeze–pump–thaw cycles, and further dried over molecular sieves (MS 4A). NMR spectra were recorded on JEOL JNM-ECX400P, JNM-ECS400, and JNM-ECA600 spectrometers (^{1}H: 400 or 600 MHz, ^{13}C: 100 MHz ^{11}B: 126 MHz). Tetramethylsilane (^{1}H, δ 0.00), $CDCl_3$ (^{13}C, δ 77.0) and $BF_3 \cdot OEt_2$ (^{11}B, δ 0.00) were employed as the external standards, respectively. CuCl (224332-25G, ≥99%) and K(O-*t*-Bu) (659878-5G, 99.99%) were purchased from Sigma-Aldrich Co. and used as received. GLC analyses were conducted with a Shimadzu GC-2014 or GC-2025 equipped with a ULBON HR-1 glass capillary column (Shinwa Chemical Industries) and an FID detector. Recycle preparative gel chromatography (GPC) was conducted with JAILC-9101 using $CHCl_3$ as an eluent. High-resolution mass spectra were recorded at the Global Facility Center, Hokkaido University and GC–MS & NMR Lab., Faculty of Agriculture, Hokkaido University. Single-crystal X-ray structural analysis was carried out on a Rigaku XtaLAB PRO MM007 diffractometer using graphite monochromated Mo-K_α and Cu-K_α radiation. The structure was solved by direct methods and expanded using Fourier techniques. Non-hydrogen atoms were refined anisotropically. Hydrogen atoms were refined using the riding model. All calculations were performed using the Olex2 crystallographic software package except for refinement, which was performed using SHELXL [87].

4.4.2 *Substrate Preparation Procedure*

All 1,1-disubstituted allenes (**1a–1m**) and an alkyl iodide **2h** were prepared according to the references and characterized as described below. Other alkyl iodides (**2a–2g** and **2i–2n**) were purchased from commercial suppliers (Tokyo Chemical Industry Co. and Sigma-Aldrich Co.). Spectroscopic data of already known compounds (**1a** [88], **1b** [89], **1g** [90], **1l** [91], **1m** [90], and **2h** [92]) were matched with those reported.

4.4.2.1 General Procedure of *gem*-Disubstituted Allenes Synthesis (GP1, Fig. 4.18) [93]

In a vacuum-dried, round-bottomed flask, CuBr (0.1 equiv) and the corresponding propargyl tosylate **S1** (1.0 equiv) were dissolved in dry THF (0.5 M). After the solution was cooled to −60 °C, the corresponding Grignard reagent **S2** (1.1 equiv) was added dropwise over 1 h under a nitrogen atmosphere. After stirred for 4 h at the same temperature, the reaction mixture was warmed up to 0 °C. Then, it was quenched by saturated NH_4Cl aqueous solution and extracted with Et_2O three times. The combined organic layer was then dried over $MgSO_4$. After filtration, the solvent was removed by evaporation. The crude mixture was purified by silica gel column chromatography (typically, Et_2O/hexane 0:100–5:95) to obtain the corresponding 1,1-disubstituted allene product **1**.

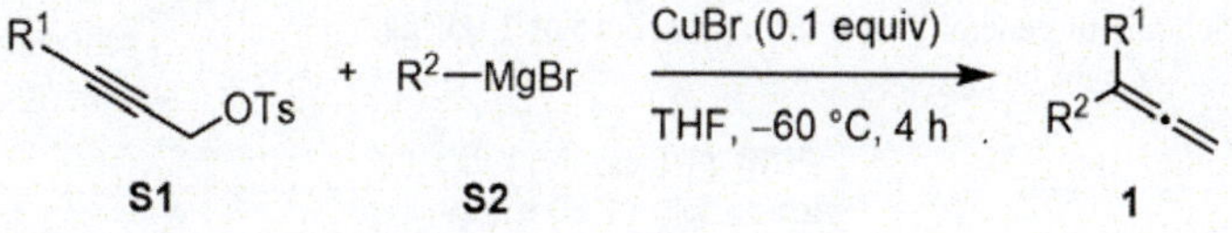

Fig. 4.18 Scheme of general preparation procedure of **1** (GP1)

4.4.2.2 General Procedure of *gem*-Disubstituted Allenes Synthesis (GP2, Fig. 4.19) [90]

In a vacuum-dried, round-bottomed flask equipped with a reflux condenser, $AlCl_3$ (0.75 equiv) was dissolved in dry THF (0.4 M against $AlCl_3$) under an argon atmosphere. After all, $AlCl_3$ was suspended, the mixture was cooled to 0 °C followed by a slow addition of a suspension of $LiAlH_4$ (2.25 equiv) in THF (1.8 M against $LiAlH_4$). After stirring for 15 min, the corresponding propargylic alcohol **S3** (1.0 equiv) was added as a solution in THF (1.6 M against propargylic alcohol). The reaction mixture was heated to 80 °C and stirred for 16 h. It was cooled to 0 °C and quenched by slow addition of H_2O (1 mL), 1M NaOH aqueous solution (2 mL), and additional H_2O (1 mL). The precipitate was filtered with Et_2O, and the crude material was purified by silica gel column chromatography (typically, Et_2O/hexane 0:100–5:95) to obtain the corresponding 1,1-disubstituted allene product **1**.

4.4.2.3 Preparation of 1-iodo-3-Methoxypropane (2h, Fig. 4.20) [92, 94]

In a vacuum-dried, round-bottomed flask, PPh_3 (3.93 g, 15.0 mmol, 1.5 equiv) and imidazole (1.27 g, 18.6 mmol, 1.86 equiv) were dissolved in dry DCM (100 mL, 0.1 M) under a nitrogen atmosphere at rt. I_2 (4.16 g, 16.4 mmol, 1.64 equiv) was then added to the reaction mixture, followed by addition of corresponding alcohol **S4** (901 mg, 10 mmol, 1.0 equiv). After stirred for 2 h at rt, the reaction mixture was quenched by saturated $Na_2S_2O_3$ aqueous solution and extracted with DCM three times. The combined organic layer was washed with saturated $NaHCO_3$ aqueous solution and brine, then dried over $MgSO_4$. After filtration, the solvent was removed by evaporation. The crude mixture was purified by silica gel column chromatography (EtOAc/hexane 0:100–5:95) to obtain the corresponding alkyl iodide **2h** in 45% yield (902.8 mg, 4.5 mmol, colorless oil).

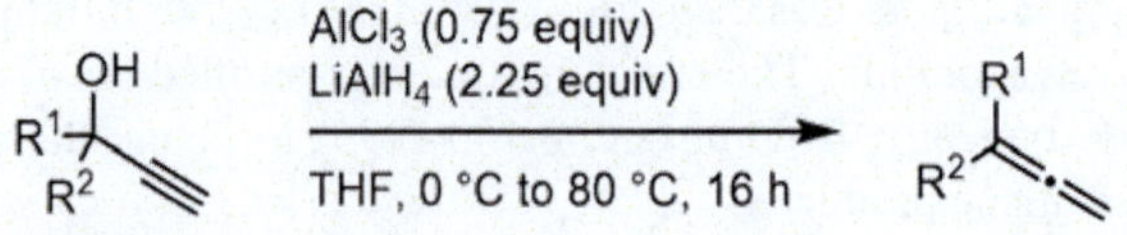

S3 **1**

Fig. 4.19 Scheme of general preparation procedure of **1** (GP2)

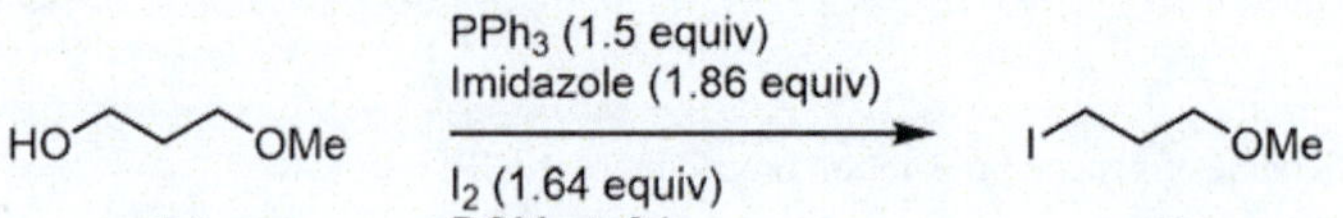

S4 **2h**
45% yield

Fig. 4.20 Scheme of preparation procedure of **2h**

^{1}H NMR (392 MHz, $CDCl_3$, δ): 1.99–2.11 (m, 2H), 3.21–3.31 (m, 2H), 3.31–3.38 (m, 3H), 3.39–3.49 (m, 2H). ^{13}C NMR (99 MHz, $CDCl_3$, δ): 3.3 (CH_2), 33.3 (CH_2), 58.7 (CH_3), 71.9 (CH_2). HRMS-EI (*m*/*z*): $[M]^+$ calcd for C_4H_9IO, 199.9698; found, 199.9701.

4.4.3 Substrate Characterization

Buta-2,3-dien-2-ylcyclohexane (1a, Fig. 4.21) [88]

1a was prepared from corresponding propargyl tosylate (6.73 g, 30.0 mmol) according to the GP1. The product **1a** was obtained in 67% yield (2.73 g, 20.0 mmol, colorless oil).

^{1}H NMR (392 MHz, $CDCl_3$, δ): 1.03–1.35 (m, 5H), 1.60–1.86 (m, 9H), 4.59 (quintet, $J = 2.9$ Hz, 2H). ^{13}C NMR (99 MHz, $CDCl_3$, δ): 17.1 (CH_3), 26.3 (CH_2), 26.5 (CH_2), 31.7 (CH_2), 40.9 (CH), 74.4 (CH_2), 103.5 (C), 205.6 (C). HRMS-EI (*m*/*z*): $[M]^+$ calcd for $C_{10}H_{16}$, 136.1252; found, 136.1253.

3,4,4-Trimethylpenta-1,2-diene (1b, Fig. 4.22) [89]

1b was prepared from corresponding propargyl tosylate (6.73 g, 30.0 mmol) according to the GP1. The product **1b** was obtained in 37% yield (1.21 g, 11.0 mmol, colorless oil).

^{1}H NMR (392 MHz, $CDCl_3$, δ): 1.05 (s, 9H), 1.69 (t, $J = 3.1$ Hz, 3H), 4.54–4.60 (m, 2H). ^{13}C NMR (99 MHz, $CDCl_3$, δ): 14.8 (CH_3), 28.9 (CH_3), 32.6 (C), 74.2 (CH_2), 107.2 (C), 205.2 (C). HRMS-EI (*m*/*z*): $[M]^+$ calcd for C_8H_{14}, 110.1096; found, 110.1092.

1-(*tert*-Butyl)-4-(2,3-dimethylpenta-3,4-dien-1-yl)-benzene (1c, Fig. 4.23)

1c was prepared from corresponding propargyl tosylate (1.92 g, 5.00 mmol) according to the GP1. The product **1c** was obtained in 81% yield (922 mg, 4.04 mmol, colorless oil).

Fig. 4.21 Scheme of **1a**

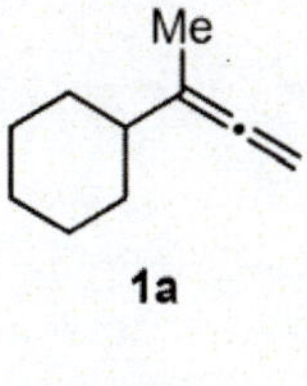

1a

Fig. 4.22 Scheme of **1b**

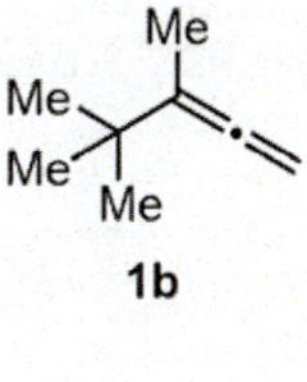

1b

Fig. 4.23 Scheme of **1c**

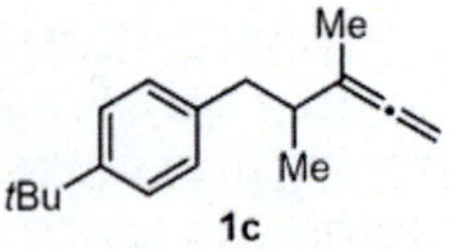

1c

^{1}H NMR (392 MHz, $CDCl_3$, δ): 0.98 (d, J = 6.7 Hz, 3H), 1.31 (s, 9H), 1.71 (t, J = 3.1 Hz, 3H), 2.18–2.28 (m, 1H), 2.36–2.44 (m, 1H), 2.79–2.86 (m, 1H), 4.56–4.62 (m, 2H), 7.06–7.11 (m, 2H), 7.24–7.31 (m, 2H). ^{13}C NMR (99 MHz, $CDCl_3$, δ): 17.1 (CH_3), 18.7 (CH_3), 31.4 (CH_3), 34.3 (C), 38.7 (CH), 41.2 (CH_2), 74.8 (CH_2), 103.0 (C), 124.9 (CH), 128.7 (CH), 138.0 (C), 148.5 (C), 205.7 (C). HRMS-EI (m/z): $[M–Me]^+$ calcd for $C_{16}H_{21}$, 213.1643; found, 213.1643.

1-(*tert*-Butyl)-4-(2-methyl-3-vinylidenehexyl)-benzene (1d, Fig. 4.24)

1d was prepared from corresponding propargyl tosylate (1.92 g, 5.00 mmol) according to the GP1. The product **1d** was obtained in 92% yield (1.18 g, 4.62 mmol, colorless oil).

^{1}H NMR (392 MHz, $CDCl_3$, δ): 0.88 (t, J = 7.4 Hz, 3H), 0.98 (d, J = 6.7 Hz, 3H), 1.30 (s, 9H), 1.37–1.49 (m, 2H), 1.82–2.00 (m, 2H), 2.09–2.26 (m, 1H), 2.31–2.47 (m, 1H), 2.75–2.90 (m, 1H), 4.63–4.71 (m, 2H), 7.04–7.14 (m, 2H), 7.23–7.31 (m, 2H). ^{13}C NMR (99 MHz, $CDCl_3$, δ): 13.9 (CH_3), 19.1 (CH_3), 20.9 (CH_2), 31.4 (CH_3), 32.9 (CH_2), 34.3 (C), 38.0 (CH), 41.7 (CH_2), 76.8 (CH_2), 108.2 (C), 124.9 (CH), 128.8 (CH), 138.2 (C), 148.4 (C), 205.3 (C). HRMS-EI (m/z): $[M–Me]^+$ calcd for $C_{18}H_{25}$, 241.1956; found, 241.1948.

1-(*tert*-Butyl)-4-(2-methyl-3-vinylideneundecyl)-benzene (1e, Fig. 4.25)

1e was prepared from corresponding propargyl tosylate (1.15 g, 3.00 mmol) according to the GP1. The product **1e** was obtained in 75% yield (587 mg, 1.79 mmol, colorless oil).

^{1}H NMR (392 MHz, $CDCl_3$, δ): 0.88 (t, J = 6.9 Hz, 3H), 0.97 (d, J = 6.4 Hz, 3H), 1.19–1.48 (m, 12H), 1.31 (s, 9H), 1.86–1.99 (m, 2H), 2.12–2.24 (m, 1H), 2.39 (dd, J = 13.3, 8.7 Hz, 1H), 2.82 (dd, J = 13.3, 5.5 Hz, 1H), 4.66–4.70 (m, 2H), 7.05–7.11 (m, 2H), 7.24–7.31 (m, 2H). ^{13}C NMR (99 MHz, $CDCl_3$, δ): 14.1 (CH_3), 19.1

Fig. 4.24 Scheme of **1d**

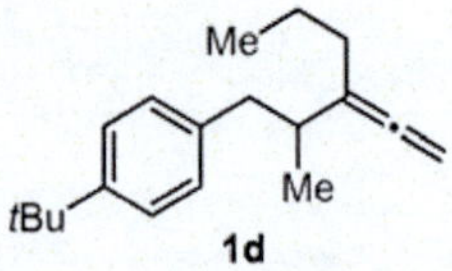

1d

Fig. 4.25 Scheme of **1e**

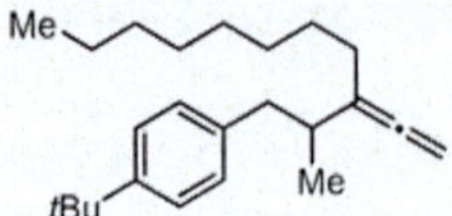

1e

(CH_3), 22.7 (CH_2), 27.6 (CH_2), 29.3 (CH_3), 29.4 (CH_3), 29.5 (CH_3), 30.7 (CH_2), 31.4 (CH_3), 31.9 (CH_2), 34.3 (*C*), 37.9 (*C*H), 41.6 (CH_2), 76.8 (CH_2), 108.5 (*C*), 124.9 (*C*H), 128.8 (*C*H), 138.2 (*C*), 148.4 (*C*), 205.2 (*C*). HRMS-EI (*m*/*z*): $[M–Me]^+$ calcd for $C_{23}H_{35}$, 311.2739; found, 311.2732.

1-(*tert*-Butyl)-4-(2,5-dimethyl-3-vinylidenehexyl)-benzene (1f, Fig. 4.26)

1f was prepared from corresponding propargyl tosylate (1.15 g, 3.00 mmol) according to the GP1. The product **1f** was obtained in 83% yield (674 mg, 2.49 mmol, colorless oil).

^1H NMR (392 MHz, $CDCl_3$, δ): 0.85 (d, $J = 6.4$ Hz, 3H), 0.88 (d, $J = 6.0$ Hz, 3H), 0.97 (d, $J = 6.9$ Hz, 3H), 1.31 (s, 9H), 1.65–1.78 (m, 1H), 1.79–1.86 (m, 2H), 2.09–2.20 (m, 1H), 2.39 (dd, $J = 13.8$, 8.7 Hz, 1H), 2.81 (dd, $J = 13.3$, 6.0 Hz, 1H), 4.65–4.71 (m, 2H), 7.05–7.12 (m, 2H), 7.24–7.31 (m, 2H). ^{13}C NMR (99 MHz, $CDCl_3$, δ): 19.1 (CH_3), 22.56 (CH_3), 22.63 (CH_3), 26.6 (*C*H), 31.4 (CH_3), 34.3 (*C*), 37.9 (*C*H), 40.5 (CH_2), 41.6 (CH_2), 76.4 (CH_2), 107.2 (*C*), 124.9 (*C*H), 128.8 (*C*H), 138.2 (*C*), 148.4 (*C*), 205.8 (*C*). HRMS-EI (*m*/*z*): $[M–Me]^+$ calcd for $C_{19}H_{27}$, 255.2113; found, 255.2115.

1-(Buta-2,3-dien-2-yl)-adamantane (1 g, Fig. 4.27) [90]

1g was prepared from corresponding propargylic (1.28 g, 6.28 mmol) alcohol according to the GP2. The product **1g** was obtained in 39% yield (456 mg, 2.42 mmol, colorless oil) via GPC purification.

^1H NMR (392 MHz, $CDCl_3$, δ): 1.55–1.77 (m, 15H), 1.99 (s, 3H), 4.55–4.62 (m, 2H). ^{13}C NMR (99 MHz, $CDCl_3$, δ): 13.4 (CH_3), 28.8 (*C*H), 34.2 (*C*), 36.9 (CH_2), 41.1 (CH_2), 74.4 (CH_2), 107.6 (*C*), 205.7 (*C*). HRMS-EI (*m*/*z*): $[M]^+$ calcd for $C_{14}H_{20}$, 188.1565; found, 188.1565.

3-(*tert*-Butyl)-hepta-1,2-diene (1h, Fig. 4.28)

1h was prepared from corresponding propargyl tosylate (7.13 g, 27.0 mmol) according to the GP1. The product **1h** was obtained in 69% yield (2.83 g, 18.6 mmol, colorless oil).

Fig. 4.26 Scheme of **1f**

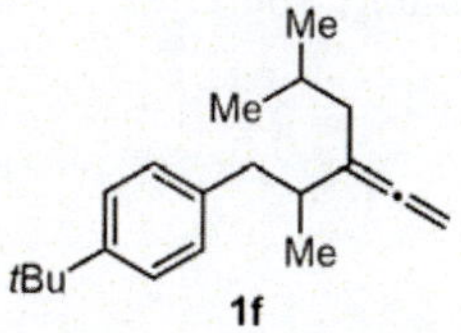

1f

Fig. 4.27 Scheme of **1g**

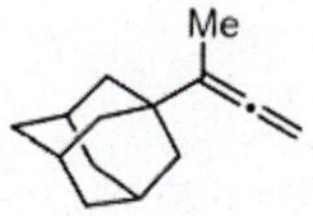

1g

Fig. 4.28 Scheme of **1h**

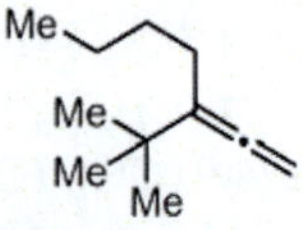

1h

^{1}H NMR (392 MHz, $CDCl_3$, δ): 0.91 (t, *J* = 7.3 Hz, 3H), 1.04 (s, 9H), 1.29–1.44 (m, 4H), 1.87–1.95 (m, 2H), 4.68 (t, *J* = 5.9 Hz, 2H). ^{13}C NMR (99 MHz, $CDCl_3$, δ): 14.1 (CH_3), 22.6 (CH_2), 26.2 (CH_2), 29.2 (CH_3), 30.5 (CH_2), 32.8 (*C*), 76.8 (CH_2), 112.6 (*C*), 204.5 (*C*). HRMS-EI (*m*/*z*): $[M–Me]^+$ calcd for $C_{10}H_{17}$, 137.1330; found, 137.1329.

(4-Methylhexa-4,5-dien-1-yl)-benzene (1i, Fig. 4.29)

1i was prepared from corresponding propargyl tosylate (6.36 g, 19.3 mmol) according to the GP1. The product **1i** was obtained in 76% yield (2.54 g, 14.7 mmol, colorless oil).

^{1}H NMR (396 MHz, $CDCl_3$, δ): 1.68 (t, *J* = 3.2 Hz, 3H), 1.76 (quintet, *J* = 7.4 Hz, 2H), 1.91–2.02 (m, 2H), 2.63 (t, *J* = 7.7 Hz, 2H), 4.61 (sextet, *J* = 3.1 Hz, 2H), 7.13–7.22 (m, 3H), 7.24–7.32 (m, 2H). ^{13}C NMR (100 MHz, $CDCl_3$, δ): 18.8 (CH_3), 29.2 (CH_2), 32.8 (CH_2), 35.4 (CH_2), 74.2 (CH_2), 98.1 (*C*), 125.6 (*C*H), 128.2 (*C*H), 128.5 (*C*H), 142.5 (*C*), 206.1 (*C*). HRMS-EI (*m*/*z*): $[M]^+$ calcd for $C_{13}H_{16}$, 172.1252; found, 172.1252.

3-Methylundeca-1,2-diene (1j, Fig. 4.30)

1j was prepared from corresponding propargyl tosylate (3.36 g, 15.0 mmol) according to the GP1. The product **1j** was obtained in 80% yield (1.99 g, 12.0 mmol, colorless oil).

^{1}H NMR (392 MHz, $CDCl_3$, δ): 0.88 (t, *J* = 6.7 Hz, 3H), 1.20–1.35 (m, 10H), 1.36–1.47 (m, 2H), 1.67 (t, *J* = 3.1 Hz, 3H), 1.87–1.96 (m, 2H), 4.57 (sextet, *J* = 3.1 Hz, 2H). ^{13}C NMR (99 MHz, $CDCl_3$, δ): 14.1 (CH_3), 18.7 (CH_3), 22.7 (CH_2), 27.4 (CH_2), 29.28 (CH_2), 29.31 (CH_2), 29.5 (CH_2), 31.9 (CH_2), 33.5 (CH_2), 73.7 (CH_2), 98.5 (*C*), 206.1 (*C*). HRMS-EI (*m*/*z*): $[M]^+$ calcd for $C_{12}H_{22}$, 166.1722; found, 166.1723.

Fig. 4.29 Scheme of **1i**

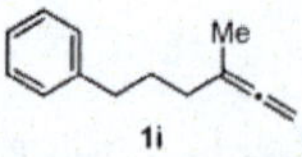

1i

Fig. 4.30 Scheme of **1j**

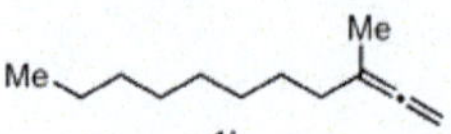

1j

(2-Ethylbuta-2,3-dien-1-yl)benzene (1k, Fig. 4.31)

1k was prepared from corresponding propargyl tosylate (6.91 g, 29.0 mmol) according to the GP1. The product **1k** was obtained in 9% yield (427 mg, 2.70 mmol, colorless oil) via GPC purification.

^{1}H NMR (401 MHz, $CDCl_3$, δ): 0.99 (t, J = 7.2 Hz, 3H), 1.85–1.95 (m, 2H), 3.31 (s, 2H), 4.66–4.73 (m, 2H), 7.16–7.24 (m, 3H), 7.24–7.32 (m, 2H). ^{13}C NMR (99 MHz, $CDCl_3$, δ): 12.1 (CH_3), 24.1 (CH_2), 39.6 (CH_2), 75.7 (CH_2), 104.4 (C), 126.1 (CH), 128.2 (CH), 128.9 (CH), 139.8 (C), 206.4 (C). HRMS-EI (m/z): $[M]^+$ calcd for $C_{12}H_{14}$, 158.1096; found, 158.1091.

(4-Vinylidenecyclohexyl)benzene (1l, Fig. 4.32) [91]

1l was prepared from corresponding propargylic alcohol (1.60 g, 8.00 mmol) according to the GP2. The product **1l** was obtained in 19% yield (274 mg, 1.49 mmol, colorless oil) via GPC purification.

^{1}H NMR (392 MHz, $CDCl_3$, δ): 1.52–1.70 (m, 2H), 1.93–2.03 (m, 2H), 2.11–2.25 (m, 2H), 2.39–2.49 (m, 2H), 2.53–2.64 (m, 1H), 4.54–4.66 (m, 2H), 7.15–7.24 (m, 3H), 7.26–7.34 (m, 2H). ^{13}C NMR (100 MHz, $CDCl_3$, δ): 31.0 (CH_2), 34.4 (CH_2), 43.9 (CH), 72.9 (CH_2), 100.0 (C), 126.0 (CH), 126.8 (CH), 128.4 (CH), 146.8 (C), 203.5 (C). HRMS-EI (m/z): $[M]^+$ calcd for $C_{14}H_{16}$, 184.1252; found, 184.1251.

(8*S*,9*S*,13*S*,14*S*)-3-Methoxy-13-methyl-17-vinylidene-7,8,9,11,12,13,14,15,16,17-decahydro-6*H*-cyclopenta[*a*]phenanthrene (1m, Fig. 4.33) [90]

1m was prepared from corresponding propargylic alcohol (1.55 g, 5.00 mmol) according to the GP2. The product **1m** was obtained in 34% yield (505 mg, 1.72 mmol, white solid).

Fig. 4.31 Scheme of **1k**

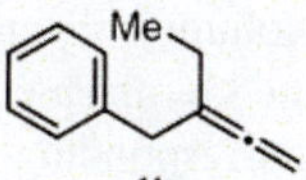

1k

Fig. 4.32 Scheme of **1l**

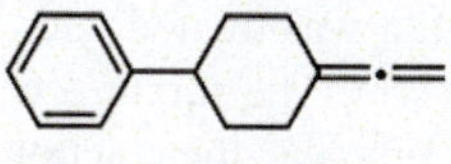

1l

Fig. 4.33 Scheme of **1m**

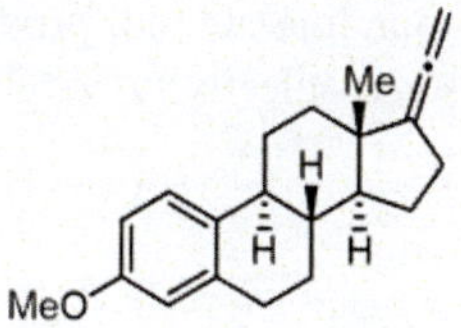

1m

^{1}H NMR (392 MHz, $CDCl_3$, δ): 0.90 (s, 3H), 1.18–1.65 (m, 6H), 1.74–2.01 (m, 3H), 2.14–2.51 (m, 3H), 2.53–2.68 (m, 1H), 2.78–2.97 (m, 2H), 3.78 (s, 3H), 4.66–4.79 (m, 2H), 6.63 (d, J = 2.7 Hz, 1H), 6.71 (dd, J = 8.4, 2.5 Hz, 1H), 7.22 (d, J = 8.6 Hz, 1H). ^{13}C NMR (99 MHz, $CDCl_3$, δ): 18.4 (CH_3), 24.5 (CH_2), 26.6 (CH_2), 27.3 (CH_2), 27.7 (CH_2), 29.8 (CH_2), 36.0 (CH_2), 38.7 (CH), 43.9 (CH), 44.4 (C), 54.4 (CH), 55.2 (CH_3), 76.8 (CH_2), 111.4 (CH), 111.9 (C), 113.7 (CH), 126.3 (CH), 132.7 (C), 137.9 (C), 157.4 (C), 200.4 (C). HRMS-EI (m/z): $[M]^+$ calcd for $C_{21}H_{26}O$, 294.1984; found, 294.1982.

4.4.4 General Borylation Procedure

4.4.4.1 Procedure for the Copper(I)-Catalyzed Alkyl Borylation of 1a (Optimized Conditions A = GP3)

SIMesCuCl was prepared according to the literature [95]. SIMesCuCl (4.1 mg, 0.010 mmol), bis(pinacolato)diboron (**3**) (152.4 mg, 0.60 mmol), and K(O-*t*-Bu) (67.3 mg, 0.60 mmol) were placed in an oven-dried reaction vial in an argon-filled glove box. After the vial was sealed with a screw cap containing a Teflon™-coated rubber septum and taken from the glove box, dry DMF (1.0 mL) was added to the vial through the rubber septum using a syringe. After stirred for 10 min, the reaction mixture was cooled to −5 °C and stirred for 10 min. Then **1a** (0.50 mmol) and **2a** (1.0 mmol) were added to the mixture. The reaction mixture was stirred at −5 °C for 24 h. The reaction monitoring and regioselectivity determination were done by GC analysis. After the reaction was completed, the reaction mixture was passed through a short silica gel column (Φ: 10 mm, the height of the silica gel column: 30 mm) eluting with Et_2O/hexane (5/95). The crude material was purified by flash column chromatography (SiO_2, Et_2O/hexane, 0:100–3:97) to give the corresponding allyl boronate (*E*)-**4aa**. In order to remove an undesired boryl substitution product of the alkyl halide and protoboration product of the allene for the further purification and the determination of the regio- and stereoselectivities, (*E*)-**4aa** was treated with $NaBO_3{\cdot}4H_2O$ as an oxidant for the boryl group in THF/H_2O (1:1). The mixture was stirred at room temperature overnight. After the reaction complete, the reaction mixture was extracted with Et_2O and dried over $MgSO_4$. The solvents were removed under reduced pressure. The crude material was purified by flash column chromatography (SiO_2, EtOAc/hexane, 0:100–12:88) to give the corresponding alcohol product (*Z*)-**6aa** as a colorless oil. Then, the regioselectivity and stereoselectivity were determined by ^{1}H and ^{13}C NMR analysis.

4.4.4.2 Procedure for the Copper(I)-Catalyzed Alkyl Borylation of 1b (Optimized Conditions B = GP4)

CuCl (2.5 mg, 0.025 mmol), Xantphos (14.5 mg, 0.025 mmol), bis(pinacolato)diboron (**3**) (152.4 mg, 0.60 mmol), and K(O-*t*-Bu) (67.3 mg, 0.60 mmol) were placed in an oven-dried reaction vial in an argon-filled glove box. After the vial was sealed with a screw cap containing a Teflon™-coated rubber septum and taken from the glove box, dry DMF (1.0 mL) was added to the vial through the rubber septum using a syringe. After stirring for 10 min, **1b** (0.50 mmol) and **2a** (1.0 mmol) were added to the mixture. The reaction mixture was stirred at 30 °C for 24 h. The reaction monitoring was done by GC analysis. After the reaction was completed, the reaction mixture was passed through a short silica gel column (Φ: 10 mm, the height of the silica gel column: 30 mm) eluting with Et_2O/hexane (5/95). The crude material was purified by flash column chromatography (SiO_2, Et_2O/hexane, 0:100–3:97) to give the corresponding allyl boronate (*E*)-**4ba**. Then, the regioselectivity and stereoselectivity were determined by ^1H NMR analysis.

4.4.5 *Borylation Product Characterization*

(Z)-2-(1-Cyclohexylethylidene)-6,6,6-trifluorohexan-1-ol [(Z)-6aa, Fig. 4.34]

The reaction was conducted with 68.0 mg (0.499 mmol) of **1a** according to the GP3. The borylation product (*E*)-**4aa** was obtained in 77% yield with E/Z = <5:95, **4**:**5** = >95:5 (determined by GC and ^{13}C NMR analysis). In order to remove an undesired boryl substitution product of the alkyl halide and protoboration product of the allene for the further purification, the oxidation of the boryl groups was performed. The product (*Z*)-**6aa** was obtained in 66% yield with E/Z = <5:95, **4**:**5** = >95:5 (86.4 mg, 0.327 mmol, colorless oil).

^1H NMR (392 MHz, $CDCl_3$, δ): 1.07–1.52 (m, 8H), 1.55–1.85 (m, 5H), 1.60 (s, 3H), 1.99–2.16 (m, 2H), 2.20 (t, J = 8.0 Hz, 2H), 2.46–2.63 (m, 1H), 4.14 (s, 2H). ^{13}C NMR (99 MHz, $CDCl_3$, δ): 13.5 (CH_3), 20.8–21.2 (m, CH_2), 26.0 (CH_2), 26.4 (CH_2), 30.1 (CH_2), 31.4 (CH_2), 33.5 (q, J = 28.6 Hz, CH_2), 41.1 (CH), 61.1 (CH_2),

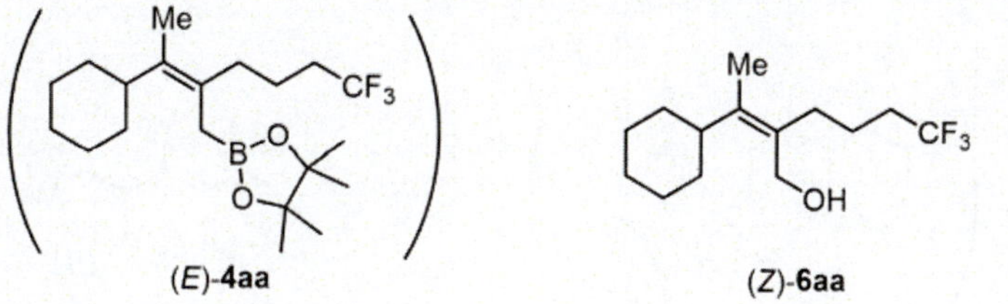

(*E*)-**4aa** (*Z*)-**6aa**

Fig. 4.34 Scheme of (*E*)-**4aa** and (*Z*)-**6aa**

127.2 (q, J = 277.6 Hz, C), 130.2 (C), 140.6 (C). HRMS-EI (m/z): $[M]^+$ calcd for $C_{14}H_{23}F_3O$, 264.1701; found, 264.1700.

(*E*)-2-[2-(3,3-Dimethylbutan-2-ylidene)-6,6,6-trifluorohexyl]-4,4,5,5-tetramethyl-1,3,2-dioxaborolane [(*E*)-4ba, Fig. 4.35]

The reaction was conducted with 55.1 mg (0.500 mmol) of **1b** according to the GP4. The product (*E*)-**4ba** was obtained in 83% yield with *E*/*Z* = >95:5, **4**:**5** = >95:5 (144.0 mg, 0.413 mmol, colorless oil).

^1H NMR (396 MHz, $CDCl_3$, δ): 1.14 (s, 9H), 1.23 (s, 12H), 1.57–1.68 (m, 5H), 1.84 (s, 2H), 1.98–2.12 (m, 4H). ^{13}C NMR (100 MHz, $CDCl_3$, δ): 17.0 (CH_3), 19.3–20.9 (m, CH_2 and br, B-CH_2), 24.6 (CH_3), 30.7 (CH_3), 33.7 (q, J = 28.4 Hz, CH_2), 35.7 (C), 36.5 (CH_2), 83.0 (C), 127.3 (q, J = 277.5 Hz, C), 128.1 (C), 135.3 (C). ^{11}B NMR (126 MHz, $CDCl_3$, δ): 33.0 (br, s). HRMS-EI (m/z): $[M]^+$ calcd for $C_{18}H_{32}{}^{11}BF_3O_2$, 348.2451; found, 348.2445.

(*Z*)-3-Cyclohexyl-2-ethylbut-2-en-1-ol [(*Z*)-6ab, Fig. 4.36]

The reaction was conducted with 68.0 mg (0.499 mmol) of **1a** according to the GP3. In order to remove an undesired boryl substitution product of the alkyl halide and protoboration product of the allene for the further purification and the determination of the regio- and stereoselectivities, the oxidation of the boryl groups was performed. The product (*Z*)-**6ab** was obtained in 62% yield with *E*/*Z* = <5:95, **4**:**5** = >95:5 (56.5 mg, 0.310 mmol, colorless oil).

^1H NMR (392 MHz, $CDCl_3$, δ): 0.97 (t, J = 7.6 Hz, 3H), 1.07–1.22 (m, 2H), 1.23–1.48 (m, 6H), 1.56 (s, 3H), 1.62–1.86 (m, 3H), 2.14 (q, J = 7.6 Hz, 2H), 2.48–2.61 (m, 1H), 4.13 (s, 2H). ^{13}C NMR (99 MHz, $CDCl_3$, δ): 13.21 (CH_3), 13.24 (CH_3),

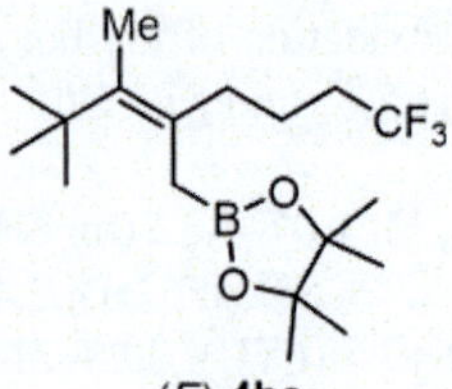

(*E*)-4ba

Fig. 4.35 Scheme of (*E*)-**4ba**

(*E*)-4ab (*Z*)-6ab

Fig. 4.36 Scheme of (*E*)-**4ab** and (*Z*)-**6ab**

24.3 (CH_2), 26.1 (CH_2), 26.5 (CH_2), 31.5 (CH_2), 41.0 (CH), 61.2 (CH_2), 133.2 (C), 138.9 (C). HRMS-EI (m/z): $[M]^+$ calcd for $C_{12}H_{22}O$, 182.1671; found, 182.1673.

(Z)-2-(1-Cyclohexylethylidene)-heptan-1-ol [(Z)-6ac, Fig. 4.37]

The reaction was conducted with 68.0 mg (0.499 mmol) of **1a** according to the GP3. In order to remove an undesired boryl substitution product of the alkyl halide and protoboration product of the allene for the further purification and the determination of the regio- and stereoselectivities, the oxidation of the boryl groups was performed. The product (*Z*)-**6ac** was obtained in 49% yield with *E*/*Z* = <5:95, **4**:**5** = >95:5 (55.2 mg, 0.246 mmol, colorless oil).

^1H NMR (396 MHz, $CDCl_3$, δ): 0.89 (t, J = 7.0 Hz, 3H), 1.06–1.20 (m, 2H), 1.23–1.48 (m, 12H), 1.59 (s, 3H), 1.63–1.84 (m, 3H), 2.05–2.15 (m, 2H), 2.49–2.60 (m, 1H), 4.12 (s, 2H). ^{13}C NMR (100 MHz, $CDCl_3$, δ): 13.5 (CH_3), 14.1 (CH_3), 22.6 (CH_2), 26.1 (CH_2), 26.5 (CH_2), 28.6 (CH_2), 31.4 (CH_2), 31.5 (CH_2), 32.0 (CH_2), 41.1 (CH), 61.6 (CH_2), 131.9 (C), 139.4 (C). HRMS-EI (m/z): $[M]^+$ calcd for $C_{15}H_{28}O$, 224.2140; found, 224.2138.

(Z)-2-(1-Cyclohexylethylidene)-5-phenylpentan-1-ol [(Z)-6ad, Fig. 4.38]

The reaction was conducted with 68.1 mg (0.500 mmol) of **1a** according to the GP3. In order to remove an undesired boryl substitution product of the alkyl halide and protoboration product of the allene for the further purification and the determination of the regio- and stereoselectivities, the oxidation of the boryl groups was performed. The product (*Z*)-**6ad** was obtained in 63% yield with *E*/*Z* = <5:95, **4**:**5** = >95:5 (86.3 mg, 0.317 mmol, colorless oil).

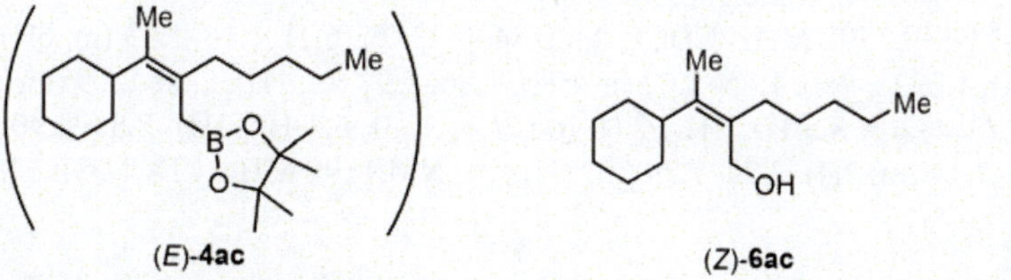

Fig. 4.37 Scheme of (*E*)-**4ac** and (*Z*)-**6ac**

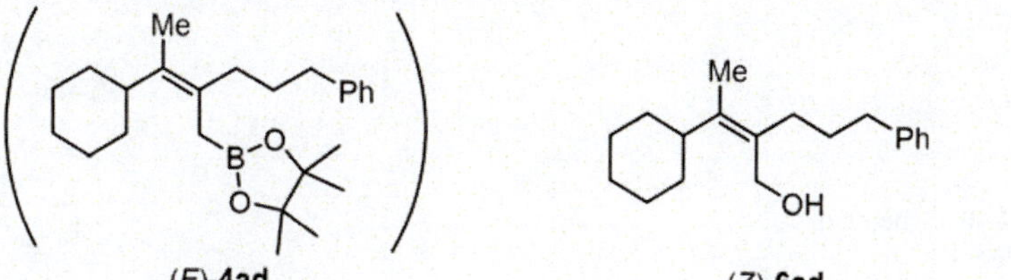

Fig. 4.38 Scheme of (*E*)-**4ad** and (*Z*)-**6ad**

^{1}H NMR (392 MHz, $CDCl_3$, δ): 1.06–1.47 (m, 8H), 1.55 (s, 3H), 1.61–1.80 (m, 5H), 2.12–2.20 (m, 2H), 2.48–2.57 (m, 1H), 2.58–2.66 (m, 2H), 4.11 (s, 2H), 7.12–7.20 (m, 3H), 7.22–7.29 (m, 2H). ^{13}C NMR (99 MHz, $CDCl_3$, δ): 13.5 (CH_3), 26.1 (CH_2), 26.5 (CH_2), 30.7 (CH_2), 31.1 (CH_2), 31.5 (CH_2), 36.0 (CH_2), 41.1 (CH), 61.4 (CH_2), 125.6 (CH), 128.2 (CH), 128.3 (CH), 131.4 (C), 139.7 (C), 142.4 (C). HRMS-EI (m/z): $[M]^+$ calcd for $C_{19}H_{28}O$, 272.2140; found, 272.2139.

(*E*)-2-{5-[4-(*tert*-Butyl)phenyl]-2-ethyl-3,4-dimethylpent-2-en-1-yl}-4,4,5,5-tetramethyl-1,3,2-dioxaborolane [(*E*)-4cb, Fig. 4.39]

The reaction was conducted with 114.0 mg (0.499 mmol) of **1c** according to the GP4. The product (*E*)-**4cb** was obtained in 88% yield with $E/Z = 88{:}12$, **4**:**5** = >95:5 (169.3 mg, 0.440 mmol, colorless oil).

^{1}H NMR (392 MHz, $CDCl_3$, δ): 0.83–0.94 (m, 6H), 1.22 (s, 12H), 1.30 (s, 9H), 1.45–1.62 (m, 5H), 1.88–2.01 (m, 1H), 2.02–2.14 (m, 1H), 2.45 (dd, J = 13.2, 8.8 Hz, 1H), 2.60 (dd, J = 13.0, 6.3 Hz, 1H), 2.79–2.88 (m, 1H), 7.06–7.14 (m, 2H), 7.22–7.28 (m, 2H). ^{13}C NMR (99 MHz, $CDCl_3$, δ): 11.9 (CH_3), 12.2 (CH_3), 16.3 (br, B-CH_2), 17.7 (CH_3), 24.66 (CH_3), 24.74 (CH_3), 27.7 (CH_2), 31.4 (CH_3), 34.2 (C), 37.8 (CH), 40.7 (CH_2), 82.8 (C), 124.6 (CH), 128.7 (CH), 129.6 (C), 129.9 (C), 138.6 (C), 148.0 (C). ^{11}B NMR (126 MHz, $CDCl_3$, δ): 32.7 (br, s). HRMS-ESI (m/z): $[M + Na]^+$ calcd for $C_{25}H_{41}{}^{11}BO_2Na$, 407.3096; found, 407.3095.

(*E*)-2-(2-{4-[4-(*tert*-Butyl)phenyl]-3-methylbutan-2-ylidene}heptyl)-4,4,5,5-tetramethyl-1,3,2-dioxaborolane [(*E*)-4 cc, Fig. 4.40]

The reaction was conducted with 114.0 mg (0.499 mmol) of **1c** according to the GP4. The product (*E*)-**4cc** was obtained in 89% yield with $E/Z = 88{:}12$, **4**:**5** = >95:5 (189.8 mg, 0.445 mmol, colorless oil).

^{1}H NMR (392 MHz, $CDCl_3$, δ): 0.84–0.93 (m, 6H), 1.16–1.33 (m, 6H), 1.22 (s, 12 H), 1.29 (s, 9H), 1.45–1.62 (m, 5H), 1.86–1.97 (m, 1H), 2.00–2.09 (m, 1H), 2.46 (dd, J = 13.5, 8.5 Hz, 1H), 2.60 (dd, J = 13.0, 6.3 Hz, 1H), 2.81–2.90 (m, 1H), 7.06–7.14 (m, 2H), 7.22–7.29 (m, 2H). ^{13}C NMR (99 MHz, $CDCl_3$, δ): 12.1 (CH_3),

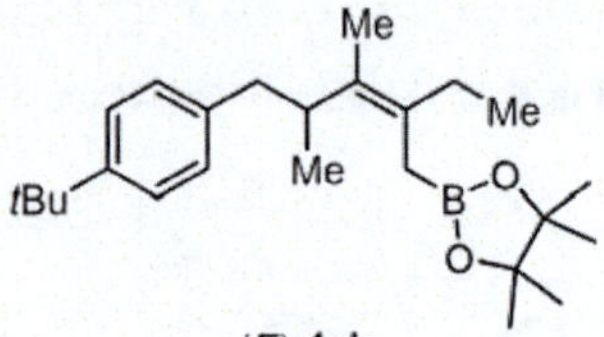

(*E*)-**4cb**

Fig. 4.39 Scheme of (*E*)-**4cb**

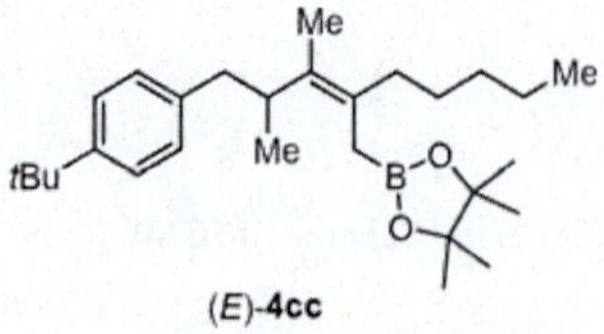

(*E*)-**4cc**

Fig. 4.40 Scheme of (*E*)-**4cc**

14.1 (CH_3), 16.8 (br, B-CH_2), 17.8 (CH_3), 22.7 (CH_2), 24.68 (CH_3), 24.74 (CH_3), 27.6 (CH_2), 31.4 (CH_3), 31.9 (CH_2), 34.2 (*C*), 34.8 (CH_2), 37.8 (*C*H), 40.8 (CH_2), 82.8 (*C*), 124.6 (*C*H), 128.6 (*C*), 128.7 (*C*H), 130.1 (*C*), 138.6 (*C*), 148.0 (*C*). ^{11}B NMR (126 MHz, $CDCl_3$, δ): 32.9 (br, s). HRMS-ESI (*m*/*z*): $[M + Na]^+$ calcd for $C_{28}H_{47}{}^{11}BO_2Na$, 449.3566; found, 449.3565.

(*E*)-2-(3-{1-[4-(*tert*-Butyl)phenyl]propan-2-yl{-2-ethylhex-2-en-1-yl)-4,4,5,5-tetramethyl-1,3,2-dioxaborolane [(*E*)-4db, Fig. 4.41]

The reaction was conducted with 128.3 mg (0.500 mmol) of **1d** according to the GP4. The product (*E*)-**4db** was obtained in 87% yield with *E*/*Z* = 89:11, **4**:**5** = >95:5 (179.9 mg, 0.436 mmol, colorless oil).

^{1}H NMR (396 MHz, $CDCl_3$, δ): 0.82–0.97 (m, 9H), 1.21 (s, 12H), 1.30 (s, 9H), 1.33–1.44 (m, 2H), 1.48–1.61 (m, 2H), 1.85–2.11 (m, 4H), 2.42 (dd, *J* = 14.7, 7.9 Hz, 1H), 2.65 (dd, *J* = 12.9, 5.7 Hz, 1H), 2.80–2.90 (m, 1H), 7.08–7.15 (m, 2H), 7.21–7.29 (m, 2H). ^{13}C NMR (100 MHz, $CDCl_3$, δ): 12.9 (CH_3), 14.9 (CH_3), 15.6 (br, B-CH_2), 18.0 (CH_3), 24.66 (CH_3 and CH_2), 24.70 (CH_3), 27.1 (CH_2), 30.2 (CH_2), 31.4 (CH_3), 34.2 (*C*), 38.2 (*C*H), 41.4 (CH_2), 82.7 (*C*), 124.6 (*C*H), 128.9 (*C*H), 130.9 (*C*), 135.0 (*C*), 138.7 (*C*), 148.0 (*C*). ^{11}B NMR (126 MHz, $CDCl_3$, δ): 32.7 (br, s). HRMS-ESI (*m*/*z*): $[M + Na]^+$ calcd for $C_{27}H_{45}{}^{11}BO_2Na$, 435.3410; found, 435.3409.

(*E*)-2-(2-{1-[4-(*tert*-Butyl)phenyl]-2-methylhexan-3-ylidene}heptyl)-4,4,5,5-tetramethyl-1,3,2-dioxaborolane [(*E*)-4dc, Fig. 4.42]

The reaction was conducted with 128.4 mg (0.501 mmol) of **1d** according to the GP4. The product (*E*)-**4dc** was obtained in 87% yield with *E*/*Z* = 89:11, **4**:**5** = >95:5 (197.4 mg, 0.434 mmol, colorless oil).

^{1}H NMR (396 MHz, $CDCl_3$, δ): 0.85–0.96 (m, 9H), 1.17–1.43 (m, 8H), 1.21 (s, 9H), 1.30 (s, 12H), 1.47–1.62 (m, 2H), 1.84–2.08 (m, 4H), 2.42 (dd, *J* = 12.9, 9.3 Hz, 1H), 2.65 (dd, *J* = 13.1, 5.4 Hz, 1H), 2.80–2.94 (m, 1H), 7.07–7.15 (m, 2H), 7.22–7.31 (m, 2H). ^{13}C NMR (100 MHz, $CDCl_3$, δ): 14.1 (CH_3), 14.9 (CH_3), 16.2

Fig. 4.41 Scheme of (*E*)-**4db**

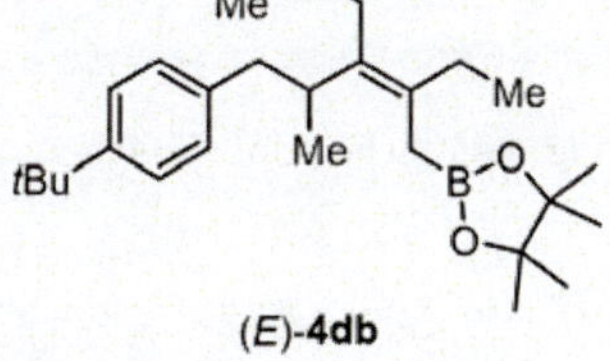

(*E*)-**4db**

Fig. 4.42 Scheme of (*E*)-**4dc**

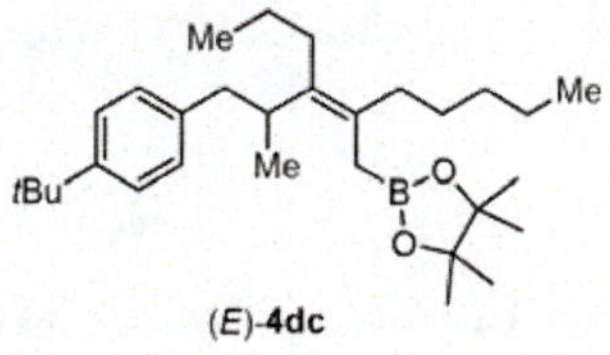

(*E*)-**4dc**

(br, B-CH_2), 18.1 (CH_3), 22.7 (CH_2), 24.6 (CH_2), 24.68 (CH_3), 24.70 (CH_3), 28.1 (CH_2), 30.3 (CH_2), 31.4 (CH_3), 32.2 (CH_2), 34.2 (*C*), 34.4 (CH_2), 38.2 (*C*H), 41.4 (CH_2), 82.7 (*C*), 124.6 (*C*H), 128.9 (*C*H), 129.8 (*C*), 135.4 (*C*), 138.7 (*C*), 148.0 (*C*). ^{11}B NMR (126 MHz, $CDCl_3$, δ): 33.3 (br, s). HRMS-ESI (*m*/*z*): $[M + Na]^+$ calcd for $C_{30}H_{51}{}^{11}BO_2Na$, 477.3880; found, 477.3877.

(*E*)-2-(3-{1-[4-(*tert*-Butyl)phenyl]propan-2-yl}-2-ethylundec-2-en-1-yl)-4,4,5,5-tetramethyl-1,3,2-dioxaborolane [(*E*)-4eb, Fig. 4.43]

The reaction was conducted with 212.0 mg (0.497 mmol) of **1e** according to the GP4. The product (*E*)-**4eb** was obtained in 82% yield with *E*/*Z* = 89:11, **4**:**5** = >95:5 (197.5 mg, 0.409 mmol, colorless oil).

^{1}H NMR (396 MHz, $CDCl_3$, δ): 0.82–0.96 (m, 9H), 1.16–1.41 (m, 12H), 1.21 (s, 12H), 1.30 (s, 9H), 1.49–1.60 (m, 2H), 1.85–2.12 (m, 4H), 2.42 (dd, *J* = 13.1, 9.5 Hz, 1H), 2.65 (dd, *J* = 12.9, 5.7 Hz, 1H), 2.79–2.92 (m, 1H), 7.08–7.15 (m, 2H), 7.23–7.29 (m, 2H). ^{13}C NMR (100 MHz, $CDCl_3$, δ): 12.9 (CH_3), 14.1 (CH_3), 15.7 (br, B-CH_2), 18.0 (CH_3), 22.7 (CH_2), 24.67 (CH_3 and CH_2), 24.70 (CH_3), 27.1 (CH_2), 27.9 (CH_2), 29.4 (CH_2), 29.5 (CH_2), 30.6 (CH_2), 31.5 (CH_3), 31.9 (CH_2), 34.2 (*C*), 38.2 (*C*H), 41.4 (CH_2), 82.7 (*C*), 124.6 (*C*H), 128.9 (*C*H), 130.8 (*C*), 135.1 (*C*), 138.7 (*C*), 148.0 (*C*). ^{11}B NMR (126 MHz, $CDCl_3$, δ): 32.6 (br, s). HRMS-EI (*m*/*z*): $[M–Me]^+$ calcd for $C_{31}H_{52}{}^{11}BO_2$, 467.4066; found, 467.4063.

(Z)-3-{1-[4-(*tert*-Butyl)phenyl]propan-2-yl}-2-pentylundec-2-en-1-ol [(Z)-6ec, Fig. 4.44]

The reaction was conducted with 213.2 mg (0.500 mmol) of **1e** according to the GP4. In order to remove an undesired boryl substitution product of the alkyl halide and protoboration product of the allene for the further purification and the determination

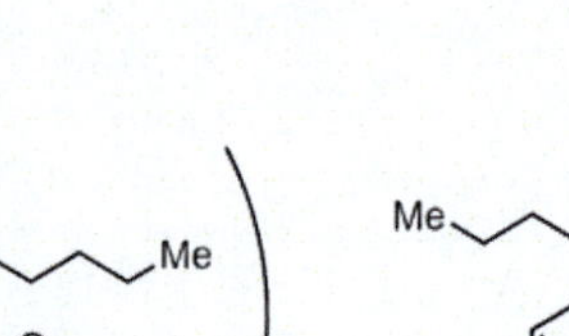

(*E*)-**4eb**

Fig. 4.43 Scheme of (*E*)-**4eb**

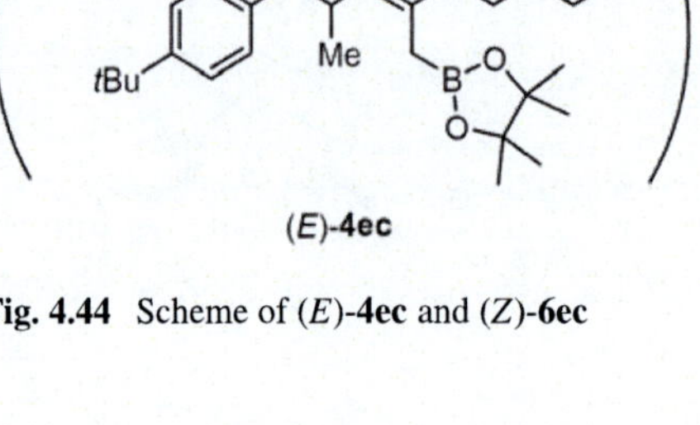

(*E*)-**4ec**

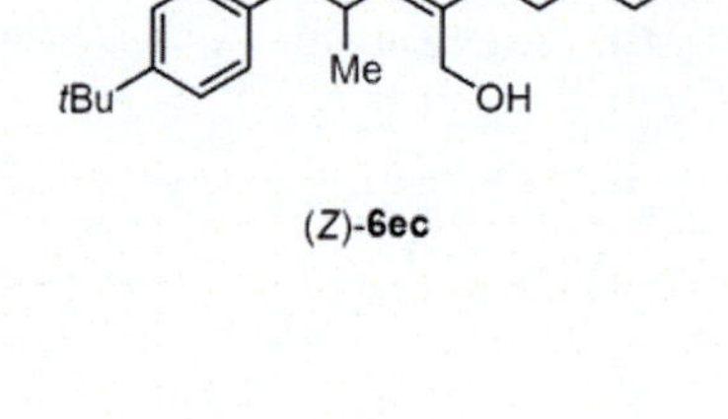

(*Z*)-**6ec**

Fig. 4.44 Scheme of (*E*)-**4ec** and (*Z*)-**6ec**

of the regio- and stereoselectivities, the oxidation of the boryl groups was performed. The corresponding alcohol (*Z*)-**6ec** was obtained in 50% yield with *E*/*Z* = 87:13, **4**:**5** = >95:5 (103.8 mg, 0.250 mmol, colorless oil).

^{1}H NMR (392 MHz, $CDCl_3$, δ): –0.13 (dd, *J* = 8.3, 3.8 Hz, 1H), 0.83–0.98 (m, 6H), 1.07 (d, *J* = 7.2 Hz, 3H), 1.17–1.46 (m, 18H), 1.28 (s, 9H), 1.86–2.15 (m, 4H), 2.47–2.64 (m, 2H), 2.85–3.03 (m, 1H), 3.66 (dd, *J* = 11.7, 8.5 Hz, 1H), 3.85 (dd, *J* = 11.4, 3.8 Hz, 1H), 6.98–7.08 (m, 2H), 7.24–7.31 (m, 2H). ^{13}C NMR (99 MHz, $CDCl_3$, δ): 14.06 (CH_3), 14.11 (CH_3), 19.8 (CH_3), 22.6 (CH_2), 22.7 (CH_2), 28.0 (CH_2), 28.6 (CH_2), 29.31 (CH_2), 29.33 (CH_2), 30.6 (CH_2), 31.0 (CH_2), 31.3 (CH_3), 31.4 (CH_2), 31.9 (CH_2), 32.3 (CH_2), 34.3 (*C*), 38.7 (*C*H), 41.7 (CH_2), 61.1 (CH_2), 124.9 (*C*H), 128.8 (*C*H), 134.7 (*C*), 138.2 (*C*), 140.3 (*C*), 148.8 (*C*). HRMS-EI (*m*/*z*): $[M]^+$ calcd for $C_{29}H_{50}O$, 414.3862; found, 414.3853.

(*E*)-2-(3-{1-[4-(*tert*-Butyl)phenyl]propan-2-yl}-2-ethyl-5-methylhex-2-en-1-yl)-4,4,5,5-tetramethyl-1,3,2-dioxaborolane [(*E*)-4fb, Fig. 4.45]

The reaction was conducted with 135.3 mg (0.500 mmol) of **1f** according to the GP4. The product (*E*)-**4fb** was obtained in 79% yield with *E*/*Z* = 89:11, **4**:**5** = >95:5 (169.6 mg, 0.398 mmol, colorless oil).

^{1}H NMR (392 MHz, $CDCl_3$, δ): 0.85–0.98 (m, 12H), 1.22 (s, 12H), 1.30 (s, 9H), 1.63–1.72 (m, 2H), 1.76–1.88 (m, 1H), 1.89–2.19 (m, 4H), 2.44 (dd, *J* = 13.0, 9.9 Hz, 1H), 2.72 (dd, *J* = 13.0, 5.4 Hz, 1H), 2.76–2.88 (m, 1H), 7.10–7.15 (m, 2H), 7.23–7.30 (m, 2H). ^{13}C NMR (99 MHz, $CDCl_3$, δ): 12.6 (CH_3), 16.3 (br, B-CH_2), 18.3 (CH_3), 22.7 (CH_3), 22.8 (CH_3), 24.7 (CH_3), 24.8 (CH_3), 27.8 (CH_2), 28.6 (*C*H), 31.4 (CH_3), 34.2 (*C*), 37.3 (CH_2), 38.5 (*C*H), 41.7 (CH_2), 82.8 (*C*), 124.7 (*C*H), 128.8 (*C*H), 132.6 (*C*), 133.2 (*C*), 138.9 (*C*), 148.0 (*C*). ^{11}B NMR (126 MHz, $CDCl_3$, δ): 32.9 (br, s). HRMS-ESI (*m*/*z*): $[M + Na]^+$ calcd for $C_{28}H_{47}{}^{11}BO_2Na$, 449.3566; found, 449.3565.

(*E*)-2-(2-{1-[4-(*tert*-Butyl)phenyl]-2,5-dimethylhexan-3-ylidene}heptyl)-4,4,5,5-tetramethyl-1,3,2-dioxaborolane [(*E*)-4fc, Fig. 4.46]

The reaction was conducted with 135.0 mg (0.499 mmol) of **1f** according to the GP4. The product (*E*)-**4fc** was obtained in 76% yield with *E*/*Z* = 91:9, **4**:**5** = >95:5 (177.4 mg, 0.379 mmol, colorless oil).

^{1}H NMR (392 MHz, $CDCl_3$, δ): 0.83–0.95 (m, 12H), 1.14–1.36 (m, 6H), 1.22 (s, 12H), 1.30 (s, 9H), 1.63–1.73 (m, 2H), 1.76–1.86 (m, 1H), 1.89–2.15 (m, 4H), 2.44 (dd, *J* = 13.2, 9.6 Hz, 1H), 2.73 (dd, *J* = 13.2, 5.2 Hz, 1H), 2.77–2.88 (m, 1H),

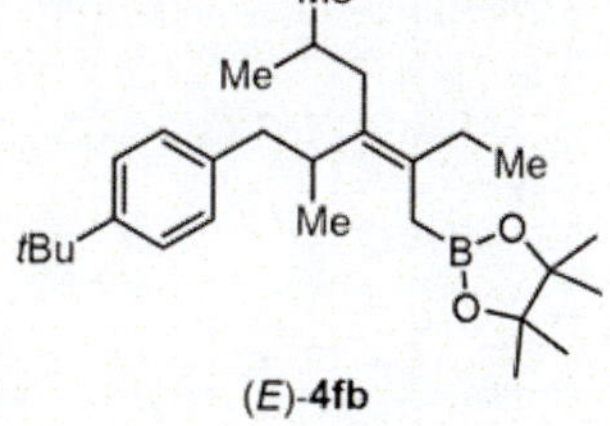

(*E*)-**4fb**

Fig. 4.45 Scheme of (*E*)-**4fb**

Fig. 4.46 Scheme of (*E*)-**4fc**

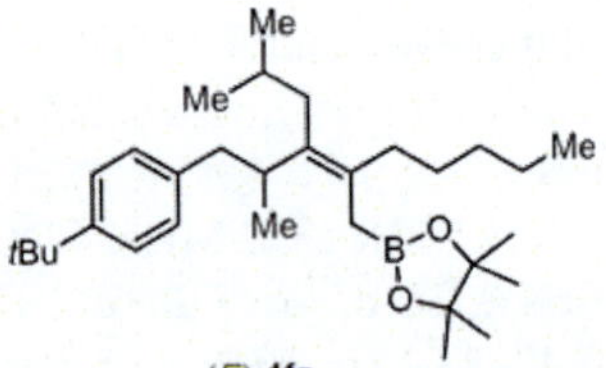

(*E*)-**4fc**

7.09–7.15 (m, 2H), 7.23–7.28 (m, 2H). ^{13}C NMR (99 MHz, $CDCl_3$, δ): 14.1 (*C*H_3), 16.8 (br, B-*C*H_2), 18.4 (*C*H_3), 22.6 (*C*H_3), 22.7 (*C*H_2), 22.8 (*C*H_3), 24.7 (*C*H_3), 24.8 (*C*H_3), 27.9 (*C*H_2), 28.7 (*C*H), 31.4 (*C*H_3), 32.2 (*C*H_2), 34.2 (*C*), 35.1 (*C*H_2), 37.4 (*C*H_2), 38.5 (*C*H), 41.7 (*C*H_2), 82.8 (*C*), 124.7 (*C*H), 128.8 (*C*H), 131.4 (*C*), 133.6 (*C*), 138.9 (*C*), 148.0 (*C*). ^{11}B NMR (126 MHz, $CDCl_3$, δ): 33.7 (br, s). HRMS-ESI (*m*/*z*): $[M + Na]^+$ calcd for $C_{31}H_{53}{}^{11}BO_2Na$, 491.4036; found, 491.4036.

2-{(*E*)-2-[1-(Adamantan-1-yl)ethylidene]-6,6,6-trifluorohexyl}-4,4,5,5-tetramethyl-1,3,2-dioxaborolane [(*E*)-4ga, Fig. 4.47]

The reaction was conducted with 94.1 mg (0.500 mmol) of **1g** according to the GP4. Then, the mixture of the product and byproducts was exposed to $NaBO_3 \cdot 4H_2O$ (300 mg, 1.94 mmol) in THF (2.0 mL) and H_2O (2.0 mL) at room temperature for 4 h. (*E*)-**4ga** was not oxidized under the conditions, and the pure product (*E*)-**4ga** was obtained in 64% yield with *E*/*Z* = >95:5, **4**:**5** = >95:5 (135.6 mg, 0.318 mmol, white solid).

^{1}H NMR (396 MHz, $CDCl_3$, δ): 1.23 (s, 12H), 1.55–1.73 (m, 11H), 1.83–1.98 (m, 11H), 2.00–2.16 (m, 4H). ^{13}C NMR (100 MHz, $CDCl_3$, δ): 16.0 (*C*H_3), 20.2 (B-*C*H_2 and *C*H_2), 24.6 (*C*H_3), 29.2 (*C*H), 33.6 (q, $J = 28.4$ Hz, CH_2), 37.0 (*C*H_2), 37.3 (*C*H_2), 38.5 (*C*), 41.0 (*C*H_2), 83.0 (*C*), 127.3 (q, $J = 277.5$ Hz, *C*), 128.1 (*C*), 135.1 (*C*). ^{11}B NMR (126 MHz, $CDCl_3$, δ): 32.9 (br, s). HRMS-EI (*m*/*z*): $[M]^+$ calcd for $C_{24}H_{38}{}^{11}BF_3O_2$, 426.2921; found, 426.2931.

(*E*)-2-[3-(*tert*-Butyl)-2-(4,4,4-trifluorobutyl)hept-2-en-1-yl]-4,4,5,5-tetramethyl-1,3,2-dioxaborolane [(*E*)-4ha, Fig. 4.48]

The reaction was conducted with 76.1 mg (0.500 mmol) of **1h** according to the GP4. The product (*E*)-**4ha** was obtained in 69% yield with *E*/*Z* = >95:5, **4**:**5** = >95:5 (134.1 mg, 0.344 mmol, colorless oil).

Fig. 4.47 Scheme of (*E*)-**4ga**

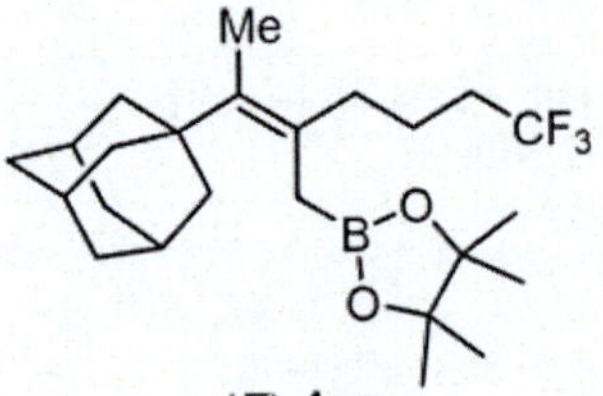

(*E*)-**4ga**

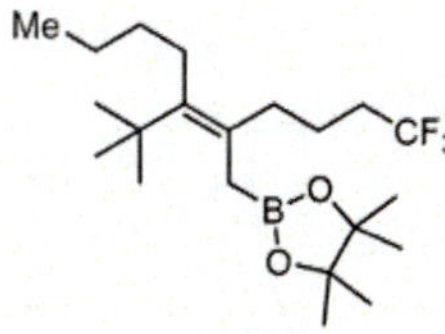

(*E*)-**4ha**

Fig. 4.48 Scheme of (*E*)-**4ha**

^{1}H NMR (396 MHz, $CDCl_3$, δ): 0.91 (t, J = 7.1 Hz, 3H), 1.16 (s, 9H), 1.22 (s, 12H), 1.27–1.39 (m, 4H), 1.60–1.71 (m, 2H), 1.85 (s, 2H), 1.96–2.14 (m, 6H). ^{13}C NMR (100 MHz, $CDCl_3$, δ): 13.9 (CH_3), 19.6 (br, B-CH_2), 21.0 (CH_2), 23.3 (CH_2), 24.6 (CH_3), 31.0 (CH_2), 31.4 (CH_3), 33.7 (CH_2), 33.8 (q, J = 28.4 Hz, CH_2), 35.5 (CH_2), 35.7 (C), 83.0 (C), 127.3 (q, J = 277.5 Hz, C), 129.1 (C), 140.1 (C). ^{11}B NMR (126 MHz, $CDCl_3$, δ): 33.0 (br, s). HRMS-EI (*m/z*): $[M]^+$ calcd for $C_{21}H_{38}{}^{11}BF_3O_2$, 390.2921; found, 390.2911.

(*E*)-2-[3-(*tert*-Butyl)-2-pentylhept-2-en-1-yl]-4,4,5,5-tetramethyl-1,3,2-dioxaborolane [(*E*)-4hc, Fig. 4.49]

The reaction was conducted with 76.1 mg (0.500 mmol) of **1h** according to the GP4. The product (*E*)-**4hc** was obtained in 92% yield with *E*/*Z* = >95:5, **4**:**5** = >95:5 (160.8 mg, 0.459 mmol, colorless oil).

^{1}H NMR (392 MHz, $CDCl_3$, δ): 0.83–0.96 (m, 6H), 1.09–1.44 (m, 10H), 1.16 (s, 9H), 1.22 (s, 12H), 1.86 (s, 2H), 1.92–2.08 (m, 4H). ^{13}C NMR (99 MHz, $CDCl_3$, δ): 14.0 (CH_3), 14.1 (CH_3), 19.7 (br, B-CH_2), 22.7 (CH_2), 23.3 (CH_2), 24.7 (CH_3), 28.6 (CH_2), 30.9 (CH_2), 31.4 (CH_3), 32.5 (CH_2), 33.9 (CH_2), 35.4 (C), 36.8 (CH_2), 82.8 (C), 130.9 (C), 138.4 (C). ^{11}B NMR (126 MHz, $CDCl_3$, δ): 33.1 (br, s). HRMS-EI (*m/z*): $[M]^+$ calcd for $C_{22}H_{43}{}^{11}BO_2$, 350.3360; found, 350.3358.

(Z)-3-Methyl-6-phenyl-2-(4,4,4-trifluorobutyl)hex-2-en-1-ol [(Z)-6ia, Fig. 4.50]

The reaction was conducted with 86.1 mg (0.500 mmol) of **1i** according to the GP3. In order to remove an undesired boryl substitution product of the alkyl halide and protoboration product of the allene for the further purification and the determination of the regio- and stereoselectivities, the oxidation of the boryl groups was performed.

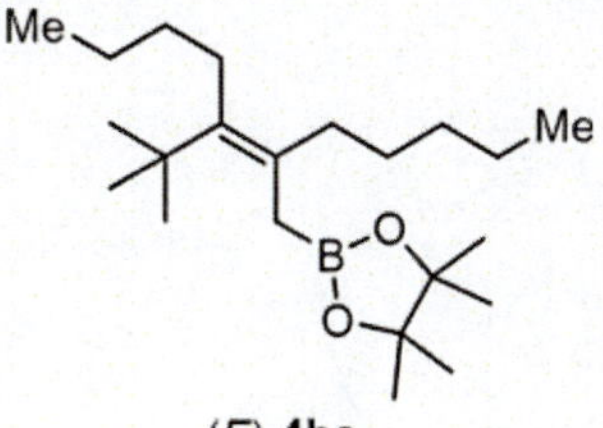

(*E*)-**4hc**

Fig. 4.49 Scheme of (*E*)-**4hc**

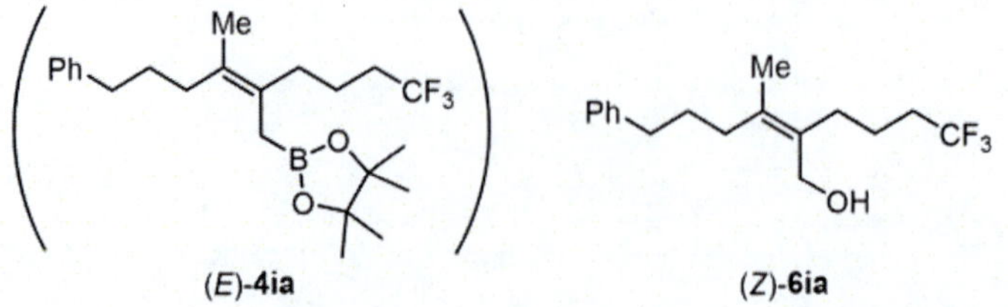

Fig. 4.50 Scheme of (*E*)-**4ia** and (Z)-**6ia**

The product (Z)-**6ia** was obtained in 59% yield with *E*/*Z* = 11:89, **4**:**5** = >95:5 (87.9 mg, 0.293 mmol, colorless oil).

^{1}H NMR (396 MHz, $CDCl_3$, δ): 1.18–1.35 (br-m, 1H), 1.52–1.81 (m, 7H), 1.91–2.32 (m, 6H), 2.59 (t, *J* = 7.7 Hz, 2H), 4.05 (s, 2H), 7.12–7.22 (m, 3H), 7.24–7.32 (m, 2H). ^{13}C NMR (100 MHz, $CDCl_3$, δ): 18.3 (CH_3), 20.9 (CH_2), 29.5 (CH_2), 30.5 (CH_2), 33.5 (q, *J* = 28.4 Hz, CH_2), 33.6 (CH_2), 35.7 (CH_2), 61.3 (CH_2), 125.8 (*C*H), 128.26 (*C*H), 128.28 (*C*H), 129.5 (q, *J* = 274.7 Hz, *C*), 131.5 (*C*), 135.2 (*C*), 142.1 (*C*). HRMS-EI (*m*/*z*): $[M]^+$ calcd for $C_{17}H_{23}F_3O$, 300.1701; found, 300.1702.

(Z)-2-(5-Phenylpentan-2-ylidene)heptan-1-ol [(Z)-6ic, Fig. 4.51]

The reaction was conducted with 86.2 mg (0.500 mmol) of **1i** according to the GP3. In order to remove an undesired boryl substitution product of the alkyl halide and protoboration product of the allene for the further purification and the determination of the regio- and stereoselectivities, the oxidation of the boryl groups was performed. The product (Z)-**6ic** was obtained in 52% yield with *E*/*Z* = 10:90, **4**:**5** = >95:5 (67.5 mg, 0.259 mmol, colorless oil).

^{1}H NMR (396 MHz, $CDCl_3$, δ): 0.89 (t, *J* = 6.9 Hz, 3H), 1.13–1.45 (m, 7H), 1.61–1.78 (m, 2H), 1.69 (s, 3H), 2.01–2.23 (m, 4H), 2.58 (t, *J* = 7.7 Hz, 2H), 4.05 (s, 2H), 7.12–7.21 (m, 3H), 7.22–7.32 (m, 2H). ^{13}C NMR (100 MHz, $CDCl_3$, δ): 14.1 (CH_3), 18.2 (CH_3), 22.6 (CH_2), 28.5 (CH_2), 30.6 (CH_2), 30.7 (CH_2), 32.0 (CH_2), 33.7 (CH_2), 35.7 (CH_2), 61.7 (CH_2), 125.7 (*C*H), 128.2 (*C*H), 133.3 (*C*), 133.8 (*C*), 142.3 (*C*). HRMS-EI (*m*/*z*): $[M]^+$ calcd for $C_{18}H_{28}O$, 260.2140; found, 260.2141.

(Z)-3-Methyl-2-(4,4,4-trifluorobutyl)undec-2-en-1-ol [(Z)-6ja, Fig. 4.52]

The reaction was conducted with 83.1 mg (0.500 mmol) of **1j** according to the GP3. In order to remove an undesired boryl substitution product of the alkyl halide and

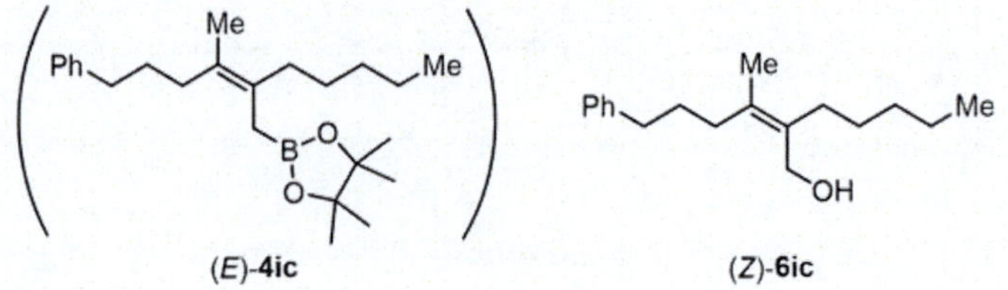

Fig. 4.51 Scheme of (*E*)-**4ic** and (Z)-**6ic**

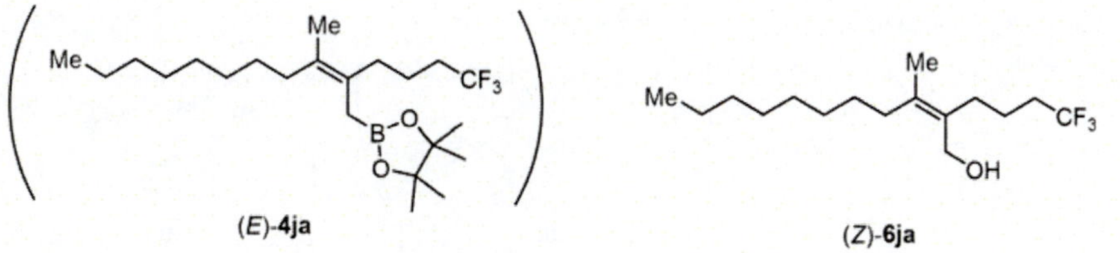

Fig. 4.52 Scheme of (*E*)-**4ja** and (*Z*)-**6ja**

protoboration product of the allene for the further purification and the determination of the regio- and stereoselectivities, the oxidation of the boryl groups was performed. The product (*Z*)-**6ja** was obtained in 64% yield with *E*/*Z* = 8:92, **4**:**5** = >95:5 (94.7 mg, 0.322 mmol, colorless oil).

^{1}H NMR (392 MHz, $CDCl_3$, δ): 0.88 (t, *J* = 6.7 Hz, 3H), 1.13–1.50 (m, 13H), 1.56–1.78 (m, 5H), 1.97–2.16 (m, 4H), 2.22 (t, *J* = 8.0 Hz, 2H), 4.12 (s, 2H). ^{13}C NMR (99 MHz, $CDCl_3$, δ): 14.0 (CH_3), 18.4 (CH_3), 21.0 (CH_2), 22.6 (CH_2), 29.0 (CH_2), 29.2 (CH_2), 29.5 (CH_2), 29.6 (CH_2), 31.8 (CH_2), 33.5 (q, *J* = 28.6 Hz, CH_2), 34.1 (CH_2), 61.4 (CH_2), 127.2 (q, *J* = 277.6 Hz, *C*), 131.1 (*C*), 135.9 (*C*). HRMS-EI (*m*/*z*): $[M]^+$ calcd for $C_{16}H_{29}F_3O$, 294.2171; found, 294.2168.

(Z)-3-Methyl-2-pentylundec-2-en-1-ol [(Z)-6jc, Fig. 4.53]

The reaction was conducted with 83.3 mg (0.501 mmol) of **1j** according to the GP3. In order to remove an undesired boryl substitution product of the alkyl halide and protoboration product of the allene for the further purification and the determination of the regio- and stereoselectivities, the oxidation of the boryl groups was performed. The product (*Z*)-**6jc** was obtained in 46% yield with *E*/*Z* = 6:94, **4**:**5** = >95:5 (58.9 mg, 0.231 mmol, colorless oil).

^{1}H NMR (392 MHz, $CDCl_3$, δ): 0.79–0.98 (m, 6H), 1.11–1.50 (m, 19H), 1.68 (s, 3H), 1.99–2.19 (m, 4H), 4.10 (s, 2H). ^{13}C NMR (99 MHz, $CDCl_3$, δ): 14.1 (CH_3), 18.2 (CH_3), 22.59 (CH_2), 22.63 (CH_2), 28.6 (CH_2), 29.0 (CH_2), 29.3 (CH_2), 29.5 (CH_2), 29.6 (CH_2), 30.8 (CH_2), 31.8 (CH_2), 32.0 (CH_2), 34.1 (CH_2), 61.8 (CH_2), 132.8 (*C*), 134.4 (*C*). HRMS-EI (*m*/*z*): $[M]^+$ calcd for $C_{17}H_{34}O$, 254.2610; found, 254.2605.

2-(1-Phenylbutan-2-ylidene)heptan-1-ol [6kc, Fig. 4.54]

The reaction was conducted with 79.1 mg (0.500 mmol) of **1k** according to the GP4. In order to remove an undesired boryl substitution product of the alkyl halide and

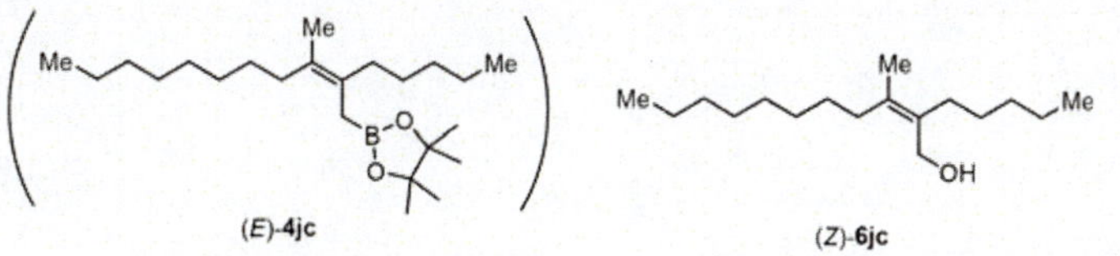

Fig. 4.53 Scheme of (*E*)-**4jc** and (*Z*)-**6jc**

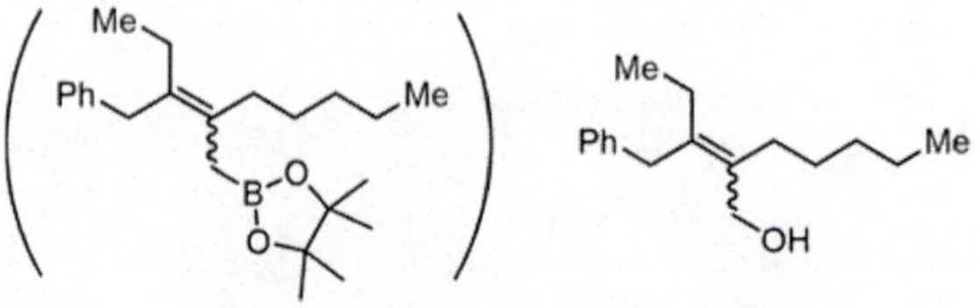

Fig. 4.54 Scheme of **4kc** and **6kc**

protoboration product of the allene for the further purification and the determination of the regio- and stereoselectivities, the oxidation of the boryl groups was performed. The corresponding alcohol product **6kc** was obtained in 67% yield with *E*/*Z* = 50:50, **4**:**5** = >95:5 (82.5 mg, 0.335 mmol, colorless oil).

^{1}H NMR (392 MHz, $CDCl_3$, δ): 0.80–1.04 (m, 6H), 1.17–1.63 (m, 7H), 1.96–2.10 (m, 2H), 2.14–2.31 (m, 2H), 3.44 and 3.50 (s, 2H, isomers), 4.18 (d, *J* = 5.9 Hz, 2H), 7.09–7.21 (m, 3H), 7.21–7.30 (m, 2H). ^{13}C NMR (99 MHz, $CDCl_3$, δ): 13.2 and 14.0 (CH_3, isomers), 14.1 and 14.2 (CH_3, isomers), 22.5 and 22.6 (CH_2, isomers), 24.3 and 24.8 (CH_2, isomers), 28.9 and 29.0 (CH_2, isomers), 30.3 and 30.7 (CH_2, isomers), 32.1 and 32.2 (CH_2, isomers), 36.4 and 36.8 (CH_2, isomers), 61.5 and 62.0 (CH_2, isomers), 125.8 (*C*H), 128.2 and 128.3 (*C*H, isomers), 128.3 and 128.5 (*C*H, isomers), 134.6 and 134.7 (*C*, isomers), 137.8 and 138.1 (*C*, isomers), 140.2 and 140.6 (*C*, isomers). HRMS-EI (*m*/*z*): $[M]^+$ calcd for $C_{17}H_{26}O$, 246.1984; found, 246.1983.

(*E*)-2-(2-Ethyl-3,4,4-trimethylpent-2-en-1-yl)-4,4,5,5-tetramethyl-1,3,2-dioxaborolane [(*E*)-4bb, Fig. 4.55]

The reaction was conducted with 55.1 mg (0.500 mmol) of **1b** according to the GP4. The product (*E*)-**4bb** was obtained in 75% yield with *E*/*Z* = >95:5, **4**:**5** = >95:5 (99.5 mg, 0.374 mmol, colorless oil).

^{1}H NMR (392 MHz, $CDCl_3$, δ): 0.93 (t, *J* = 7.4 Hz, 3H), 1.13 (s, 9H), 1.23 (s, 12H), 1.63 (s, 3H), 1.83 (s, 2H), 2.01 (q, *J* = 7.5 Hz, 2H). ^{13}C NMR (99 MHz, $CDCl_3$, δ): 12.3 (CH_3), 16.6 (CH_3), 19.4 (br, B-CH_2), 24.7 (CH_3), 30.4 (CH_2), 30.7 (CH_3), 35.6 (*C*), 82.9 (*C*), 130.8 (*C*), 133.3 (*C*). ^{11}B NMR (126 MHz, $CDCl_3$, δ): 33.1 (br, s). HRMS-EI (*m*/*z*): $[M]^+$ calcd for $C_{16}H_{31}{}^{11}BO_2$, 266.2420; found, 266.2419.

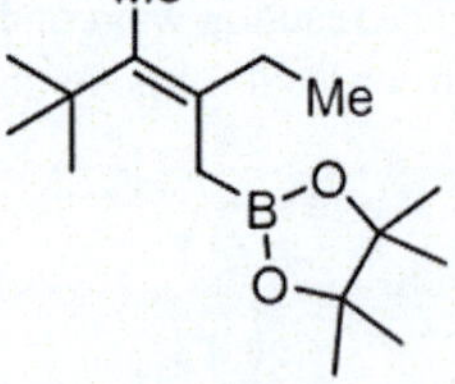

Fig. 4.55 Scheme of (*E*)-**4bb**

(*E*)-2-[2-(3,3-Dimethylbutan-2-ylidene)heptyl]-4,4,5,5-tetramethyl-1,3,2-dioxaborolane [(*E*)-4bc, Fig. 4.56]

The reaction was conducted with 55.3 mg (0.502 mmol) of **1b** according to the GP4. The product (*E*)-**4bc** was obtained in 83% yield with *E*/*Z* = >95:5, **4**:**5** = >95:5 (128.3 mg, 0.416 mmol, colorless oil).

^{1}H NMR (396 MHz, $CDCl_3$, δ): 0.89 (t, *J* = 6.9 Hz, 3H), 1.13 (s, 9H), 1.19–1.41 (m, 6H), 1.23 (s, 12H), 1.62 (s, 3H), 1.84 (s, 2H), 1.93–2.01 (m, 2H). ^{13}C NMR (100 MHz, $CDCl_3$, δ): 14.1 (CH_3), 16.9 (CH_3), 20.0 (br, B-CH_2), 22.7 (CH_2), 24.7 (CH_3), 27.8 (CH_2), 30.7 (CH_3), 32.3 (CH_2), 35.6 (*C*), 37.7 (CH_2), 82.9 (*C*), 129.7 (*C*), 133.6 (*C*). ^{11}B NMR (126 MHz, $CDCl_3$, δ): 33.1 (br, s). HRMS-EI (*m*/*z*): $[M]^+$ calcd for $C_{19}H_{37}{}^{11}BO_2$, 308.2890; found, 308.2886.

(*E*)-2-[2-(3,3-Dimethylbutan-2-ylidene)dodecyl]-4,4,5,5-tetramethyl-1,3,2-dioxaborolane [(*E*)-4be, Fig. 4.57]

The reaction was conducted with 55.0 mg (0.499 mmol) of **1b** according to the GP4. The product (*E*)-**4be** was obtained in 88% yield with *E*/*Z* = >95:5, **4**:**5** = >95:5 (166.9 mg, 0.441 mmol, colorless oil).

^{1}H NMR (392 MHz, $CDCl_3$, δ): 0.88 (t, *J* = 7.1 Hz, 3H), 1.13 (s, 9H), 1.18–1.44 (m, 16H), 1.23 (s, 12H), 1.62 (s, 3H), 1.83 (s, 2H), 1.96 (t, *J* = 7.8 Hz, 2H). ^{13}C NMR (99 MHz, $CDCl_3$, δ): 14.1 (CH_3), 16.9 (CH_3), 20.0 (br, B-CH_2), 22.7 (CH_2), 24.7 (CH_3), 28.1 (CH_2), 29.3 (CH_2), 29.6 (CH_2), 29.7 (CH_2), 30.0 (CH_2), 30.7 (CH_3), 31.9 (CH_2), 35.6 (*C*), 37.7 (*C*), 82.9 (*C*), 129.7 (*C*), 133.6 (*C*). ^{11}B NMR (126 MHz, $CDCl_3$, δ): 33.3 (br, s). HRMS-EI (*m*/*z*): $[M]^+$ calcd for $C_{24}H_{47}{}^{11}BO_2$, 378.3674; found, 378.3666.

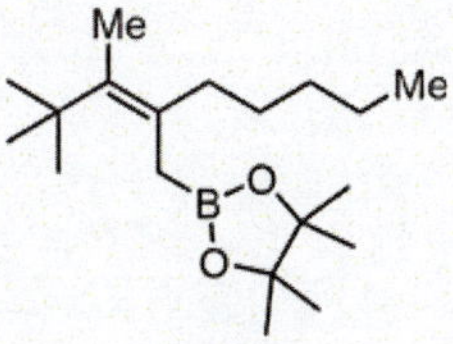

(*E*)-**4bc**

Fig. 4.56 Scheme of (*E*)-**4bc**

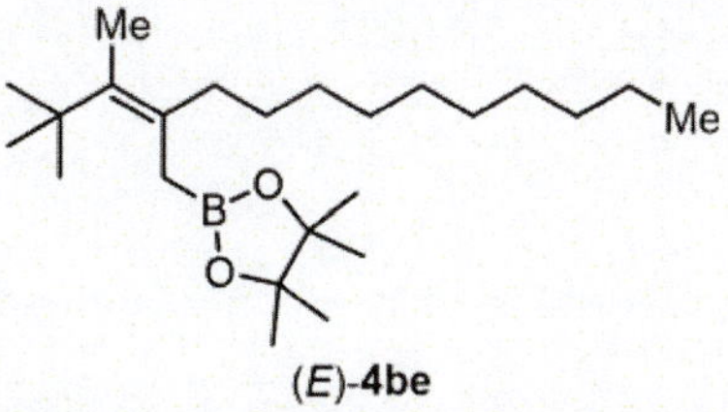

(*E*)-**4be**

Fig. 4.57 Scheme of (*E*)-**4be**

(*E*)-2-(2-Isobutyl-3,4,4-trimethylpent-2-en-1-yl)-4,4,5,5-tetramethyl-1,3,2-dioxaborolane [(*E*)-4bf, Fig. 4.58]

The reaction was conducted with 55.2 mg (0.501 mmol) of **1b** according to the GP4. The product (*E*)-**4bf** was obtained in 81% yield with *E*/*Z* = 91:9, **4**:**5** = >95:5 (119.7 mg, 0.406 mmol, colorless oil).

^{1}H NMR (392 MHz, $CDCl_3$, δ): 0.86 (d, *J* = 6.3 Hz, 6H), 1.15 (s, 9H), 1.23 (s, 12H), 1.64 (s, 3H), 1.69–1.83 (m, 1H), 1.87 (s, 2H), 1.95 (d, *J* = 7.4 Hz, 2H). ^{13}C NMR (99 MHz, $CDCl_3$, δ): 17.6 (CH_3), 20.3 (br, B-CH_2), 22.5 (CH_3), 24.7 (CH_3), 27.6 (*C*H), 30.9 (CH_3), 35.8 (*C*), 45.8 (CH_2), 82.8 (*C*), 128.8 (*C*), 134.8 (*C*). ^{11}B NMR (126 MHz, $CDCl_3$, δ): 33.1 (br, s). HRMS-EI (*m*/*z*): $[M]^+$ calcd for $C_{18}H_{35}{}^{11}BO_2$, 294.2733; found, 294.2732.

(*E*)-2-(2-Benzyl-3,4,4-trimethylpent-2-en-1-yl)-4,4,5,5-tetramethyl-1,3,2-dioxaborolane [(*E*)-4bg, Fig. 4.59]

The reaction was conducted with 55.0 mg (0.500 mmol) of **1b** according to the GP4. The product (*E*)-**4bg** was obtained in 55% yield with *E*/*Z* = >95:5, **4**:**5** = >95:5 (90.3 mg, 0.275 mmol, colorless oil).

^{1}H NMR (399 MHz, $CDCl_3$, δ): 1.21 (s, 9H), 1.22 (s, 12H), 1.69 (s, 3H), 1.81 (s, 2H), 3.46 (s, 2H), 7.10–7.18 (m, 3H), 7.23–7.28 (m, 2H). ^{13}C NMR (100 MHz, $CDCl_3$, δ): 17.7 (CH_3), 20.2 (br, B–CH_2), 24.7 (CH_3), 30.7 (CH_3), 35.8 (*C*), 42.2 (CH_2), 82.9 (*C*), 125.5 (*C*H), 127.5 (*C*), 128.2 (*C*H), 128.2 (*C*H), 136.6 (*C*), 140.7 (*C*). ^{11}B NMR (126 MHz, $CDCl_3$, δ): 32.9 (br, s). HRMS-EI (*m*/*z*): $[M]^+$ calcd for $C_{21}H_{33}{}^{11}BO_2$, 328.2577; found, 328.2578.

(*E*)-2-[2-(3-Methoxypropyl)-3,4,4-trimethylpent-2-en-1-yl]-4,4,5,5-tetramethyl-1,3,2-dioxaborolane [(*E*)-4bh, Fig. 4.60]

The reaction was conducted with 55.1 mg (0.500 mmol) of **1b** according to the GP4. The product (*E*)-**4bh** was obtained in 74% yield with *E*/*Z* = >95:5, **4**:**5** = >95:5 (114.8 mg, 0.370 mmol, colorless oil).

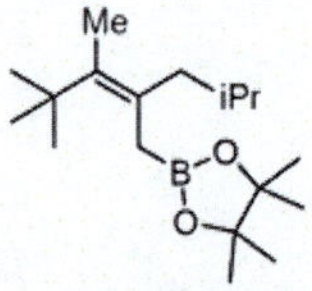

(*E*)-**4bf**

Fig. 4.58 Scheme of (*E*)-**4bf**

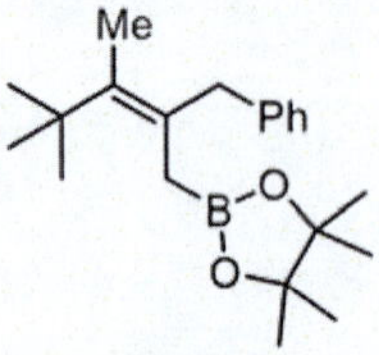

(*E*)-**4bg**

Fig. 4.59 Scheme of (*E*)-**4bg**

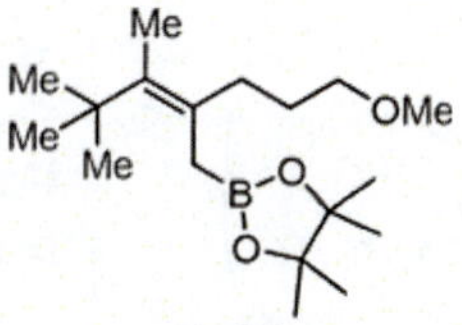

(E)-4bh

Fig. 4.60 Scheme of (*E*)-**4bh**

^{1}H NMR (392 MHz, $CDCl_3$, δ): 1.13 (s, 9H), 1.23 (s, 12H), 1.57–1.71 (m, 2H), 1.63 (s, 3H), 1.83 (s, 2H), 2.05 (t, $J = 7.8$ Hz, 2H), 3.32 (s, 3H), 3.35 (t, $J = 6.5$ Hz, 2H). ^{13}C NMR (99 MHz, $CDCl_3$, δ): 16.8 (CH_3), 19.9 (br, B-CH_2), 24.6 (CH_3), 27.9 (CH_2), 30.7 (CH_3), 33.7 (CH_2), 35.6 (C), 58.3 (CH_3), 72.7 (CH_2), 82.9 (C), 128.7 (C), 134.3 (C). ^{11}B NMR (126 MHz, $CDCl_3$, δ): 33.1 (br, s). HRMS-ESI (*m*/*z*): [M + Na]$^+$ calcd for $C_{18}H_{35}{}^{11}BO_3Na$, 333.2575; found, 333.2569.

(Z)-6-[(*tert*-Butyldimethylsilyl)oxy]-2-(3,3-dimethylbutan-2-ylidene)hexan-1-ol [(Z)-6bi, Fig. 4.61]

The reaction was conducted with 55.3 mg (0.502 mmol) of **1b** according to the GP4. In order to remove an undesired boryl substitution product of the alkyl halide and protoboration product of the allene for the further purification and the determination of the regio- and stereoselectivities, the oxidation of the boryl groups was performed. The corresponding alcohol product (*Z*)-**6bi** was obtained in 44% yield with *E*/*Z* = <5:95, **4**:**5** = >95:5 (69.6 mg, 0.221 mmol, colorless oil).

^{1}H NMR (396 MHz, $CDCl_3$, δ): 0.05 (s, 6H), 0.89 (s, 9H), 1.18 (s, 9H), 1.35–1.47 (m, 2H), 1.48–1.62 (m, 3H), 1.68 (s, 3H), 2.13 (t, $J = 8.1$ Hz, 2H), 3.62 (t, $J = 6.3$ Hz, 2H), 4.23 (d, $J = 5.9$ Hz, 2H). ^{13}C NMR (99 MHz, $CDCl_3$, δ): –5.3 (CH_3), 17.1 (CH_3), 18.3 (C), 25.0 (CH_2), 25.9 (CH_3), 31.7 (CH_3), 32.88 (CH_2), 32.95 (CH_2), 36.0 (C), 62.5 (CH_2), 63.0 (CH_2), 133.1 (C), 142.1 (C). HRMS-EI (*m*/*z*): [M –*t*Bu]$^+$ calcd for $C_{14}H_{29}O_2Si$, 257.1937; found, 257.1938.

(Z)-2-(3-Chloropropyl)-3,4,4-trimethylpent-2-en-1-ol [(Z)-6bj, Fig. 4.62]

The reaction was conducted with 55.1 mg (0.500 mmol) of **1b** according to the GP4. In order to remove an undesired boryl substitution product of the alkyl halide and protoboration product of the allene for the further purification and the determination

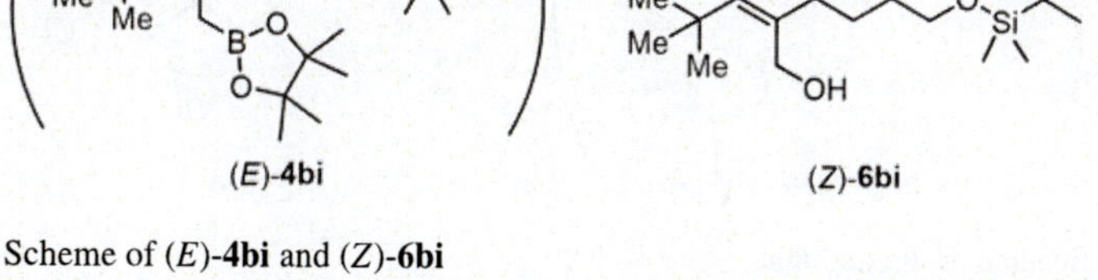

(E)-4bi (Z)-6bi

Fig. 4.61 Scheme of (*E*)-**4bi** and (*Z*)-**6bi**

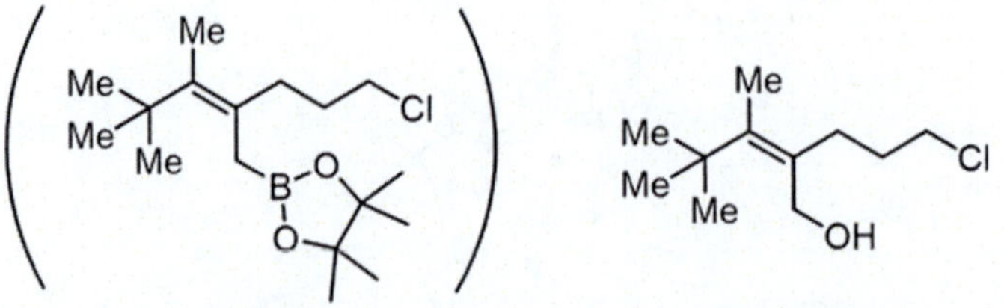

(*E*)-**4bj** (*Z*)-**6bj**

Fig. 4.62 Scheme of (*E*)-**4bj** and (*Z*)-**6bj**

of the regio- and stereoselectivities, the oxidation of the boryl groups was performed. The corresponding alcohol product (*Z*)-**6bj** was obtained in 51% yield with E/Z = <5:95, **4**:**5** = >95:5 (52.3 mg, 0.255 mmol, colorless oil). The selectivity of C–C bond formation at chloride and iodate was 86:14 (determined by ^{1}H NMR analysis).

^{1}H NMR (396 MHz, $CDCl_3$, δ): 1.19 (s, 9H), 1.30 (br-s, 1H), 1.71 (s, 3H), 1.79–1.98 (m, 2H), 2.19–2.35 (m, 2H), 3.54 (t, J = 6.5 Hz, 2H), 4.25 (s, 2H). ^{13}C NMR (100 MHz, $CDCl_3$, δ): 17.2 (CH_3), 30.5 (CH_2), 31.7 (CH_3), 31.9 (CH_2), 36.1 (C), 45.2 (CH_2), 62.6 (CH_2), 131.6 (C), 143.2 (C). HRMS-EI (m/z): $[M]^+$ calcd for $C_{11}H_{21}ClO$, 204.1281; found, 204.1281.

6,6,6-Trifluoro-2-(4-phenylcyclohexylidene)hexan-1-ol (6la, Fig. 4.63)

The reaction was conducted with 92.1 mg (0.500 mmol) of **1l** according to the GP4. In order to remove an undesired boryl substitution product of the alkyl halide and protoboration product of the allene for the further purification and the determination of the regio- and stereoselectivities, the oxidation of the boryl groups was performed. The corresponding alcohol product **6la** was obtained in 69% yield with **4**:**5** = >95:5 (107.3 mg, 0.343 mmol, colorless oil).

^{1}H NMR (392 MHz, $CDCl_3$, δ): 1.05–1.20 (m, 1H), 1.41–1.60 (m, 2H), 1.62–1.74 (m, 2H), 1.91–2.18 (m, 6H), 2.28 (t, J = 8.1 Hz, 2H), 2.68–2.80 (m, 2H), 2.84–2.93 (m, 1H), 4.14–4.24 (m, 2H), 7.16–7.23 (m, 3H), 7.25–7.33 (m, 2H). ^{13}C NMR (99 MHz, $CDCl_3$, δ): 21.7 (CH_2), 29.5 (CH_2), 30.1 (CH_2), 30.3 (CH_2), 33.5 (q, J = 28.6 Hz, CH_2), 35.4 (CH_2), 35.6 (CH_2), 44.5 (CH), 61.6 (CH_2), 126.0 (CH), 126.7 (CH), 127.2 (q, J = 277.9 Hz, C), 128.3 (CH), 128.6 (C), 138.7 (C), 146.4 (C). HRMS-EI (m/z): $[M]^+$ calcd for $C_{18}H_{23}F_3O$, 312.1701; found, 312.1703.

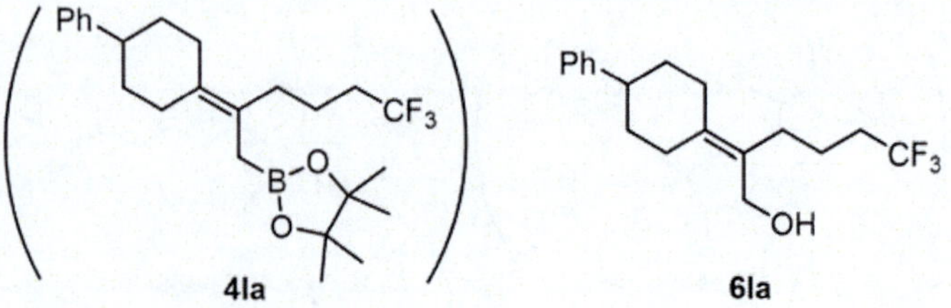

4la **6la**

Fig. 4.63 Scheme of **4la** and **6la**

4,4,5,5-Tetramethyl-2-[(*E*)-6,6,6-trifluoro-2-{(8*S*,9*S*,13*S*,14*S*)-3-methoxy-13-methyl-6,7,8,9,11,12,13,14,15,16-decahydro-17*H*-cyclopenta[a]phenanthren-17-ylidene}hexyl]-1,3,2-dioxaborolane [(*E*)-4ma, Fig. 4.64]

The reaction was conducted with 147.2 mg (0.500 mmol) of **1m** according to the GP4. The product (*E*)-**4ma** was obtained in 90% yield with *E*/*Z* = >95:5, **4**:**5** = >95:5 (240.6 mg, 0.452 mmol, colorless oil).

^{1}H NMR (396 MHz, $CDCl_3$, δ): 0.87 (s, 3H), 1.02–1.82 (m, 10H), 1.24 (s, 12H), 1.83–2.39 (m, 11H), 2.78–2.93 (m, 2H), 3.76 (s, 3H), 6.62 (d, *J* = 2.8 Hz, 1H), 6.70 (dd, *J* = 8.7 Hz, 2.8 Hz, 1H), 7.20 (d, *J* = 8.3 Hz, 1H). ^{13}C NMR (100 MHz, $CDCl_3$, δ): 16.1 (br, B-*C*H_2), 16.4 (*C*H_3), 19.6 (*C*H_2), 24.0 (*C*H_2), 24.6 (*C*H_3), 24.7 (*C*H_3), 27.0 (*C*H_2), 27.5 (*C*H_2), 29.5 (*C*H_2), 29.8 (*C*H_2), 33.4 (q, *J* = 28.1 Hz, *C*H_2), 35.3 (*C*H_2), 37.7 (*C*H_2), 38.2 (*C*H), 43.5 (*C*H), 44.5 (*C*), 55.0 (*C*H), 55.1 (*C*H), 83.0 (*C*), 111.3 (*C*H), 113.6 (*C*H), 123.6 (*C*), 126.1 (*C*H), 127.4 (q, *J* = 277.5 Hz, *C*), 132.8 (*C*), 137.9 (*C*), 143.3 (*C*), 157.3 (*C*). ^{11}B NMR (126 MHz, $CDCl_3$, δ): 33.0 (br, s). HRMS-EI (*m*/*z*): $[M]^+$ calcd for $C_{31}H_{44}{}^{11}BF_3O_3$, 532.3341; found, 532.3322.

2-{(*E*)-2-[(8*S*,9*S*,13*S*,14*S*)-3-Methoxy-13-methyl-6,7,8,9,11,12,13,14,15,16-decahydro-17H-cyclopenta[a]phenanthren-17-ylidene]butyl}-4,4,5,5-tetramethyl-1,3,2-dioxaborolane [(*E*)-4mb, Fig. 4.65]

The reaction was conducted with 147.2 mg (0.500 mmol) of **1m** according to the GP4. The product (*E*)-**4mb** was obtained in 78% yield with *E*/*Z* = >95:5, **4**:**5** = >95:5 (176.2 mg, 0.391 mmol, colorless oil).

^{1}H NMR (392 MHz, $CDCl_3$, δ): 0.86 (s, 3H), 0.94 (t, *J* = 7.4 Hz, 3H), 1.16–1.59 (m, 5H), 1.25 (s, 12H), 1.62–1.80 (m, 3H), 1.84–2.05 (m, 4H), 2.14–2.39 (m, 5H), 2.77–2.97 (m, 2H), 3.77 (s, 3H), 6.63 (d, *J* = 2.7 Hz, 1H), 6.70 (dd, *J* = 8.2, 2.7 Hz, 1H), 7.21 (d, *J* = 9.0 Hz, 1H). ^{13}C NMR (99 MHz, $CDCl_3$, δ): 11.8 (*C*H_3), 15.8 (br, B–*C*H_2), 16.5 (*C*H_3), 24.1 (*C*H_2), 24.6 (*C*H_3), 24.8 (*C*H_3), 27.1 (*C*H_2), 27.6 (*C*H_2), 29.06 (*C*H_2), 29.09 (*C*H_2), 29.9 (*C*H_2), 37.8 (*C*H_2), 38.2 (*C*H), 43.6 (*C*H), 44.2 (*C*), 55.1 (*C*H), 55.2 (*C*H_3), 82.9 (*C*), 111.4 (*C*H), 113.6 (*C*H), 126.2 (*C*H), 126.7 (*C*), 133.0 (*C*), 138.0 (*C*), 141.1 (*C*), 157.3 (*C*). ^{11}B NMR (126 MHz, $CDCl_3$, δ): 33.0 (br, s). HRMS-EI (*m*/*z*): $[M]^+$ calcd for $C_{29}H_{43}{}^{11}BO_3$, 450.3310; found, 450.3306.

Fig. 4.64 Scheme of (*E*)-**4ma**

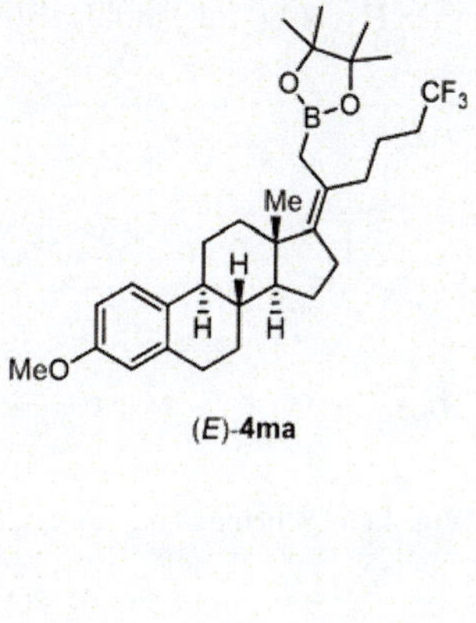

(*E*)-**4ma**

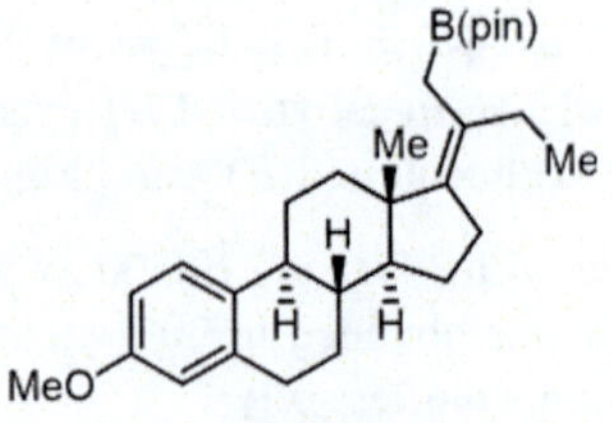

(*E*)-**4mb**

Fig. 4.65 Scheme of (*E*)-**4mb**

4.4.6 Large Scale Reaction and Synthetic Applications

4.4.6.1 Gram-Scale Reaction (Fig. 4.66)

CuCl (29.7 mg, 0.30 mmol), bis(pinacolato)diboron (**3**) (1.83 g, 7.20 mmol), Xantphos (174 mg, 0.300 mmol) and K(O-*t*-Bu) (808 mg, 7.20 mmol) were placed in a round-bottomed flask under a nitrogen atmosphere. Dry DMF (12.0 mL) was added to the flask through the rubber septum using a syringe. After stirring for 30 min at room temperature, **1h** (914 mg, 6.00 mmol) and **2c** (2.38 g, 12.0 mmol) were added to the mixture. Then, the reaction mixture was heated to 30 °C. After the reaction was completed, the reaction mixture was quenched with water and extracted with EtOAc three times. The combined organic layer was dried over $MgSO_4$. After filtration, the solvents were removed under reduced pressure. The crude material was purified by flash column chromatography (SiO_2, Et_2O/hexane, 0:100–3:97) to give the corresponding borylation product (*E*)-**4hc** in 88% yield with *E*/*Z* = >95:5, **4**:**5** = >95:5 (1.86 g, 5.29 mmol, colorless oil).

4.4.6.2 Synthesis of (Z)-3-(*tert*-Butyl)-2-Pentylhept-2-en-1-ol [(Z)-6hc, Fig. 4.67]

In a reaction vial, (*E*)-**4hc** (70.1 mg, 0.200 mmol) was dissolved in THF (2.0 mL), then H_2O (2.0 mL) and $NaBO_3{\cdot}4H_2O$ (300 mg, 0.80 mmol) were subsequently added

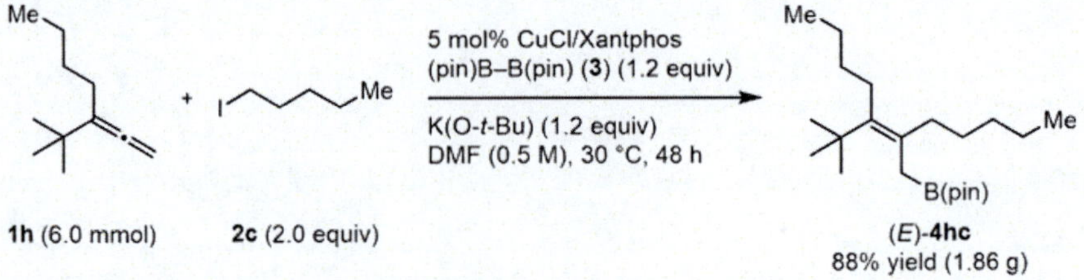

Fig. 4.66 Scheme of a gram-scale reaction

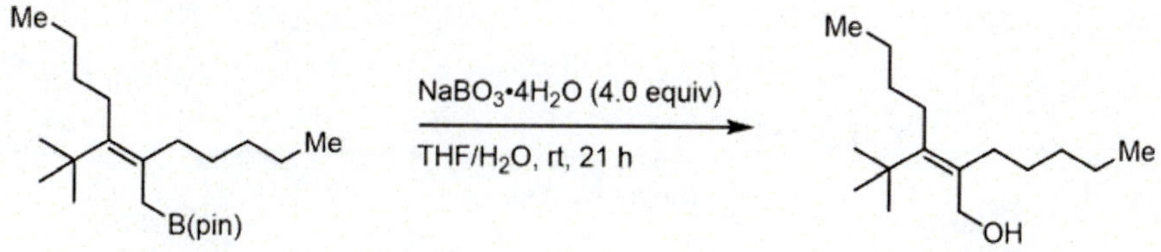

(*E*)-**4hc**

(*Z*)-**6hc**
89% yield

Fig. 4.67 Oxidation of borylation product

to the reaction mixture at rt under air. The reaction mixture was stirred at rt for 21 h. The reaction mixture was then quenched by H_2O and extracted three times with EtOAc. The combined organic layer was dried over $MgSO_4$ followed by filtration. The crude material was purified by silica gel chromatography (EtOAc/hexane, 0:100–8:92) to give the corresponding alcohol (Z)-**6hc** in 89% yield with $E/Z = <5:95$ (42.9 mg, 0.178 mmol, colorless oil).

^{1}H NMR (392 MHz, $CDCl_3$, δ): 0.86–0.97 (m, 6H), 1.20 (s, 9H), 1.24–1.45 (m, 11H), 2.01–2.12 (m, 4H), 4.26 (s, 2H). ^{13}C NMR (99 MHz, $CDCl_3$, δ): 13.9 (*C*H_3), 14.1 (*C*H_3), 22.6 (*C*H_2), 23.4 (*C*H_2), 29.2 (*C*H_2), 30.8 (*C*H_2), 32.1 (*C*H_2), 32.4 (*C*H_3 and *C*H_2), 33.4 (*C*H_2), 35.8 (*C*), 62.3 (*C*H_2), 134.2 (*C*), 146.2 (*C*). HRMS-EI (*m*/*z*): $[M]^+$ calcd for $C_{16}H_{32}O$, 240.2453; found, 240.2454.

4.4.6.3 Synthesis of *tert*-Butyl (Z)-(3-(*tert*-Butyl)-2-Pentylhept-2-en-1-yl)Carbamate [(Z)-7hc, Fig. 4.68]

According to the literature procedure, the amination of (*E*)-**4hc** and preparation of $MeONH_2$ solution in THF (0.80 M) were conducted [12]. In an oven-dried reaction vial, $MeONH_2$ (0.80 M in THF, 751 μL, 0.60 mmol) was diluted with THF (600 μL). After the mixture was cooled to −78 °C, *n*-BuLi (1.57 M in hexane, 375 μL, 0.60 mmol) was added dropwise under a nitrogen atmosphere. Then (*E*)-**4hc** (70.1 mg, 0.20 mmol) in THF (400 μL) was added dropwise to the solution and stirred at 60 °C. After 48 h, $(Boc)_2O$ (138 μL, 0.60 mmol) was added to the mixture and stirred for 2 h at rt. The mixture was then quenched by addition of H_2O and extracted three times with EtOAc. The combined organic layer was dried over $MgSO_4$ followed by filtration and evaporation. After purification by flash silica gel column chromatography (Et_2O/hexane, 0:100–4:96), (*Z*)-**7hc** was further purified by oxidation reaction with $NaBO_3 \cdot 4H_2O$ (300 mg, 0.80 mmol) in THF (2.0 mL) and H_2O (2.0 mL) to remove unreacted starting material. After the oxidation reaction, the crude material was purified by silica gel chromatography (Et_2O/hexane, 0:100–4:96) to give the pure (Z)-**7hc** in 46% yield with $E/Z = <5:95$ (30.8 mg, 0.091 mmol, colorless oil).

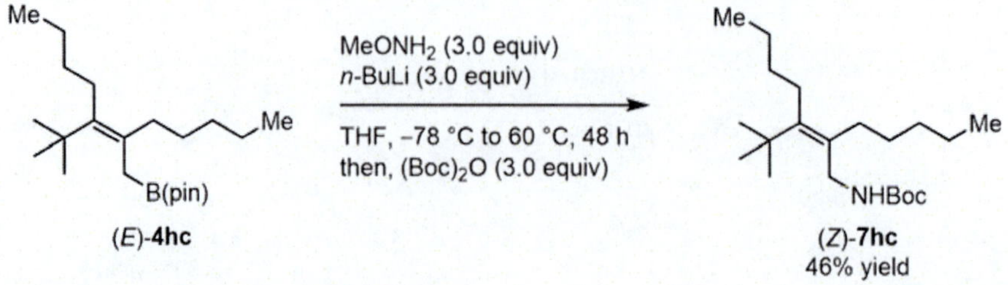

Fig. 4.68 Amination of borylation product

^{1}H NMR (392 MHz, $CDCl_3$, δ): 0.84–0.95 (m, 6H), 1.17 (s, 9H), 1.22–1.41 (m, 10H), 1.45 (s, 9H), 1.89–1.97 (m, 2H), 1.98–2.08 (m, 2H), 3.88 (d, J = 5.4 Hz, 2H), 4.22–4.34 (m, 1H). ^{13}C NMR (99 MHz, $CDCl_3$, δ): 13.9 (CH_3), 14.1 (CH_3), 22.5 (CH_2), 23.4 (CH_2), 28.4 (CH_3), 29.1 (CH_2), 30.8 (CH_2), 32.1 (CH_3), 32.3 (CH_2), 32.4 (CH_2), 33.5 (CH_2), 35.6 (C), 42.1 (CH_2), 79.1 (C), 131.3 (C), 146.1 (C), 156.0 (C). HRMS-ESI (m/z): $[M + Na]^+$ calcd for $C_{21}H_{41}O_2NNa$, 362.3030; found, 362.3029.

4.4.6.4 Synthesis of (*E*)-2-(4-(*tert*-Butyl)-3-Pentyloct-3-en-1-yl)-4,4,5,5-Tetramethyl-1,3,2-Dioxaborolane [(*E*)-8hc, Fig. 4.69]

Homologation reaction of (*E*)-**4hc** was performed according to the literature procedure [96]. In an oven-dried reaction vial, (*E*)-**4hc** (70.1 mg, 0.200 mmol) and $BrCH_2Cl$ (53.6 μL, 0.800 mmol) were dissolved in THF (1.2 mL) in a nitrogen atmosphere, and the mixture was cooled to –78 °C. *n*-BuLi (1.6 M in hexane, 375 μL, 0.600 mmol) was then added dropwise to the reaction mixture, and the mixture was stirred at rt for 21 h. The mixture was then quenched by the addition of a saturated aqueous NH_4Cl solution and extracted three times with Et_2O. Next, the combined organic layer was washed with brine, dried over $MgSO_4$ followed by filtration. After evaporation, the crude material was purified by silica gel chromatography (Et_2O/hexane, 0:100–3:97) to give the corresponding boronate (*E*)-**8hc** in 73% yield with E/Z = >95:5 (52.8 mg, 0.145 mmol, colorless oil).

^{1}H NMR (392 MHz, $CDCl_3$, δ): 0.81–0.94 (m, 8H), 1.16 (s, 9H), 1.20–1.42 (m, 10H), 1.25 (s, 12H), 1.87–1.94 (m, 2H), 1.94–2.02 (m, 2H), 2.19–2.27 (m, 2H). ^{13}C NMR (99 MHz, $CDCl_3$, δ): 11.1 (br, B-CH_2), 14.0 (CH_3), 14.1 (CH_3), 22.7 (CH_2), 23.4 (CH_2), 24.8 (CH_3), 26.9 (CH_2), 29.2 (CH_2), 30.7 (CH_2), 31.9 (CH_3), 32.5 (CH_2), 32.7 (CH_2), 33.8 (CH_2), 35.5 (C), 82.8 (C), 137.1 (C), 139.4 (C). ^{11}B NMR (126 MHz, $CDCl_3$, δ): 33.7 (br, s). HRMS-EI (m/z): $[M]^+$ calcd for $C_{23}H_{45}{}^{11}BO_2$, 364.3517; found, 364.3516.

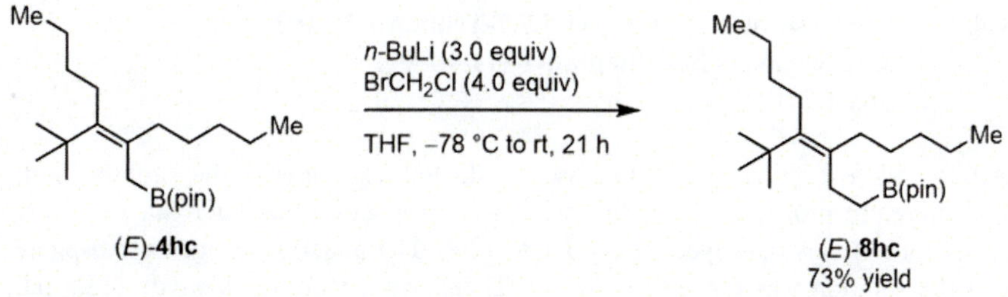

Fig. 4.69 Homologation of borylation product

4.4.7 Procedure of Allylboration Reactions

4.4.7.1 Synthesis of 2-(*tert*-Butyl)-2-Butyl-3-Methyleneoctan-1-ol (9a, Fig. 4.70)

Allylboration reaction of (*E*)-**4hc** was conducted according to the slightly modified literature procedure [65]. In an oven-dried reaction vial, (*E*)-**4hc** (70.1 mg, 0.200 mmol) and paraformaldehyde (30.0 mg, 1.00 mmol) were dissolved in THF (1.0 mL) under a nitrogen atmosphere. The reaction mixture was stirred at 65 °C for 48 h. The mixture was then quenched by the addition of triethanolamine (10% solution in DCM) and extracted three times with DCM. The combined organic layer was washed with brine, dried over $MgSO_4$ followed by filtration. After evaporation, the crude material was purified by silica gel chromatography (Et_2O/hexane, 0:100–5:95) to give the corresponding homoallyl alcohol **9a** in 90% yield (45.8 mg, 0.180 mmol, colorless oil).

^{1}H NMR (392 MHz, $CDCl_3$, δ): 0.92 (s, 9H), 0.80–0.98 (m, 6H), 1.15–1.40 (m, 9H), 1.41–1.56 (m, 2H), 1.60–1.72 (m, 2H), 1.98 (t, $J = 7.8$ Hz, 2H), 3.76–3.93 (m, 2H), 4.83 (s, 1H), 5.16 (s, 1H). ^{13}C NMR (99 MHz, $CDCl_3$, δ): 14.1 (CH_3), 14.2 (CH_3), 22.7 (CH_2), 24.1 (CH_2), 27.5 (CH_3), 27.9 (CH_2), 29.2 (CH_2), 30.3 (CH_2), 32.2 (CH_2), 33.5 (CH_2), 36.4 (*C*), 51.7 (*C*), 64.8 (CH_2), 112.5 (CH_2), 150.9 (*C*). HRMS-FD (*m*/*z*): $[M]^+$ calcd for $C_{17}H_{34}O$, 254.2610; found, 254.2605.

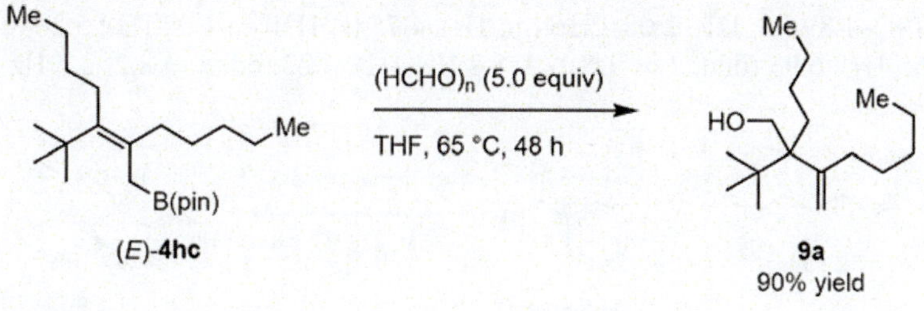

Fig. 4.70 Scheme of arylboration using formaldehyde

4.4.7.2 Synthesis of 2-Cyclohexyl-7,7,7-Trifluoro-2-Methyl-3-Methylene-1-Phenylheptan-1-ol (9b, Fig. 4.71)

Allylboration reaction of (*E*)-**4aa** was conducted according to the slightly modified literature procedure [68]. In an oven-dried reaction vial, (*E*)-**4aa** (38.3 mg, 0.102 mmol) was dissolved in dry DCM (200 μL) under a nitrogen atmosphere, and the solution was cooled to –78 °C. Then, corresponding aldehyde (15.5 μL, 0.15 mmol) was added to the reaction mixture, followed by the addition of $BF_3 \cdot OEt_2$ (12.8 μL, 0.10 mmol). After stirring for 3 h at –78 °C, the reaction mixture was quenched by the addition of triethanolamine (10% solution in DCM) and extracted three times with DCM. Next, the combined organic layer was washed with brine, dried over $MgSO_4$ followed by filtration. After evaporation, the crude material was purified by silica gel chromatography (EtOAc/hexane, 0:100–3:97) to give the corresponding homoallyl alcohol **9b** in 75% yield with dr = >95:5 (27.1 mg, 0.0765 mmol, colorless oil).

^1H NMR (600 MHz, $CDCl_3$, δ): 1.08–1.44 (m, 6H), 1.16 (s, 3H), 1.46–1.61 (m, 3H), 1.62–1.92 (m, 8H), 2.02–2.10 (m, 1H), 4.88–4.98 (m, 3H), 7.19–7.31 (m, 5H). ^{13}C NMR (99 MHz, $CDCl_3$, δ): 15.4 (CH_3), 20.1 (CH_2), 26.9 (CH_2), 27.19 (CH_2), 27.22 (CH_2), 28.7 (CH_2), 29.0 (CH_2), 32.2 (CH_2), 33.5 (q, J = 28.6 Hz, CH_2), 44.3 (CH), 51.3 (C), 77.0 (CH), 112.6 (CH_2), 127.1 (q, J = 277.6 Hz, C), 127.2 (CH), 127.4 (CH), 127.5 (CH), 142.6 (C), 149.8 (C). HRMS-ESI (*m*/*z*): $[M–H]^-$ calcd for $C_{21}H_{28}OF_3$, 353.2098; found, 353.2099.

4.4.7.3 Synthesis of 1-(2-Bromo-4-Fluorophenyl)-2-Cyclohexyl-2-Methyl-3-Methylenepentan-1-ol (9c, Fig. 4.72)

The allylboration reaction was conducted with 58.6 mg (0.200 mmol) of (*E*)-**4ab** according to the procedure described above. The product **9c** was obtained in 60% yield with dr = >95:5 (44.5 mg, 0.120 mmol, colorless oil).

^1H NMR (399 MHz, $CDCl_3$, δ): 0.67 (t, J = 7.3 Hz, 3H), 1.01 (qd, J = 12.3, 3.3 Hz, 1H), 1.11–1.40 (m, 4H), 1.28 (s, 3H), 1.53–1.60 (m, 1H), 1.63–1.78 (m, 6H), 1.82–1.89 (m, 1H), 2.08–2.16 (m, 1H), 4.75 (s, 1H), 4.91 (s, 1H), 5.26 (d, J = 4.1 Hz, 1H), 6.95 (ddd, J = 9.6, 6.9, 1.8 Hz, 1H), 7.22 (dd, J = 8.2, 2.7 Hz, 1H),

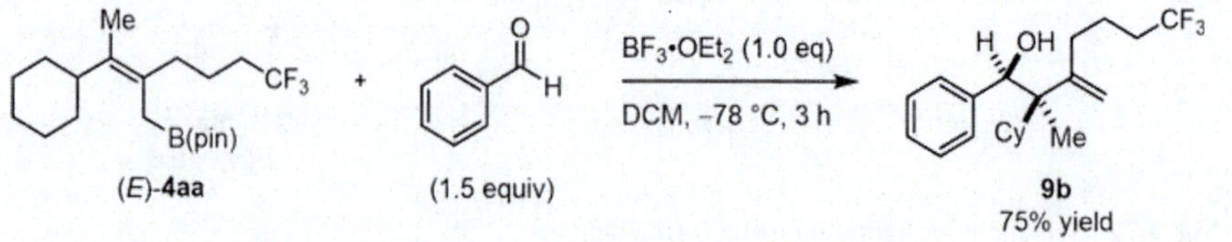

Fig. 4.71 Scheme of arylboration using an aryl aldehyde

Fig. 4.72 Scheme of **9c**

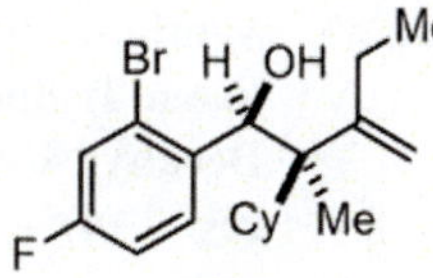

9c

7.56 (dd, $J = 8.7, 6.4$ Hz, 1H). ^{13}C NMR (100 MHz, $CDCl_3$, δ): 12.2 (CH_3), 13.7 (CH_3), 25.0 (CH_2), 26.8 (CH_2), 27.3 (CH_2), 27.5 (CH_2), 28.4 (CH_2), 28.8 (CH_2), 46.8 (*C*H), 51.5 (*C*), 76.6 (*C*H), 110.3 (CH_2), 114.3 (d, $J = 20.1$ Hz, *C*H), 119.5 (d, $J = 24.0$ Hz, *C*H), 124.1 (d, $J = 8.6$ Hz, *C*), 130.9 (d, $J = 8.6$ Hz, *C*H), 138.6 (d, $J = 3.8$ Hz, *C*), 153.2 (*C*), 161.1 (d, $J = 251.1$ Hz, *C*). HRMS-ESI (*m*/*z*): $[M–H]^-$ calcd for $C_{19}H_{25}OBrF$, 367.1078; found, 367.1082.

4.4.7.4 Synthesis of 2-Cyclohexyl-2-Methyl-3-Methylene-1-(Naphthalen-2-yl)Pentan-1-ol (9d, Fig. 4.73)

The allylboration reaction was conducted with 120.8 mg (0.413 mmol) of (*E*)-**4ab** according to the procedure described above. The product **9d** was obtained in 52% yield with dr = >95:5 (69.7 mg, 0.216 mmol, white solid).

^{1}H NMR (392 MHz, $CDCl_3$, δ): 0.81 (t, $J = 7.1$ Hz, 3H), 1.10–1.42 (m, 5H), 1.19 (s, 3H), 1.51–1.81 (m, 6H), 1.82–1.93 (m, 1H), 2.01–2.21 (m, 2H), 4.88 (s, 1H), 4.96 (s, 1H), 5.12 (s, 1H), 7.39–7.49 (m, 3H), 7.69–7.76 (m, 2H), 7.76–7.84 (m, 2H). ^{13}C NMR (99 MHz, $CDCl_3$, δ): 12.1 (CH_3), 15.8 (CH_3), 25.8 (CH_2), 27.0 (CH_2), 27.2 (CH_2), 27.3 (CH_2), 28.8 (CH_2), 29.0 (CH_2), 44.2 (*C*H), 51.6 (*C*), 76.7 (*C*H), 111.4 (CH_2), 125.5 (*C*H), 125.7 (*C*H), 125.9 (*C*H), 126.1 (*C*H), 126.5 (*C*H), 127.4 (*C*H), 128.0 (*C*H), 132.5 (*C*), 132.6 (*C*), 140.4 (*C*), 152.5 (*C*). HRMS-ESI (*m*/*z*): $[M + Na]^+$ calcd for $C_{23}H_{30}ONa$, 345.2189; found, 345.2191.

Fig. 4.73 Scheme of **9d**

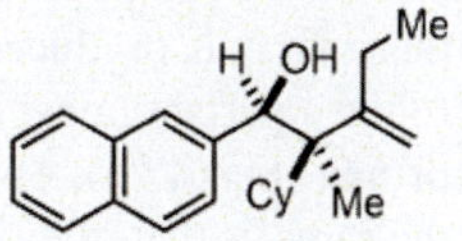

9d

4.4.7.5 Synthesis of {(8*S*,9*S*,13*S*,14*S*,17*S*)-3-Methoxy-13-Methyl-17-(6,6,6-Trifluorohex-1-en-2-yl)-7,8,9,11,12,13,14,15,16,17-Decahydro-6H-Cyclopenta[a]phenanthren-17-yl}Methanol (9e, Fig. 4.74)

The allylboration reaction was conducted with 240.6 mg (0.452 mmol) of (*E*)-**4ma** according to the procedure described in chapter 7.1. The product **9e** was obtained in 84% yield with dr = >95:5 (165.1 mg, 0.378 mmol, white solid).

^{1}H NMR (392 MHz, $CDCl_3$, δ): 0.72 (s, 3H), 1.18–1.58 (m, 5H), 1.59–2.03 (m, 9H), 2.04–2.40 (m, 6H), 2.76–2.95 (m, 2H), 3.45 (d, *J* = 10.2 Hz, 1H), 3.71 (s, 3H), 3.71–3.85 (m, 1H), 5.09 (d, *J* = 18.8 Hz, 2H), 6.62 (d, *J* = 2.7 Hz, 1H), 6.69 (dd, *J* = 8.6 Hz, 2.7 Hz, 1H), 7.16 (d, *J* = 8.6 Hz, 1H). ^{13}C NMR (99 MHz, $CDCl_3$, δ): 15.7 (CH_3), 20.5 (CH_2), 23.4 (CH_2), 26.3 (CH_2), 27.7 (CH_2), 27.8 (CH_2), 29.7 (CH_2), 32.4 (br, CH_2), 33.1 (CH_2), 33.4 (q, *J* = 28.6 Hz, CH_2), 39.2 (*C*H), 43.3 (*C*H), 45.9 (*C*), 50.9 (*C*H), 55.0 (CH_3), 59.5 (*C*), 63.7 (CH_2), 111.3 (*C*H), 112.9 (CH_2), 113.6 (*C*H), 126.0 (*C*H), 127.0 (q, *J* = 277.6 Hz, *C*), 132.3 (*C*), 137.8 (*C*), 150.6 (*C*), 157.3 (*C*). HRMS-EI (*m*/*z*): $[M]^+$ calcd for $C_{26}H_{35}F_3O_2$, 436.2589; found, 436.2582.

4.4.8 *Evaluation of Accuracy of ^{13}C NMR Analysis*

During the optimization of reaction conditions (Table 4.1), the stereoselectivity of allyl boron compound **4aa** was determined by ^{13}C NMR analysis because the other analyses, including a quantitative ^{1}H and ^{19}F NMR, GC, HPLC analysis, were unfortunately not effective for determining the stereoselectivity of **4aa** due to lack of the peak separation between the regio- and stereoisomers. As the accuracy of ^{13}C NMR analysis is not thoroughly investigated [97], I evaluated the accuracy of ^{13}C NMR for determining the stereoselectivity of **4aa** by comparing the selectivity determined by ^{13}C NMR analysis of **4aa** against that determined by quantitative-^{1}H NMR (q-^{1}H NMR) analysis of corresponding oxidation product **6aa** obtained via stereospecific oxidation reaction of **4aa** (Fig. 4.75). For this purpose, I chose Xantphos, IMes, and IAd ligand (Table 4.1, entries 1, 6, and 7, respectively), because the reactions with those ligands produced **4aa** with high regioselectivity (**4aa/5aa** = >94:6). Then, I found a perfect linear correlation between (*E*)-selectivity of **4aa** and (*Z*)-selectivity of **6aa** (Fig. 4.76). Thus, I could confirm that the accuracy of ^{13}C NMR analysis is the same as that of quantitative-^{1}H NMR analysis, and I assured the accuracy of the stereoselectivity of **4aa** described in Table 4.1.

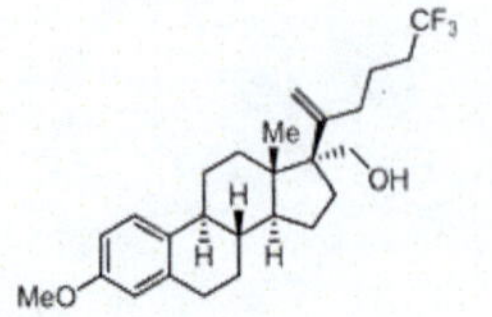

9e

Fig. 4.74 Scheme of **9e**

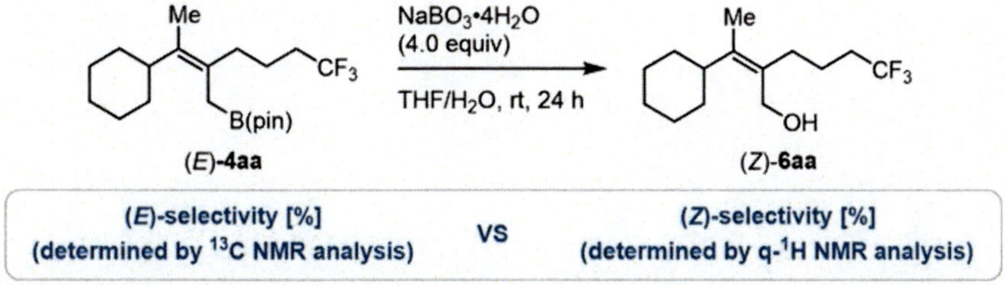

Fig. 4.75 Method for evaluation of accuracy of ^{13}C NMR analysis

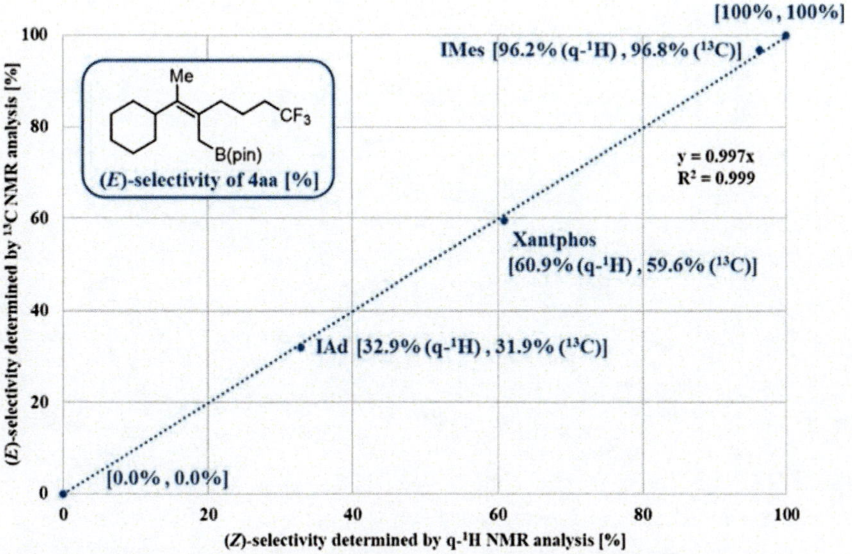

Fig. 4.76 Relationship between (*E*)-selectivity of **4aa** and (*Z*)-selectivity of **6aa**

4.4.9 2D NMR Study for Determination of Product Stereochemistry

4.4.9.1 Stereochemistry of (*E*)-4aa

The stereochemistry of (*E*)-**4aa** was determined by ^{1}H NMR NOESY experiment after the stereospecific oxidation of the boryl group in (*E*)-**4aa**. After the oxidation reaction of (*E*)-**4aa** (Fig. 4.77), the crude material was purified by silica gel column chromatography to give the (*Z*)-**6aa** as a single isomer (*E*/*Z* = <5:95). The results were summarized as the following schemes (Figs. 4.78 and 4.79). A solid curved arrow indicates the selected nOe around the alkene moiety.

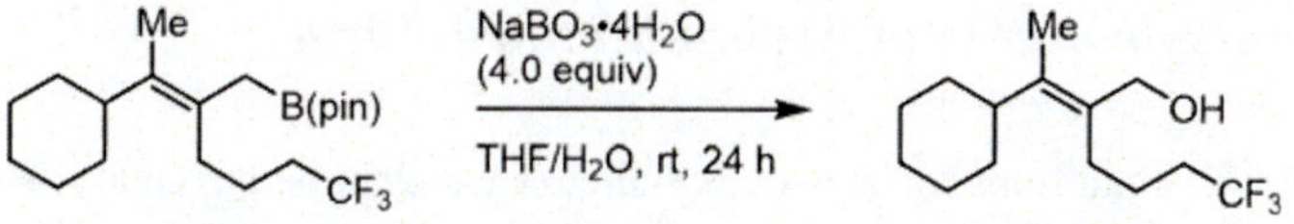

(*Z*)-**4aa** (*E*)-**6aa**

Fig. 4.80 Oxidation of borylation product (*E*)-**4aa**

gel column chromatography to give the (*E*)-**6aa** as a single isomer (E/Z = >95:5). The results were summarized as the following schemes (Figs. 4.81 and 4.82). A solid curved arrow indicates the selected nOe around the alkene moiety.

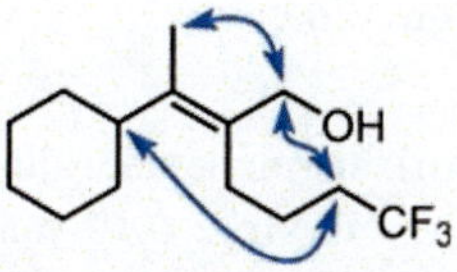

Fig. 4.81 Summary of NOESY analysis of the alcohol product (*E*)-**6aa**

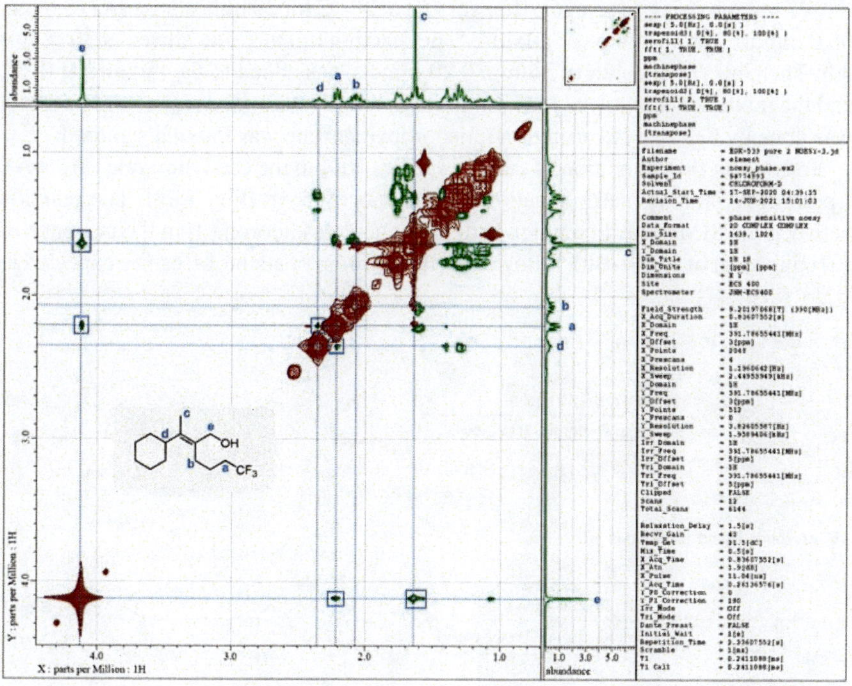

Fig. 4.82 NOESY spectrum of the alcohol product (*E*)-**6aa** and the selected correlation peaks

4.4.10 Alkylboration of Aryl-Substituted Allene

The optimized conditions for *gem*-dialkylallenes were applied to an aryl-substituted allene **1n** (Fig. 4.83). Under Conditions **A**, the reaction afforded a complex mixture (Fig. 4.83A). Allylic boronate **4nb** and alkenyl boronate **5nb** were found in the mixture, albeit with no regioselectivity (**4/5** = 51:49). Conversely, **4nb** was obtained as the major product under Conditions **B**, although the regioselectivity was not sufficient (Fig. 4.83B). Thus, I conclude that the optimized conditions for *gem*-dialkylallenes are not suitable for aryl-substituted allenes.

4.4.11 Stoichiometric Reaction

SIMesCuCl (40.5 mg, 0.10 mmol), bis(pinacolato)diboron (**3**) (25.4 mg, 0.10 mmol), and K(O-*t*-Bu) (for Fig. 4.84A; 11.2 mg, 0.10 mmol, for Fig. 4.84B; 22.4 mg, 0.10 mmol) were placed in an oven-dried reaction vial in an argon-filled glove box. After the vial was sealed with a screw cap containing a Teflon™-coated rubber septum and taken from the glove box, dry DMF (2.0 mL) was added to the vial through the rubber septum using a syringe. After stirring at 30 °C for 30 min, allene **1a** (13.6 mg, 0.10 mmol) was added to the mixture. The reaction mixture was stirred at 30 °C for 3 h. Then, alkyl halide **2a** (47.6 mg, 0.20 mmol) was added to the mixture at 0 °C, and the mixture was allowed to warm to room temperature. The reaction monitoring was done by GC analysis using 1,4-diisopropylbenzene was internal standard.

To confirm the generation of cuprate species **Int4** in the catalytic cycle (Fig. 6A), I performed stoichiometric reactions as described above (Fig. 4.84). The reaction between the SIMesCuCl complex and bis(pinacolato)diboron (**3**) in the presence of 1.0 equivalent of K(O-*t*-Bu), followed by the addition of allene **1a,** can be expected to

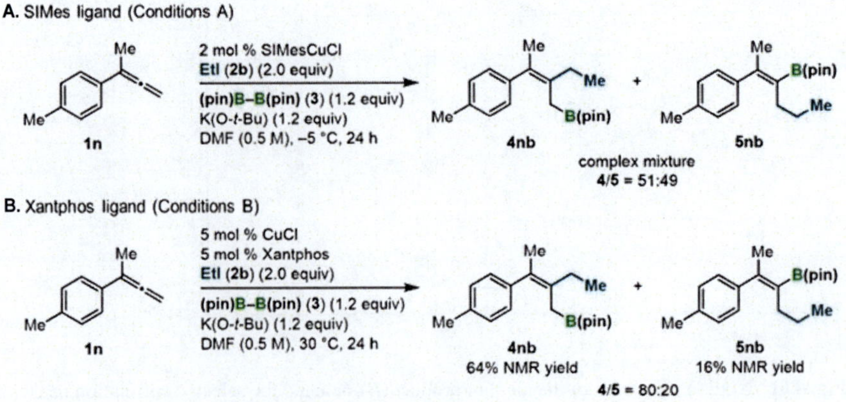

Fig. 4.83 Alkylboration of aryl-substituted allene **1n** using Conditions **A** and **B**

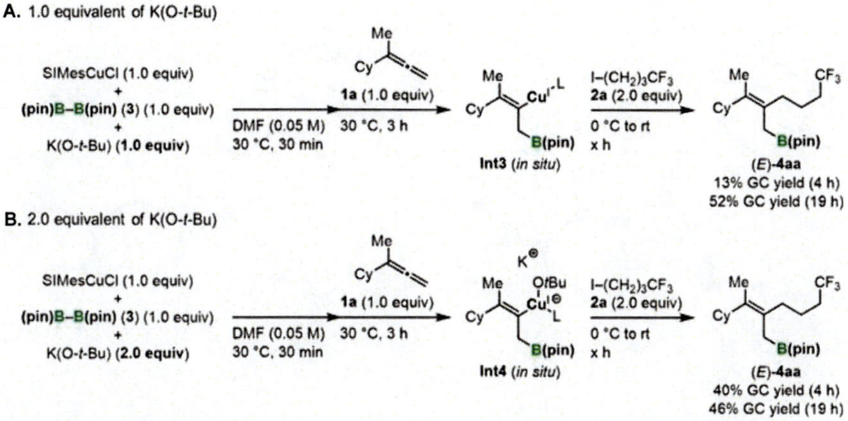

Fig. 4.84 Stoichiometric reactions using 1.0 and 2.0 equivalents of the alkoxide base

form boryl copper(I) species **Int3** in situ (Fig. 4.84A). After stirring the mixture for 3 h, the consumption of allene **1a** was 66%. The following reaction between **Int3** and alkyl halide **2a** gave (*E*)-**4aa** in low yield at the initial stage of the reaction (4 h: 13% GC yield). Thereafter, the yield reached moderate levels (19 h: 52% GC yield). On the other hand, the reaction with 2.0 equivalent of K(O-*t*-Bu) can be expected to form cuprate species **Int4** (Fig. 4.84B). After stirring the mixture for 3 h, the consumption of allene **1a** was 65%. After the adding **2a**, (*E*)-**4aa** was immediately generated in moderate yield (4 h: 40% GC yield). Finally, the yield increased to almost the same level as that with 1.0 equivalent of the base (19 h: 46% GC yield). These results indicate that the consumption of the allene is not affected by the amount of alkoxide base. In contrast, an excess of alkoxide can be expected to promote the alkylation step to produce the alkylboration product (*E*)-**4aa** faster in the initial stage of the reaction. However, further studies are required to propose a conclusive mechanism for the alkylation step in situ.

4.4.12 Single-Crystal X-ray Structural Analysis

4.4.12.1 Molecular Structure and X-ray Crystallographic Data of (*E*)-4ga

The stereochemistry of (*E*)-**4ga** was determined by X-ray crystallographic analysis. The details were summarized in Fig. 4.85 and Table 4.5.

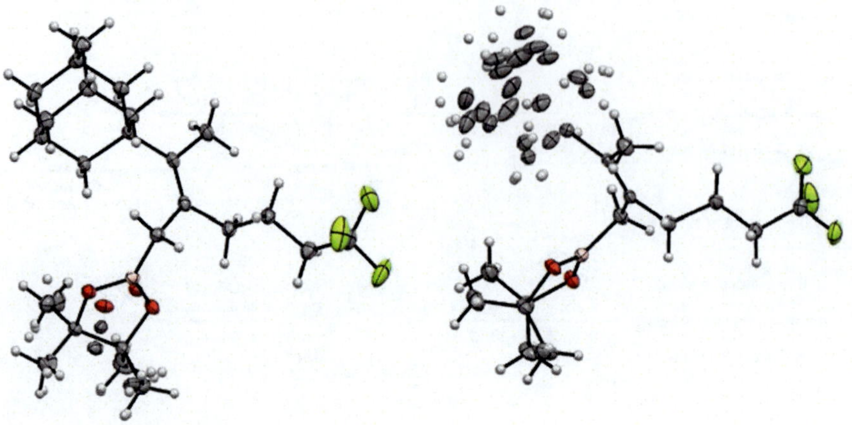

Fig. 4.85 Molecular structure of (*E*)-**4ga**

Table 4.5 Summary of X-ray crystallographic data of (*E*)-4ga

CCDC name	2,062,673
Empirical formula	$C_{24}H_{38}BF_3O_2$
Formula weight	426.35
Crystal system	Monoclinic
Crystal size (mm)	0.100 × 0.100 × 0.100
a (Å)	11.5507(2)
b (Å)	10.4530(2)
c (Å)	38.0880(8)
α (°)	90.0000
β (°)	90.206(2)
γ (°)	90.0000
V (Å^3)	4598.70(15)
Space group	$P2_1/n$
Z value	8
Dcalc (g cm^{-3})	1.232
Temperature (K)	123
$2\theta_{max}$ (°)	58.84
μ (CuKα) (mm^{-1})	–
μ (MoKα) (mm^{-1})	0.091
No. of reflections measured	Total: 60,487 Unique: 11,431 (R_{int} = 0.0402)
No. of observations (all reflections)	11,431
Residuals: R_1 (I > 2.00σ (I))	0.0579
Residuals: wR_2 (all reflections)	0.1396

(continued)

Table 4.5 (continued)

CCDC name	2,062,673
Goodness of fit indicator (GOF)	1.126
Maximum peak in final diff. map (Å^3)	0.31
Minimum peak in final diff. map (Å^3)	−0.32
Flack parameter	–

4.4.12.2 Molecular Structure and X-ray Crystallographic Data of 9d

The stereochemistry of **9d** was determined by X-ray crystallographic analysis. The details were summarized in Fig. 4.86 and Table 4.6.

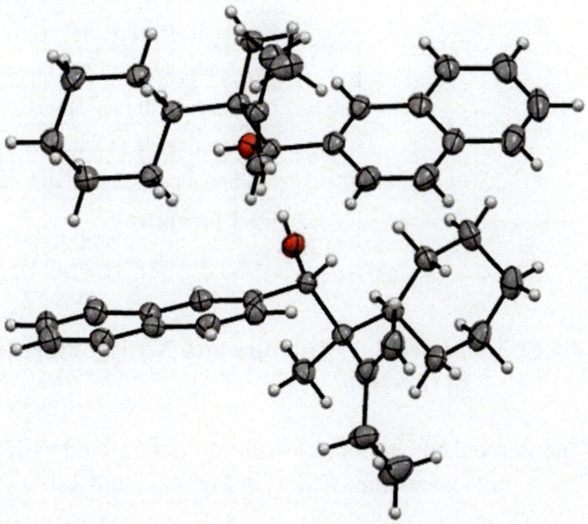

Fig. 4.86 Molecular structure of **9d**

Table 4.6 Summary of X-ray crystallographic data of 9d

CCDC name	2,067,460
Empirical formula	$C_{23}H_{30}O$
Formula weight	322.47
Crystal system	Monoclinic
Crystal size (mm)	0.200 × 0.100 × 0.100
a (Å)	10.56870(10)
b (Å)	15.6326(2)
c (Å)	22.9525(3)
α (°)	90.0000
β (°)	101.9420(10)
γ (°)	90.0000
V (Å^3)	3710.05(8)

(continued)

Table 4.6 (continued)

CCDC name	2,067,460
Space group	$P2_1/c$
Z value	8
Dcalc (g cm^{-}3)	1.155
Temperature (K)	123
$2\theta_{max}$ (°)	150.372
μ (CuKα) (mm^{-1})	0.516
μ (MoKα) (mm^{-1})	–
No. of reflections measured	Total: 18,532 Unique: 18,532 (R_{int} = N/A[a])
No. of observations (all reflections)	18,532
Residuals: R_1 (I > 2.00σ (I))	0.0708
Residuals: wR_2 (all reflections)	0.2115
Goodness of fit indicator (GOF)	1.111
Maximum peak in final diff. map (Å^3)	0.43
Minimum peak in final diff. map (Å^3)	−0.31
Flack parameter	–

[a] Not available because of the twin analyses

4.4.12.3 Molecular Structure and X-ray Crystallographic Data of (*E*)-4mb

The stereochemistry of (*E*)-**4mb** was determined by X-ray crystallographic analysis. The details were summarized in Fig. 4.87 and Table 4.7.

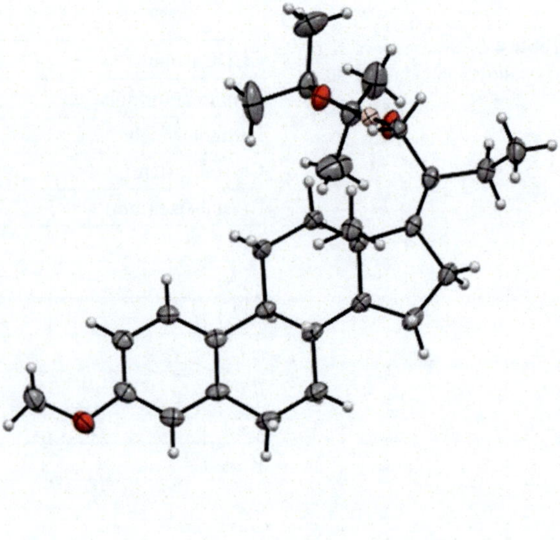

Fig. 4.87 Molecular structure of (*E*)-**4mb**

Table 4.7 Summary of X-ray crystallographic data of (*E*)-4mb

CCDC name	2,070,912
Empirical formula	$C_{29}H_{43}BO_3$
Formula weight	450.44
Crystal system	Orthorhombic
Crystal size (mm)	0.200 × 0.200 × 0.200
a (Å)	7.41170(10)
b (Å)	12.02580(10)
c (Å)	29.9155(2)
α (°)	90.0000
β (°)	90.0000
γ (°)	90.0000
V (Å^3)	2666.42(5)
Space group	$P2_12_12_1$
Z value	4
Dcalc (g cm^{-3})	1.122
Temperature (K)	123
$2\theta_{max}$ (°)	153.09
μ (CuKα) (mm^{-1})	0.538
μ (MoKα) (mm^{-1})	–
No. of reflections measured	Total: 42,609 Unique: 5506 (R_{int} = N/A^a)
No. of observations (all reflections)	5506
Residuals: R_1 (I > 2.00σ (I))	0.0407
Residuals: wR_2 (all reflections)	0.1061
Goodness of fit indicator (GOF)	1.079
Maximum peak in final diff. map (Å^3)	0.29
Minimum peak in final diff. map (Å^3)	−0.22
Flack parameter	−0.10(7)

4.4.12.4 Molecular Structure and X-ray Crystallographic Data of 9e

The stereochemistry of **9e** was determined by X-ray crystallographic analysis. The details were summarized in Fig. 4.88 and Table 4.8.

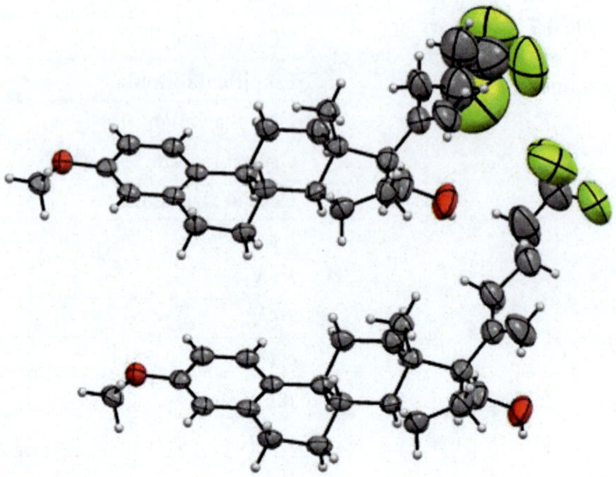

Fig. 4.88 Molecular structure of **9e**

Table 4.8 Summary of X-ray crystallographic data of 9e

CCDC name	2,067,461
Empirical formula	$C_{26}H_{35}F_3O_2$
Formula weight	436.54
Crystal system	Monoclinic
Crystal size (mm)	1.50 × 1.50 × 0.200
a (Å)	12.6735(2)
b (Å)	6.86020(10)
c (Å)	26.7831(5)
α (°)	90.0000
β (°)	97.082(2)
γ (°)	90.0000
V (Å^3)	2310.83(7)
Space group	$P2_1$
Z value	4
Dcalc (g cm^{-3})	1.255
Temperature (K)	293
$2\theta_{max}$ (°)	150.316
μ (CuKα) (mm^{-1})	0.772
μ (MoKα) (mm^{-1})	–
No. of reflections measured	Total: 23,607 Unique: 8055 (R_{int} = 0.0427)
No. of observations (all reflections)	8055
Residuals: R_1 (I > 2.00σ (I))	0.0920
Residuals: wR_2 (All reflections)	0.3124
Goodness of fit indicator (GOF)	1.221

(continued)

Table 4.8 (continued)

CCDC name	2,067,461
Maximum peak in final diff. map (Å^3)	0.58
Minimum peak in final diff. map (Å^3)	−0.41
Flack parameter	−0.3(2)

4.5 Computational Details

4.5.1 Calculation Method Details

All geometry optimizations and thermal energy correction calculations (frequency analyses) using density functional theory (DFT) were performed with the Gaussian 16 (revision C.01) [98] suite of programs. The thresholds defined in Gaussian 16 were used. The geometry optimizations were carried out at ωB97X-D [99] level of theory with Def2-SVP [100, 101] basis set because this combination of the DFT model can reproduce experimental reactivities and selectivities of a various type of borylation reaction with copper(I)/diboron catalyst system [43, 49, 102, 103]. The solvation effect of DMF was included in the calculations using CPCM solvation model [104, 105]. Harmonic frequency calculations were conducted at the same level of theory on the optimized geometries to check all the stationary points as either minima or first-order saddle points. Intrinsic reaction coordinate (IRC) [106] calculations were carried out to confirm the transition states connecting the correct reactants and products on the potential energy surface.

Then, the self-consistent field (SCF) energies of the optimized molecular systems were corrected at a higher level of theory. For this purpose, I chose ωB97M-V [107] functional with Def2-TZVPP [100, 101] basis set because I need a method with high accuracy to compare two different types of the transition state (TS), coordination of boryl copper(I) species to allenes (**TS1**), and borylcupration of allenes (**TS2**). ωB97M-V is one of the state-of-the-art DFT methods and showed excellent performance on many benchmarks involving non-covalent interactions, isomerization energies, thermochemical properties, barrier height, etc., of organic systems including transition metals [108–110]. The single-point calculations were performed with the Orca 4.1.2 program [111, 112]. The solvation effect of DMF was included in the calculations using SMD solvation model [113]. The calculated structures were visualized with VMD program [114].

Summary: ωB97M-V/Def2-TZVPP/SMD(DMF)//ωB97X-D/Def2-SVP/CPCM(DMF).

4.5.2 DFT Study for the Case of Xantphos Ligand Conditions

The DFT study on the regio- and stereoselectivity-determining steps in the case of the substrate **1b** and Xantphos ligand was conducted (Fig. 4.89), as well as the substrate **1a** and SIMes ligand described in the main text (Figs. 4.13 and 4.14).

In the major path to the allylic boronate (*E*)-**4by**, the barrierless coordination of the boryl copper(I) species **Int2** to allene **1b** forms the slightly unstable π-complex **EQ2**$^{\text{allyl-}E}$ relative to the precursor state **EQ1** (**EQ1**: $\Delta G = 0.0$ kcal/mol, **EQ2**$^{\text{allyl-}E}$: $\Delta G = 7.3$ kcal/mol). The subsequent borylcupration generates the highly stable alkenyl copper(I) species **Int3**$^{\text{allyl-}E}$ via the small transition state **TS2**$^{\text{allyl-}E}$ (**TS2**$^{\text{allyl-}E}$: $\Delta G^{\ddagger} = 12.0$ kcal/mol, **Int3**$^{\text{allyl-}E}$: $\Delta G = -24.2$ kcal/mol). Thus, this borylcupration step is irreversible and the selectivity-determining step of this major path. For a path to the stereoisomer (*Z*)-**4by**, the corresponding π-complex **EQ2**$^{\text{allyl-}Z}$ was not found on the IRC from the transition state of the coordination step **TS1**$^{\text{allyl-}Z}$

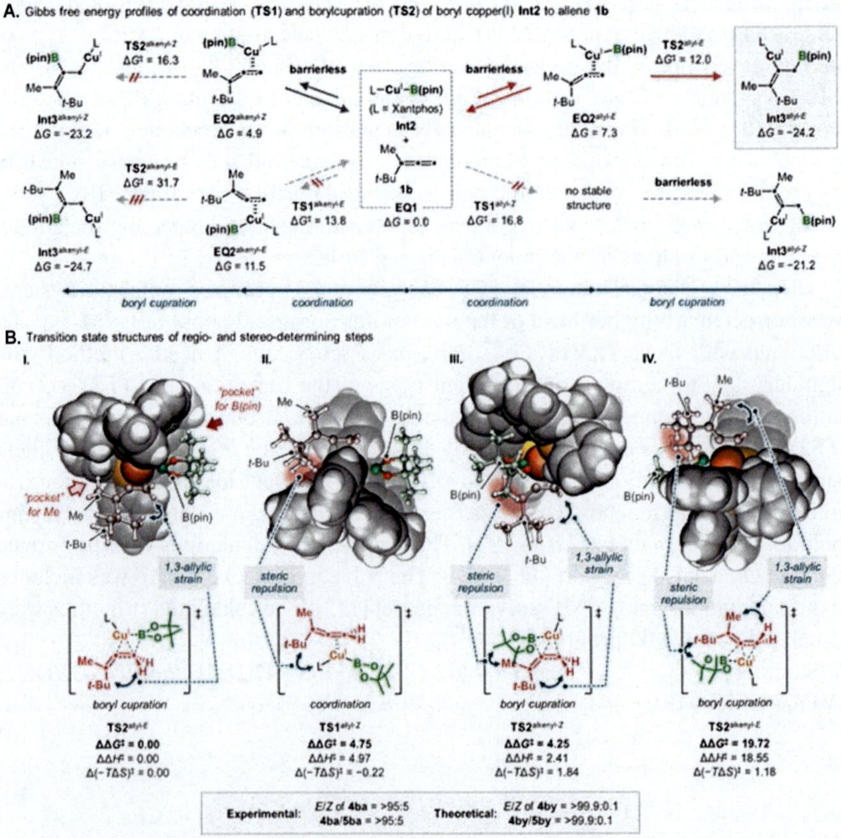

Fig. 4.89 Gibbs free energy profiles and transition state structure with Xantphos ligand; energy values are given in kcal/mol

(**TS1**$^{\text{allyl-}Z}$: $\Delta G^{\ddagger}$ = 16.8 kcal/mol). The IRC is directly connected to the alkenyl copper(I) species **Int3**$^{\text{allyl-}Z}$ (**Int3**$^{\text{allyl-}Z}$: $\Delta G = -21.2$ kcal/mol). A path to the regioisomer (*Z*)-**5by**, as well as the major path, has no **TS1** corresponding to **TS1**$^{\text{alkenyl-}Z}$. Also, the borylcupration generates the alkenyl copper(I) **Int3**$^{\text{alkenyl-}Z}$ via the transition state **TS2**$^{\text{alkenyl-}Z}$ (**TS2**$^{\text{alkenyl-}Z}$: $\Delta G^{\ddagger}$ = 16.3 kcal/mol, **Int3**$^{\text{alkenyl-}Z}$: $\Delta G = -$23.2 kcal/mol). Finally, a path to another regioisomer (*E*)-**5by** has an extremely high transition state of the borylcupration **TS2**$^{\text{alkenyl-}E}$. However, the prior transition state of the coordination **TS1**$^{\text{alkenyl-}E}$ is also higher than the selectivity-determining TS of the major path **TS2**$^{\text{allyl-}E}$ (**TS1**$^{\text{alkenyl-}E}$: $\Delta G^{\ddagger}$ = 13.8 kcal/mol, **TS2**$^{\text{alkenyl-}E}$: $\Delta G^{\ddagger}$ = 31.7 kcal/mol). In summary, **TS2**$^{\text{allyl-}E}$, **TS1**$^{\text{allyl-}Z}$, **TS2**$^{\text{alkenyl-}Z}$, and **TS2**$^{\text{alkenyl-}E}$ are selectivity-determining TSs.

Structures of the respective selectivity-determining TSs were shown in Fig. 4.89B. The predicted selectivities using relative Gibbs free energies are in good agreement with the experimental values (for **4ba**, experimentally *E*/*Z* = >95:5, **4**:**5** = >95:5; theoretically *E*/*Z* = >99.9:0.1, **4**:**5** = >99.9:0.1). In the major TS **TS2**$^{\text{allyl-}E}$, the methyl (Me) group of allene **1b** and the boryl [B(pin)] group incline into the pocket-like space of the catalyst. Therefore, the **TS2**$^{\text{allyl-}E}$ can be formed without large steric repulsions, although the 1,3-allylic strain is found in the substrate **1b** [$\Delta\Delta G^{\ddagger}$ = 0.00 kcal/mol, $\Delta\Delta H^{\ddagger}$ = 0.00 kcal/mol, $\Delta(-T\Delta S^{\ddagger})$ = 0.00 kcal/mol]. For the minor TS **TS1**$^{\text{allyl-}Z}$, the steric repulsion between the *tert*-butyl (*t*-Bu) group and the catalyst causes a large enthalpic destabilization of the structure. However, this minor TS is slightly entropically favored than the major TS **TS2**$^{\text{allyl-}E}$ due to the absence of the 1,3-allylic repulsion [$\Delta\Delta G^{\ddagger}$ = 4.75 kcal/mol, $\Delta\Delta H^{\ddagger}$ = 4.97 kcal/mol, $\Delta(-T\Delta S^{\ddagger})$ = −0.22 kcal/mol]. In the TS of one of the minor regioisomers, **TS2**$^{\text{alkenyl-}Z}$, the combination of the steric repulsion between the substrate and the B(pin) group, and the 1,3-allylic strain destabilizes the structure both enthalpically and entropically [$\Delta\Delta G^{\ddagger}$ = 4.25 kcal/mol, $\Delta\Delta H^{\ddagger}$ = 2.41 kcal/mol, $\Delta(-T\Delta S^{\ddagger})$ = 1.84 kcal/mol]. Finally, for the TS of another regioisomer, **TS2**$^{\text{alkenyl-}E}$, the tremendous steric repulsion between the *t*-Bu group and the B(pin) group induces the huge enthalpic destabilization of this structure [$\Delta\Delta G^{\ddagger}$ = 19.72 kcal/mol, $\Delta\Delta H^{\ddagger}$ = 18.55 kcal/mol, $\Delta(-T\Delta S^{\ddagger})$ = 1.18 kcal/mol]. In conclusion, those minor paths are destabilized by predominantly enthalpic effect as described above; meanwhile, the desired isomer (*E*)-**4by** can be produced via the major TS **TS2**$^{\text{allyl-}E}$ without significant steric repulsion.

As shown in Figs. 4.13 and 4.89A, the barrier of the selectivity-determining major TS for SIMes (Fig. 4.13, **TS1**$^{\text{allyl-}E}$, $\Delta G^{\ddagger}$ = 17.0 kcal/mol) is larger than that for Xantphos (Fig. 4.89A, **TS2**$^{\text{allyl-}E}$, $\Delta G^{\ddagger}$ = 12.0 kcal/mol). However, the reaction with SIMes ligand proceeds under a lower catalyst loading and reaction temperature than that with Xantphos (Table 4.1). I speculate that this conflict between the reaction barrier and the reactivity of those ligands is because the alkylation step (**Int4** to **Int5**) rather than the coordination **TS1** and borylcupration **TS2** is the turnover-limiting step (Fig. 4.12). The alkylation with Xantphos ligand should be disfavored compared to SIMes ligand as the bidentate of Xantphos to the copper(I) reaction center sterically and electronically suppress the additional coordination of alkoxide base to make the following alkylation disadvantageous. Thus, Xantphos ligand showed lower reactivity than SIMes ligand.

4.5.3 Discussion About Ligand Effect

As discussed in Tables 4.1 and 4.2, SIMes is the optimal ligand for the allene substrates bearing a small R^1 and a *medium* R^2 substituent in this alkylboration reaction. In contrast, Xantphos ligand is optimal for the substrate bearing a small R^1 and a *large* R^2 substituent. However, the product was obtained with moderate stereoselectivity when Xantphos ligand was applied to the substrates bearing a small R^1 and a *medium* R^2 substituent (e.g., **1a**; Table 4.1, entry 1, *E*/*Z* of **4aa** = 60:40). Also, the combination of SIMes and the substrate bearing a small R^1 and a *large* R^2 resulted in the isomer mixture (e.g., **1b**; Table 4.1, entry 13, 77:23 mixture of isomers, unfortunately, the structure of the minor isomer could not be identified).

As shown in Fig. 4.89B-II, Xantphos suppresses the $\mathbf{TS1}^{\text{allyl-}Z}$ by steric repulsion between the ligand and the R^2 substituent. Thus, the smaller R^2 substituent is, the more stable the $\mathbf{TS1}^{\text{allyl-}E}$ becomes. Although the smaller R^2 substituent should also make the 1,3-allylic strain in the $\mathbf{TS2}^{\text{allyl-}E}$ smaller, the degree of stabilization of $\mathbf{TS1}^{\text{allyl-}Z}$ would be larger than that of $\mathbf{TS2}^{\text{allyl-}Z}$. Consequently, the stereoselectivity of the substrate bearing a *medium* R^2 is decreased in the reaction with Xantphos ligand.

On the other hand, I then focus on combining SIMes ligand and the substrate bearing a *large* R^2. As shown in Figs. 4.13 and 4.14-I, $\mathbf{TS1}^{\text{allyl-}E}$ is the selectivity-determining step because $\mathbf{TS1}^{\text{allyl-}E}$ is larger than $\mathbf{TS2}^{\text{allyl-}E}$ in the case of SIMes ligand and the substrate **1a**. However, I speculate that replacing the Cy group in **1a** with larger substituent cause destabilization of $\mathbf{TS2}^{\text{allyl-}E}$ to make this TS higher than $\mathbf{TS1}^{\text{allyl-}E}$ because the increasing 1,3-allylic strain along the borylcupration via $\mathbf{TS2}^{\text{allyl-}E}$ become significant. Also, this large destabilization of $\mathbf{TS2}^{\text{allyl-}E}$ makes this major path more disfavored to cause the decrease of the selectivity.

4.5.4 Calculated Properties of All Structures (Table 4.9)

See Table 4.9.

Table 4.9 Calculated energies and thermochemical parameters of the optimized structures

Structure	*E* [Hartree]	*H* [Hartree]	*TS* [Hartree]	*G* [Hartree]
Ligand: SIMes, substrate: 1a				
Catalyst Int2	−2897.899856	−2897.898912	−0.102980	−2898.001892
Substrate 1a	−390.355657	−390.354713	−0.048561	−390.403274
$TS1^{\text{allyl-}E}$	−3288.257399	−3288.256454	−0.121639	−3288.378093
$TS1^{\text{allyl-}Z}$	−3288.254904	−3288.253960	−0.121093	−3288.375053
$TS1^{\text{alkenyl-}E}$	−3288.256106	−3288.255162	−0.124582	−3288.379744
$TS1^{\text{alkenyl-}Z}$	−3288.260313	−3288.259369	−0.122300	−3288.381669

(continued)

Table 4.9 (continued)

Structure	*E* [Hartree]	*H* [Hartree]	*TS* [Hartree]	*G* [Hartree]
EQ2$^{allyl\text{-}E}$	−3288.261781	−3288.260837	−0.122604	−3288.383441
EQ2$^{allyl\text{-}Z}$	−3288.261602	−3288.260658	−0.121819	−3288.382477
EQ2$^{alkenyl\text{-}E}$	−3288.266323	−3288.265379	−0.122734	−3288.388113
EQ2$^{alkenyl\text{-}Z}$	−3288.267039	−3288.266095	−0.121835	−3288.387930
TS2$^{allyl\text{-}E}$	−3288.256618	−3288.255674	−0.123364	−3288.379038
TS2$^{allyl\text{-}Z}$	−3288.259006	−3288.258062	−0.120780	−3288.378842
TS2$^{alkenyl\text{-}E}$	−3288.253543	−3288.252598	−0.120554	−3288.373152
TS2$^{alkenyl\text{-}Z}$	−3288.254710	−3288.253766	−0.120446	−3288.374212
Int3$^{allyl\text{-}E}$	−3288.322829	−3288.321885	−0.121583	−3288.443468
Int3$^{allyl\text{-}Z}$	−3288.325142	−3288.324198	−0.123632	−3288.447830
Int3$^{alkenyl\text{-}E}$	−3288.327429	−3288.326485	−0.119278	−3288.445763
Int3$^{alkenyl\text{-}Z}$	−3288.328378	−3288.327434	−0.119909	−3288.447343
Ligand: Xantphos, substrate: 1b				
Catalyst Int2	−4314.300915	−4314.299970	−0.132457	−4314.432427
Substrate 1b	−312.971634	−312.970690	−0.045390	−313.016080
TS1$^{allyl\text{-}Z}$	−4627.272228	−4627.271284	−0.150487	−4627.421771
TS1$^{alkenyl\text{-}Z}$	−4627.277299	−4627.276354	−0.150181	−4627.426535
EQ2$^{allyl\text{-}E}$	−4627.284956	−4627.284012	−0.152872	−4627.436884
EQ2$^{alkenyl\text{-}E}$	−4627.290122	−4627.289177	−0.151570	−4627.440747
EQ2$^{alkenyl\text{-}Z}$	−4627.281502	−4627.280557	−0.149602	−4627.430159
TS2$^{allyl\text{-}E}$	−4627.280154	−4627.279210	−0.150130	−4627.429340
TS2$^{alkenyl\text{-}E}$	−4627.276318	−4627.275374	−0.147191	−4627.422565
TS2$^{alkenyl\text{-}Z}$	−4627.250600	−4627.249656	−0.148257	−4627.397913
Int3$^{allyl\text{-}E}$	−4627.337341	−4627.336397	−0.149024	−4627.485421
Int3$^{allyl\text{-}Z}$	−4627.339475	−4627.338530	−0.149307	−4627.487837
Int3$^{alkenyl\text{-}E}$	−4627.337375	−4627.336430	−0.150686	−4627.487116
Int3$^{alkenyl\text{-}Z}$	−4627.332616	−4627.331671	−0.150579	−4627.482250

References

1. Hall DG (ed) (2011) Boronic acids: preparation, applications in organic synthesis and medicine, vols 1–2. Wiley-VCH, Weinheim
2. Diner C, Szabó KJ (2017) Recent advances in the preparation and application of allylboron species in organic synthesis. J Am Chem Soc 139:2–14
3. Hoffmann RW (1982) Diastereogenic addition of crotylmetal compounds to aldehydes. Angew Chem Int Ed 21:555–566
4. Kennedy JWJ, Hall DG (2003) Recent advances in the activation of boron and silicon reagents for stereocontrolled allylation reactions. Angew Chem Int Ed 42:4732–4739

5. Marek I, Sklute G (2007) Creation of quaternary stereocenters in carbonyl allylation reactions. Chem Commun 1683–1691
6. Hall DG (2008) New preparative methods for allylic boronates and their application in stereoselective catalytic allylborations. Pure Appl Chem 80:913–927
7. Yus M, González-Gómez JC, Foubelo F (2013) Diastereoselective allylation of carbonyl compounds and imines: application to the synthesis of natural products. Chem Rev 113:5595–5698
8. Huo H-X, Duvall JR, Huang M-Y, Hong R (2014) Catalytic asymmetric allylation of carbonyl compounds and imines with allylic boronates. Org Chem Front 1:303–320
9. Scott HK, Aggarwal VK (2011) Highly enantioselective synthesis of tertiary boronic esters and their stereospecific conversion to other functional groups and quaternary stereocentres. Chem Eur J 17:13124–13132
10. Leonori D, Aggarwal VK (2014) Lithiation-borylation methodology and its application in synthesis. Acc Chem Res 47:3174–3183
11. Leonori D, Aggarwal VK (2015) Stereospecific couplings of secondary and tertiary boronic esters. Angew Chem Int Ed 54:1082–1096
12. Mlynarski SN, Karns AS, Morken JP (2012) Direct stereospecific amination of alkyl and aryl pinacol boronates. J Am Chem Soc 134:16449–16451
13. Alam R, Vollgraff T, Eriksson L, Szabó KJ (2015) Synthesis of adjacent quaternary stereocenters by catalytic asymmetric allylboration. J Am Chem Soc 137:11262–11265
14. Alam R, Diner C, Jonker S, Eriksson L, Szabó KJ (2016) Catalytic asymmetric allylboration of indoles and dihydroisoquinolines with allylboronic acids: stereodivergent synthesis of up to three contiguous stereocenters. Angew Chem Int Ed 55:14417–14421
15. García-Ruiz C, Chen JL-Y, Sandford C, Feeney K, Lorenzo P, Berionni G, Mayr H, Aggarwal VK (2017) Stereospecific allylic functionalization: the reactions of allylboronate complexes with electrophiles. J Am Chem Soc 139:15324–15327
16. Jonker SJT, Diner C, Schulz G, Iwamoto H, Eriksson L, Szabó KJ (2018) Catalytic asymmetric propargyl- and allylboration of hydrazonoesters: a metal-free approach to sterically encumbered chiral α-amino acid derivatives. Chem Commun 54:12852–12855
17. Kiyokawa K, Hata S, Kainuma S, Minakata S (2019) Electrophilic cyanation of allylic boranes: synthesis of β, γ-unsaturated nitriles containing allylic quaternary carbon centers. Chem Commun 55:458–461
18. Hari DP, Madhavachary R, Fasano V, Haire J, Aggarwal VK (2021) Highly diastereoselective strain-increase allylborations: rapid access to alkylidenecyclopropanes and alkylidenecyclobutanes. J Am Chem Soc 143:7462–7470
19. Elford TG, Hall DG (2010) Total synthesis of (+)-chinensiolide B via tandem allylboration/lactonization. J Am Chem Soc 132:1488–1489
20. Wohlfahrt M, Harms K, Koert U (2011) Asymmetric allylboration of *vic*-tricarbonyl compounds: total synthesis of (+)-awajanomycin. Angew Chem Int Ed 50:8404–8406
21. Beveridge RE, Batey RA (2014) An organotrifluoroborate-based convergent total synthesis of the potent cancer cell growth inhibitory depsipeptides kitastatin and respirantin. Org Lett 16:2322–2325
22. Allais C, Roush WR (2017) Enantio- and diastereoselective synthesis of 1,5-*syn*-(Z)-amino alcohols via imine double allylboration: synthesis of *trans*-1,2,3,6-tetrahydropyridines and total synthesis of andrachcine. Org Lett 19:2646–2649
23. Kennedy JWJ, Hall DG (2002) Novel isomerically pure tetrasubstituted allylboronates: stereocontrolled synthesis of α-exomethylene γ-lactones as aldol-like adducts with a stereogenic quaternary carbon center. J Am Chem Soc 124:898–899
24. Zhu N, Hall DG (2003) Effect of additives on the stereochemical integrity and reactivity of α-alkoxycarbonyl alkenylcopper intermediates. Optimal conditions for the synthesis of isomerically pure tetrasubstituted alkenes. J Org Chem 68:6066–6069
25. Kennedy JWJ, Hall DG (2004) Lewis acid catalyzed allylboration: discovery, optimization, and application to the formation of stereogenic quaternary carbon centers. J Org Chem 69:4412–4428

Z

26. Baumann AN, Music A, Karaghiosoff K, Didier D (2016) Highly diastereoselective approach to methylenecyclopropanes via boron-homologation/allylboration sequences. Chem Commun 52:2529–2532
27. Eisold M, Kiefl GM, Didier D (2016) Single-pot asymmetric approach toward enantioenriched quaternary stereocenter-containing alkylidenecyclobutanes. Org Lett 18:3022–3025
28. Eisold M, Baumann AN, Kiefl GM, Emmerling ST, Didier D (2017) Unsaturated four-membered rings: efficient strategies for the construction of cyclobutenes and alkylidenecyclobutanes. Chem Eur J 23:1634–1644
29. Parsutkar MM, Pagar VV, RajanBabu TV (2019) Catalytic enantioselective synthesis of cyclobutenes from alkynes and alkenyl derivatives. J Am Chem Soc 141:15367–15377
30. The platinum-catalyzed diboration of *gem*-disubstituted allenes is an alternative strategy for the preparation of differentially 2,3,3-trisubstituted allylic boronates that have an alkenyl boronate structure instead of an all-carbon tetrasubstituted alkene structure; for details, see Ref. [31]
31. Guo X, Nelson AK, Slebodnick C, Santos WL (2015) Regio- and chemoselective diboration of allenes with unsymmetrical diboron: formation of vinyl and allyl boronic acid derivatives. ACS Catal 5:2172–2176
32. Pulis AP, Yeung K, Procter DJ (2017) Enantioselective copper catalysed, direct functionalisation of allenes via allyl copper intermediates. Chem Sci 8:5240–5247
33. Fujihara T, Tsuji Y (2018) Cu-catalyzed borylative and silylative transformations of allenes: use of β-functionalized allyl copper intermediates in organic synthesis. Synthesis (Stuttg) 50:1737–1749
34. Whyte A, Torelli A, Mirabi B, Zhang A, Lautens M (2020) Copper-catalyzed borylative difunctionalization of π-systems. ACS Catal 10:11578–11622
35. Talbot FJT, Dherbassy Q, Manna S, Shi C, Zhang S, Howell GP, Perry GJP, Procter DJ (2020) Copper-catalyzed borylative couplings with C–N electrophiles. Angew Chem Int Ed 59:20278–20289
36. Zhou Y, You W, Smith KB, Brown MK (2014) Copper-catalyzed cross-coupling of boronic esters with aryl iodides and application to the carboboration of alkynes and allenes. Angew Chem Int Ed 53:3475–3479
37. Semba K, Bessho N, Fujihara T, Terao J, Tsuji Y (2014) Copper-catalyzed borylative allyl–allyl coupling reaction. Angew Chem Int Ed 53:9007–9011
38. Meng F, McGrath KP, Hoveyda AH (2014) Multifunctional organoboron compounds for scalable natural product synthesis. Nature 513:367–374
39. Xiong M, Xie X, Liu Y (2017) Copper-catalyzed borylative cyclization of in situ generated *o*-allenylaryl nitriles with bis(pinacolato)diboron. Org Lett 19:3398–3401
40. Meng F, Jang H, Jung B, Hoveyda AH (2013) Cu-catalyzed chemoselective preparation of 2-(pinacolato)boron-substituted allylcopper complexes and their in situ site-, diastereo-, and enantioselective additions to aldehydes and ketones. Angew Chem Int Ed 52:5046–5051
41. Rae J, Yeung K, McDouall JJW, Procter DJ (2016) Copper-catalyzed borylative cross-coupling of allenes and imines: selective three-component assembly of branched homoallyl amines. Angew Chem Int Ed 55:1102–1107
42. Yeung K, Ruscoe RE, Rae J, Pulis AP, Procter DJ (2016) Enantioselective generation of adjacent stereocenters in a copper-catalyzed three-component coupling of imines, allenes, and diboranes. Angew Chem Int Ed 55:11912–11916
43. Meng F, Li X, Torker S, Shi Y, Shen X, Hoveyda AH (2016) Catalytic enantioselective 1,6-conjugate additions of propargyl and allyl groups. Nature 537:387–393
44. Zhao Y-S, Tang X-Q, Tao J-C, Tian P, Lin G-Q (2016) Efficient access to *cis*-decalinol frameworks: copper(I)-catalyzed borylative cyclization of allene cyclohexanediones. Org Biomol Chem 14:4400–4404
45. Zhao W, Montgomery J (2016) Cascade copper-catalyzed 1,2,3-trifunctionalization of terminal allenes. J Am Chem Soc 138:9763–9766
46. Fujihara T, Sawada A, Yamaguchi T, Tani Y, Terao J, Tsuji Y (2017) Boraformylation and silaformylation of allenes. Angew Chem Int Ed 56:1539–1543

Z

68. Ishiyama T, Ahiko T-A, Miyaura N (2002) Acceleration effect of lewis acid in allylboration of aldehydes: catalytic, regiospecific, diastereospecific, and enantioselective synthesis of homoallyl alcohols. J Am Chem Soc 124:12414–12415
69. Sakata K, Fujimoto H (2008) Quantum chemical study of lewis acid catalyzed allylboration of aldehydes. J Am Chem Soc 130:12519–12526
70. Kubota K, Iwamoto H, Ito H (2017) Formal nucleophilic borylation and borylative cyclization of organic halides. Org Biomol Chem 15:285–300
71. Royes J, Ni S, Farré A, La Cascia E, Carbó JJ, Cuenca AB, Maseras F, Fernández E (2018) Copper-catalyzed borylative ring closing C-C coupling toward spiro- and dispiroheterocycles. ACS Catal 8:2833–2838
72. In the catalytic cycle, the detailed mechanism of the alkylation step from **Int3** to **Int1** is still unclear at present. However, I would like to propose a mechanism that involves the formation of cuprate species **Int4** and a subsequent S_N2-type oxidative addition to alkyl halides based on the references Refs. [34, 49, 70, 71, and 73–80], and stoichiometric reactions as a mechanistic study (for details, see the Supporting Information). Alternatively, the formation of borates rather than cuprates was also reported in Refs. [81 and 82]. A direct alkylation (methylation) of **Int3** has been proposed in Ref. [83]
73. Nakamura E, Mori S (2000) Wherefore art thou copper? Structures and reaction mechanisms of organocuprate clusters in organic chemistry. Angew Chem Int Ed 39:3750–3771
74. Mori S, Nakamura E, Morokuma K (2000) Mechanism of S_N2 alkylation reactions of lithium organocuprate clusters with alkyl halides and epoxides. Solvent effects, BF_3 effects, and trans-diaxial epoxide opening. J Am Chem Soc 122:7294–7307
75. Yoshikai N, Nakamura E (2012) Mechanisms of nucleophilic organocopper(I) reactions. Chem Rev 112:2339–2372
76. Posner GH, Sterling JJ (1973) Reaction of α,α'-dibromo ketones with organocopper reagents. New method for α alkylation of ketones. J Am Chem Soc 95:3076–3077
77. Posner GH, Whitten CE, Sterling JJ (1973) New class of mixed cuprate(I) reagents, lithium hetero(alkyl) cuprate(I), which allow selective alkyl group transfer. J Am Chem Soc 95:7788–7800
78. Posner GH, Whitten CE (1973) Substitution and conjugate addition reactions using new lithium -butoxy-alkylcuprate(I) reagents. Tetrahedron Lett 14:1815–1818
79. Please see comments on a mechanistic study in the reference section of one of the previous reports of the Ito group (Ref. [80])
80. Ito H, Toyoda T, Sawamura M (2010) Stereospecific synthesis of cyclobutylboronates through copper(I)-catalyzed reaction of homoallylic sulfonates and a diboron derivative. J Am Chem Soc 132:5990–5992
81. Kim-Lee SH, Alonso I, Mauleón P, Arrayás RG, Carretero JC (2018) Rationalizing the role of NaOtBu in copper-catalyzed carboboration of alkynes: assembly of allylic all-carbon quaternary stereocenters. ACS Catal 8:8993–9005
82. Lin S, Lin Z (2019) DFT studies on the mechanism of copper-catalyzed boracarboxylation of alkene with CO_2 and diboron. Organometallics 38:240–247
83. Alfaro R, Parra A, Alemán J, García Ruano JL, Tortosa M (2012) Copper(I)-catalyzed formal carboboration of alkynes: synthesis of tri- and tetrasubstituted vinylboronates. J Am Chem Soc 134:15165–15168
84. Zhao H, Dang L, Marder TB, Lin Z (2008) DFT studies on the mechanism of the diboration of aldehydes catalyzed by copper(I) boryl complexes. J Am Chem Soc 130:5586–5594
85. Han L, Xu B, Liu T (2018) Mechanisms of the synthesis of trialkylsubstituted alkenylboronates from unactivated internal alkynes catalyzed by copper: a theoretical study. J Organomet Chem 864:154–159
86. Falivene L, Cao Z, Petta A, Serra L, Poater A, Oliva R, Scarano V, Cavallo L (2019) Towards the online computer-aided design of catalytic pockets. Nat Chem 11:872–879
87. Sheldrick GM (2013) SHELXL-2013, program for the refinement of crystal structures. University of Göttingen, Göttingen, Germany

Z

88. Liu J, Nie M, Zhou Q, Gao S, Jiang W, Chung LW, Tang W, Ding K (2017) Enantioselective palladium-catalyzed diboration of 1,1-disubstituted allenes. Chem Sci 8:5161–5165
89. Pasto DJ, Warren SE, Morrison MA (1981) Radical-chain addition of benzenethiol to aIienes. Analysis of steric effects and reversibility. J Org Chem 46:2837–2841
90. Köpfer A, Breit B (2015) Rhodium-catalyzed hydroformylation of 1,1-disubstituted allenes employing the self-assembling 6-DPPon system. Angew Chem Int Ed 54:6913–6917
91. Takaya J, Iwasawa N (2008) Hydrocarboxylation of allenes with CO_2 catalyzed by silyl pincer-type palladium complex. J Am Chem Soc 130:15254–15255
92. Wang B, Gandamana DA, Gagosz F, Chiba S (2019) Diastereoselective intramolecular hydride transfer under brønsted acid catalysis. Org Lett 21:2298–2301
93. Wang C, Teo WJ, Ge S (2017) Access to stereodefined (Z)-allylsilanes and (Z)-allylic alcohols via cobalt-catalyzed regioselective hydrosilylation of allenes. Nat Commun 8:1–9
94. Liu X, Liu B, Liu Q (2020) Migratory hydrogenation of terminal alkynes by base/cobalt relay catalysis. Angew Chem Int Ed 59:6750–6755
95. Santoro O, Collado A, Slawin AMZ, Nolan SP, Cazin CSJ (2013) A general synthetic route to [Cu(X)(NHC)] (NHC = N-heterocyclic carbene, X = Cl, Br, I) complexes. Chem Commun 49:10483–10485
96. Sadhu KM, Matteson DS (1985) (Chloromethyl)Llthium: efficient generation and capture by boronic esters and a simple preparation of diisopropyl (chloromethyl)boronate. Organometallics 4:1687–1689
97. Otte DAL, Borchmann DE, Lin C, Weck M, Woerpel KA (2014) ^{13}C NMR spectroscopy for the quantitative determination of compound ratios and polymer end groups. Org Lett 16:1566–1569
98. Gaussian 16, Revision C.01, Frisch MJ, Trucks GW, Schlegel HB, Scuseria GE, Robb MA, Cheeseman JR, Scalmani G, Barone V, Petersson GA, Nakatsuji H, Li X, Caricato M, Marenich AV, Bloino J, Janesko BG, Gomperts R, Mennucci B, Hratchian HP, Ortiz JV, Izmaylov AF, Sonnenberg JL, Williams-Young D, Ding F, Lipparini F, Egidi F, Goings J, Peng B, Petrone A, Henderson T, Ranasinghe D, Zakrzewski VG, Gao J, Rega N, Zheng G, Liang W, Hada M, Ehara M, Toyota K, Fukuda R, Hasegawa J, Ishida M, Nakajima T, Honda Y, Kitao O, Nakai H, Vreven T, Throssell K, Montgomery JA Jr, Peralta JE, Ogliaro F, Bearpark MJ, Heyd JJ, Brothers EN, Kudin KN, Staroverov VN, Keith TA, Kobayashi R, Normand J, Raghavachari K, Rendell AP, Burant JC, Iyengar SS, Tomasi J, Cossi M, Millam JM, Klene M, Adamo C, Cammi R, Ochterski JW, Martin RL, Morokuma K, Farkas O, Foresman JB, Fox DJ (2016) Gaussian, Inc., Wallingford CT
99. Chai JD, Head-Gordon M (2008) Long-range corrected hybrid density functionals with damped atom-atom dispersion corrections. Phys Chem Chem Phys 10:6615–6620
100. Weigend F, Ahlrichs R (2005) Balanced basis sets of split valence, triple zeta valence and quadruple zeta valence quality for H to Rn: design and assessment of accuracy. Phys Chem Chem Phys 7:3297–3305
101. Weigend F (2006) Accurate coulomb-fitting basis sets for H to Rn. Phys Chem Chem Phys 8:1057–1065
102. Lee J, Radomkit S, Torker S, Pozo J, Hoveyda AH (2018) Mechanism-based enhancement of scope and enantioselectivity for reactions involving a copper-substituted stereogenic carbon centre. Nat Chem 10:99–108
103. Iwamoto H, Ozawa Y, Takenouchi Y, Imamoto T, Ito H (2021) Backbone-modified c_2-symmetrical chiral bisphosphine TMS-QuinoxP*: asymmetric borylation of racemic allyl electrophiles. J Am Chem Soc 143:6413–6422
104. Barone V, Cossi M (1998) Quantum calculation of molecular energies and energy gradients in solution by a conductor solvent model. J Phys Chem A 102:1995–2001
105. Cossi M, Rega N, Scalmani G, Barone V (2003) Energies, structures, and electronic properties of molecules in solution with the C-PCM solvation model. J Comput Chem 24:669–681
106. Fukui K (1981) The path of chemical reactions - the IRC approach. ACC Chem Res 14:363–368

Z

107. Mardirossian N, Head-Gordon M (2016) ωB97M-V: a combinatorially optimized, range-separated hybrid, meta-GGA density functional with VV10 nonlocal correlation. J Chem Phys 144
108. Mardirossian N, Head-Gordon M (2017) Thirty years of density functional theory in computational chemistry: an overview and extensive assessment of 200 density functionals. Mol Phys 115:2315–2372
109. Najibi A, Goerigk L (2018) The Nonlocal Kernel in van der waals density functionals as an additive correction: an extensive analysis with special emphasis on the B97M-V and ωB97M-V Approaches. J Chem Theory Comput 14:5725–5738
110. Vogiatzis KD, Polynski MV, Kirkland JK, Townsend J, Hashemi A, Liu C, Pidko EA (2019) Computational approach to molecular catalysis by 3d transition metals: challenges and opportunities. Chem Rev 119:2453–2523
111. Neese F (2012) The ORCA program system. Wiley Interdiscip Rev Comput Mol 2:73
112. Neese F (2017) Software update: the ORCA program system, version 4.0. Wiley Interdiscip Rev Comput Mol Sci 8:e1327
113. Marenich AV, Cramer CJ, Truhlar DG (2009) Universal solvation model based on solute electron density and on a continuum model of the solvent defined by the bulk dielectric constant and atomic surface tensions. J Phys Chem B 113:6378–6396
114. Humphrey W, Dalke A, Schulten K (1996) VMD: visual molecular dynamics. J Mol Graph 14:33–38

Z

Chapter 5
Computational Investigation on Copper(I)-Catalyzed Enantioselective Radical Borylation of Benzyl Halides

5.1 Introduction

α-Chiral alkyl boron compounds have multifaceted applications in organic synthesis and pharmaceuticals [1–4]. To date, various catalytic and non-catalytic methods for the enantioselective synthesis have been developed [3–10]. In 1961, the Brown group reported the first synthesis of α-chiral alkyl boranes through asymmetric hydroboration of alkenes (Fig. 5.1A) [11]. Subsequently, in 1980, the Matteson group developed the new methodology for the synthesis of α-chiral alkyl boronates using a homologation reaction of achiral boron compounds (Fig. 5.1B) [12]. However, these classical procedures require multi-steps and stoichiometric amount of chiral auxiliary groups. Catalytic asymmetric borylation of achiral double bonds is one of the reliable routes to the chiral alkyl boron compounds. In 1989, the Hayashi and Ito group achieved the first enantioselective hydroboration of styrene substrates (Fig. 5.1C) [13]. Afterward, various catalytic methods have been reported. However, the enantioconvergent boryl substitution of racemic substrates has not been reported up to date (Fig. 5.1D).

In 2012, the Steel, Marder, Liu group and the Ito group independently reported copper(I)-catalyzed boryl substitution reactions of alkyl halide substrates (Fig. 5.2A) [14–17]. The mechanistic study revealed the reaction proceeds via alkyl radical intermediate (Fig. 5.2B). Thus, if this achiral radical intermediate is asymmetrically trapped with the chiral catalyst and borylated, the enantioenriched compounds can be obtained from the racemic alkyl halides [18–23]. Based on this hypothesis, my co-workers in the Ito group developed the first copper(I)-catalyzed enantioconvergent boryl substitution of racemic benzyl chlorides (Fig. 5.2C) [24]. However, the mechanism of the enantioface recognition of the radical intermediate could not be proved experimentally. In this study, I performed density functional theory (DFT) calculations as the computational mechanistic study and analyzed the enantiodetermining transition states (TSs). Interaction analyses based on the non-covalent interaction (NCI) plot revealed that the combination of an attractive C–H/π interaction in the favored TS and steric repulsion in the disfavored TS contributes to the high enantioselectivity.

Y. Ozawa, *Copper(I)-Catalyzed Stereoselective Borylation Reactions*, Springer Theses,
https://doi.org/10.1007/978-981-99-1098-4_5

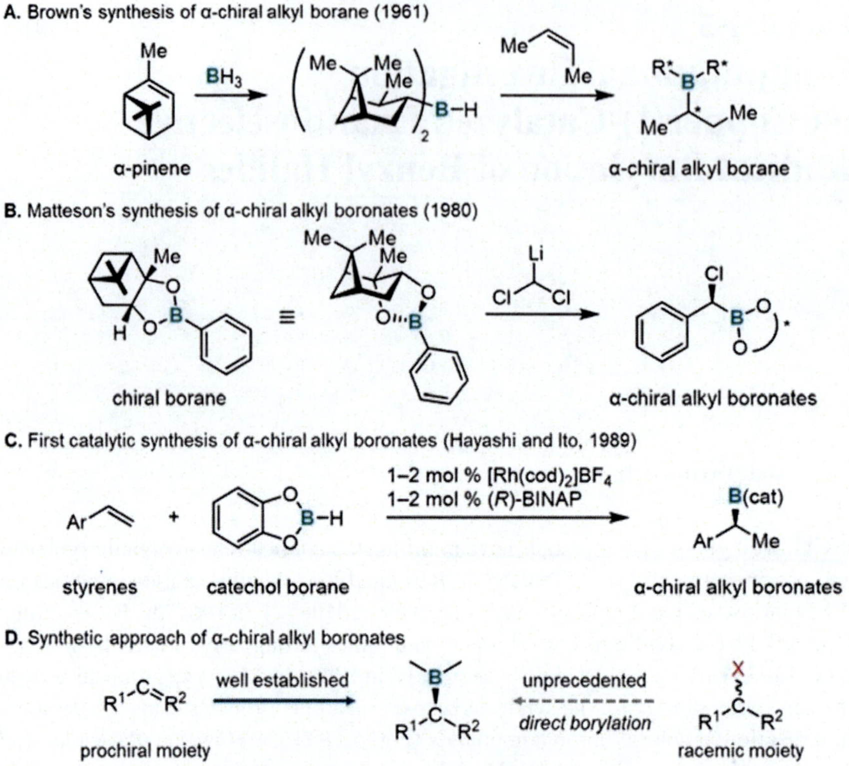

Fig. 5.1 Synthetic procedure of α-chiral alkyl boron compounds

5.2 Results and Discussion

The proposed catalytic cycle is shown in Fig. 5.3. The σ-bond metathesis between a copper(I) alkoxide **I** and a diboron reagent **2** generates the boryl copper(I) species **II**. After the coordination of potassium methoxide to the copper(I) center to form the cuprate species [17], a benzyl radical should be afforded via the reaction between the cuprate and benzyl halide. The generation of the cuprate species was experimentally probed by my co-workers [25]. Then, the radical species is borylated via recombination with copper(II) intermediate, which produce the benzyl boronate (*S*)-**3** and the copper(I) alkoxide **I**, which is assumed to be the enantiodetermining step.

To investigate the enantiodetermining step involving **IV** in Fig. 5.3, I first optimized the structure of the copper(II) species (Fig. 5.4). Although some copper(II) complexes are reported to be a square planer geometry [26–28], the geometry optimization from the square planer boryl copper(II)/bisphosphine complex resulted in the distorted tetrahedral geometry. It is considered that the steric repulsion between the boryl group and substituents on the phosphorous atoms destabilizes the square planer geometry and forces the copper(I) center in the distorted tetrahedral geometry.

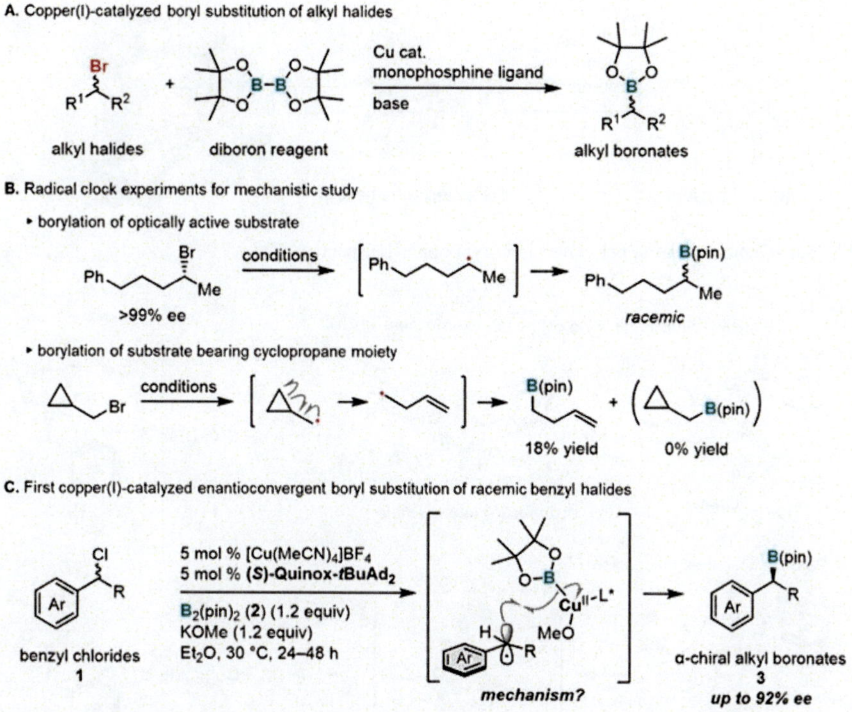

Fig. 5.2 Copper(I)-catalyzed boryl substitution of alkyl halides via radical intermediate

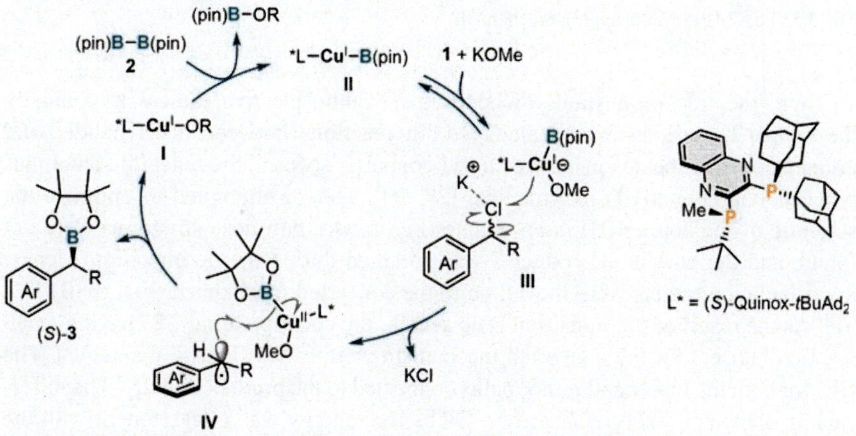

Fig. 5.3 Proposed catalytic cycle of copper(I)-catalyzed enantioconvergent boryl substitution of alkyl halides

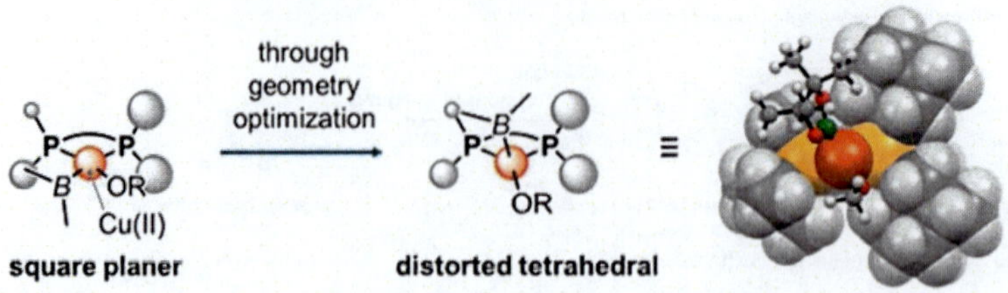

Fig. 5.4 Geometry of copper center in boryl copper(II) species

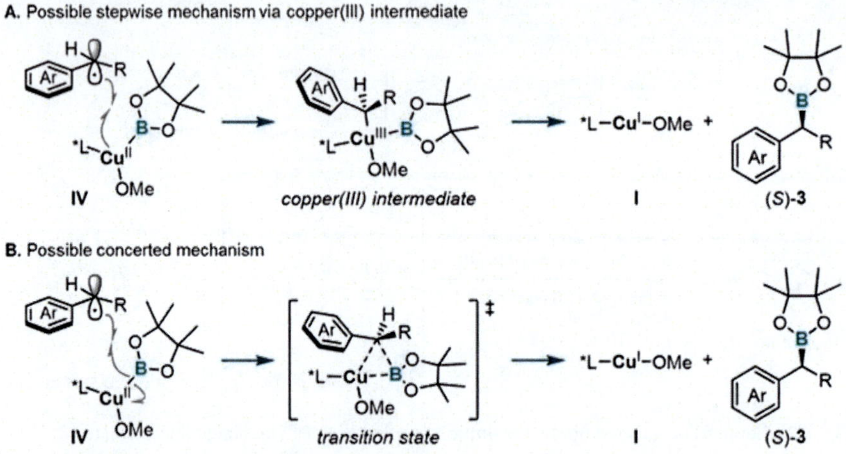

Fig. 5.5 Possible radical trapping mechanism

Then, the transition states (TSs) of the enantioselective radical trapping by the copper(II) species were calculated. In reactions between alkyl halides and organocopper(I) species instead of boryl copper(I) species, the reaction sometimes proceeds via copper(III) intermediate [29, 30]. Thus, I attempted to optimize the structure of the copper(III) intermediate (Fig. 5.5A), but those structures were not found, and the borylated product **3** was obtained during the optimization. Hence, my calculation targets were moved on to the concerted mechanism (Fig. 5.5B) [29, 31]. As the result of the transition state search, the corresponding TSs having small reaction barriers for the concerted mechanism were found (Figs. 5.6 and 5.7). The IRC for both the major and minor paths connected to the product directly. The difference in activation energies based on Gibbs free energy was in agreement with the experimental results (experiment: 67% ee; calculation: 68% ee).

Finally, the enantiodetermining TSs were analyzed using NCIPLOT (Fig. 5.7) [32, 33]. The strong attractive interaction, which resembles hydrogen bonding, was found between the benzylic hydrogen atom and the methoxide oxygen atom coordinating to the copper center, which was confirmed by existence of the blue isosurface between those atoms [34]. Also, the boryl group inclines into the sterically less hindered upper-left space for avoiding steric repulsion of the bulky alkyl groups on the phosphorous atoms of the ligand, thus creating a rigid chiral reaction pocket for the enantioselective

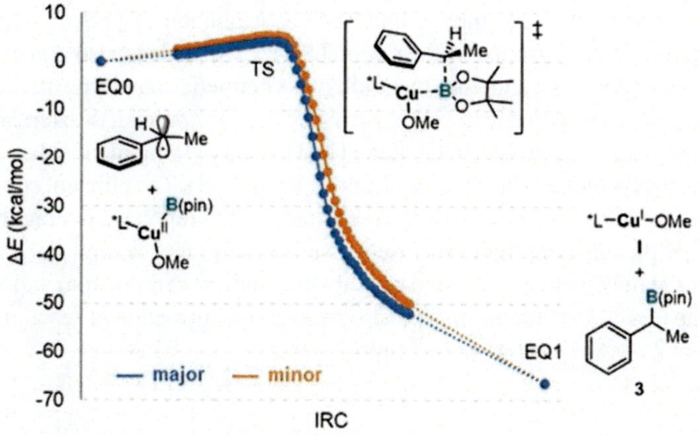

Fig. 5.6 IRC of major and minor paths

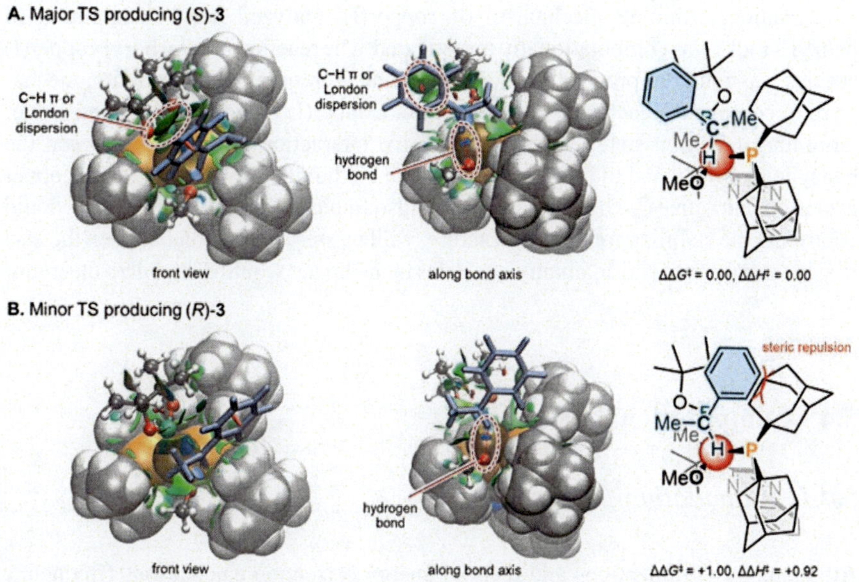

Fig. 5.7 Structure of major and minor TSs of radical trapping

reaction with the radical species. A strong C–H/π interaction between the phenyl ring in the substrate and the methyl group in the B(pin) moiety was found in the major TS, as evident from the large green isosurface in the NCIPLOT analysis (Fig. 5.7A). This analysis on the C–H/π interactions is in agreement with the experimental results, where larger aryl groups exhibiting stronger attractive C–H/π interactions with the B(pin) group than a phenyl group have a positive effect on the enantioselectivity (for 1-naphtyl: 86% ee; for phenyl: 71% ee). The adamantyl groups induce not only repulsive but also attractive interactions such as a London dispersion between the adamantyl and methyl groups in the major TS [35, 36]. This is also consistent with the experimental results, showing that bulky alkyl moieties have a positive effect on the enantioselectivity (for Me: 67% ee; *i*-Bu: 80% ee). In contrast, the steric repulsion between the adamantyl groups in the ligand and the phenyl ring in the substrate found in the minor TS should destabilize the structure (Fig. 5.7B). The efficient combination of the rigid conformation of ligand and the attractive and repulsive interactions in the TSs realizes this challenging enantioselective recognition of achiral radical species in spite of the long distance between the substrate radical carbon atom and the boron atom compared to that found in typical transition-metal-mediated reactions (major TS: B–C = 2.89 Å, Cu–C = 3.30 Å; minor TS: B–C = 2.93 Å, Cu–C = 3.22 Å).

5.3 Conclusions

An enantiodetermining mechanism of copper(I)-catalyzed boryl substitution of benzyl halides was computationally investigated. The reaction using a boryl copper(I) species was found to proceed via concerted radical trapping and borylation mechanism in contrast to reactions involving organocopper(I) reagents. In the enantiodetermining transition states, a strong attractive interaction was found between the benzylic hydrogen and the oxygen atom of the methoxide coordinated to the copper center. An attractive C–H/π interaction was also found in the major TS, which should contribute the stabilization of the structure, while a steric repulsion between the aryl ring in the substrate and adamantyl moieties in the ligand would destabilize the minor TS.

5.4 Computational Details

5.4.1 Calculation Method Details

All geometry optimizations and thermal energy correction calculations (frequency analyses) using density functional theory (DFT) were performed with the Gaussian 09 (revision D.01) [37] and Gaussian 16 (revision B.01) [38] suite of programs. The geometry optimizations were carried out at unrestricted (U) B3PW91 [39] level of theory in gas phase with a mixed basis set; Def2-SVPD for the atoms around the reaction center, B, P, Cu and a carbon atom to form C–B bond and Def2-SVP for the other atoms [40, 41]. The wave function stabilities were tested with "Stable" or "Stable = Opt" keyword [42, 43] in Gaussian program. Harmonic frequency calculations were conducted at the same level of theory on the optimized geometries to check all the stationary points as either minima or first-order saddle points. Intrinsic reaction coordinate (IRC) [44] calculations were carried out to confirm the transition states connecting the correct reactants and products on the potential energy surface. The zero-point energy (ZPE) and thermal energy corrections were calculated using vibrational frequencies with *GoodVibes* Python script [45] employing a quasi-harmonic approximation for entropy calculation at the solution of 0.25 M and temperature of 303.15 K with free-rotor description below 100 cm^{-1}, as proposed by Grimme [46]. Frequency scaling was not applied to the *GoodVibes* calculations. The self-consistent field energies were corrected at unrestricted (U) ωB97X-D [47] level of theory in diethyl ether using IEF-PCM [48] solvation model with a mixed basis set; Def2-TZVPD for B, P, Cu and a carbon atom to form C–B bond and Def2-TZVP for the other atoms [40, 41].

Summary: UωB97X-D/ Def2-TZVPD, Def2-TZVP/ IEF-PCM (Et_2O)// UB3PW91 / Def2-SVPD, Def2-SVP/ gas phase.

5.4.2 *Non-Covalent Interaction (NCI)-Plot*

Non-covalent interactions (NCI) in the transition states were computed using the non-covalent interaction index from the optimized electron density at the same level of theory as the geometry optimizations (Figs. 5.8 and 5.9). The wave function

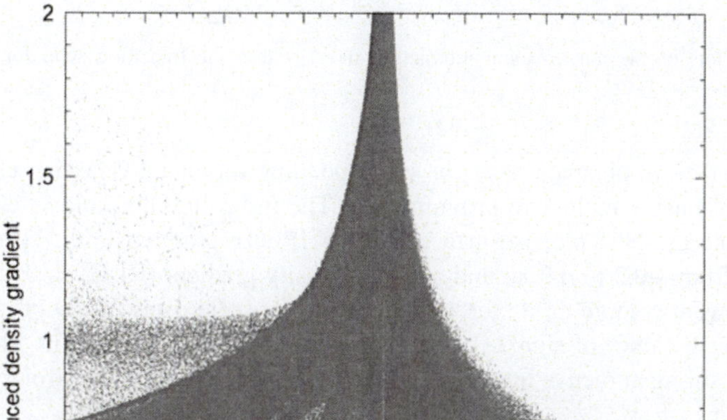

Fig. 5.8 Plots of the non-covalent interaction index within the transition state for the major enantiomer

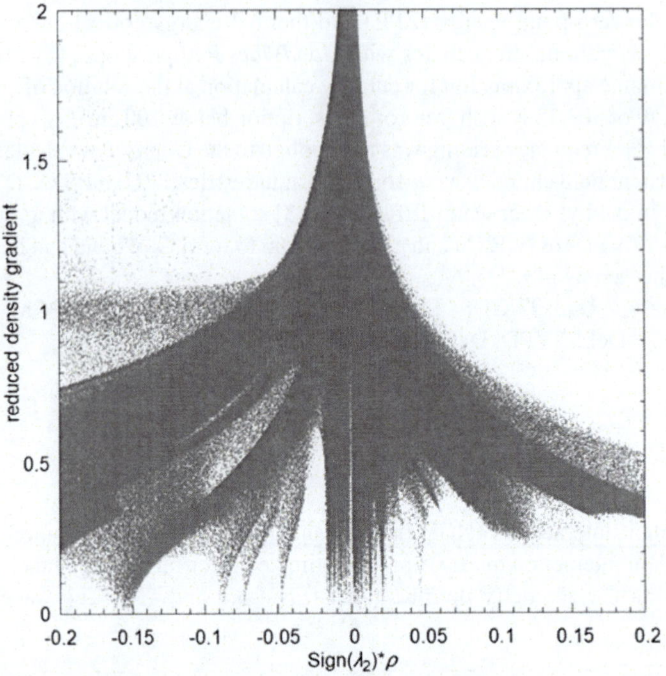

Fig. 5.9 Plots of the non-covalent interaction index within the transition state for the minor enantiomer

files (.wfn) were obtained from the corresponding formatted Gaussian checkpoint files (.fchk) using multi-wfn program [49]. The following thresholds were applied to generate the NCI plot isosurface with NCIPLOT program [50, 51]; sign(λ_2)ρ ranging from −0.2 to 0.2 au and reduced density gradient (RDG) = 0.45 au. The surfaces were colored on a blue-green-red (BGR) scale using VMD program [52] according to values of sign(λ_2)ρ ranging from −0.02 to 0.02 au. The blue region indicates strong attractive interactions, and the red region indicates strong repulsive interactions.

5.4.3 Calculated Properties and Geometries (Table 5.1)

Table 5.1 Calculated energies and thermochemical parameters of the optimized structures

Structure	*E* [Hartree]	*H* [Hartree]	−*TS* [Hartree]	*G* [Hartree]
Benzyl radical	−310.232767	−310.081337	−0.039575	−310.120912
Cu(II) complex	−4244.629849	−4243.622168	−0.127366	−4243.749534
Major TS	−4554.877122	−4553.716580	−0.141170	−4553.857750
Minor TS	−4554.875916	−4553.715121	−0.141032	−4553.856153
Cu(I) complex	−3833.364832	−3832.549433	−0.104286	−3832.653719
Product	−721.605168	−721.258231	−0.060780	−721.319011

References

1. Hall DG (ed) (2011) Boronic acids: preparation, applications in organic synthesis and medicine, vols 1–2. Wiley-VCH, Weinheim
2. Sandford C, Aggarwal VK (2017) Stereospecific functionalizations and transformations of secondary and tertiary boronic esters. Chem Commun 53:5481–5494
3. Crudden CM, Glasspoole BW, Lata CJ (2009) Expanding the scope of transformations of organoboron species: carbon-carbon bond formation with retention of configuration. Chem Commun 6704–6716
4. Leonori D, Aggarwal VK (2015) Stereospecific couplings of secondary and tertiary boronic esters. Angew Chem Int Ed 54:1082–1096
5. Brown HC, Singaram B (1988) Development of a simple general procedure for synthesis of pure enantiomers via chiral organoboranes. Acc Chem Res 21:287–293
6. Matteson DS (1989) α-Halo boronic esters: intermediates for stereodirected synthesis. Chem Rev 89:1535–1551
7. Carroll AM, O'Sullivan TP, Guiry PJ (2005) The development of enantioselective rhodium-catalysed hydroboration of olefins. Adv Synth Catal 347:609–631
8. Leonori D, Aggarwal VK (2014) Lithiation-borylation methodology and its application in synthesis. Acc Chem Res 47:3174–3183
9. Collins BSL, Wilson CM, Myers EL, Aggarwal VK (2017) Asymmetric synthesis of secondary and tertiary boronic esters. Angew Chem Int Ed 56:11700–11733
10. Hu J, Ferger M, Shi Z, Marder TB (2021) Recent advances in asymmetric borylation by transition metal catalysis. Chem Soc Rev 50:13129–13188
11. Brown HC, Zweifel G (1961) Hydroboration as a convenient procedure for the asymmetric synthesis of alcohols of high optical purity. J Am Chem Soc 83:486–487
12. Matteson DS, Ray R (1980) Directed chiral synthesis with pinanediol boronic esters. J Am Chem Soc 102:7590–7591
13. Hayashi T, Matsumoto Y, Ito Y (1989) Catalytic asymmetric hydroboration of styrenes. J Am Chem Soc 111:3426–3428
14. Yang CT, Zhang ZQ, Tajuddin H, Wu CC, Liang J, Liu JH, Fu Y, Czyzewska M, Steel PG, Marder TB, Liu L (2012) Alkylboronic esters from copper-catalyzed borylation of primary and secondary alkyl halides and pseudohalides. Angew Chem Int Ed 51:528–532
15. Ito H, Kubota K (2012) Copper(I)-catalyzed boryl substitution of unactivated alkyl halides. Org Lett 14:890–893

16. Iwamoto H, Kubota K, Yamamoto E, Ito H (2015) Copper(I)-catalyzed carbon-halogen bond-selective boryl substitution of alkyl halides bearing terminal alkene moieties. Chem Commun 51:9655–9658
17. Iwamoto H, Akiyama S, Hayama K, Ito H (2017) Copper(I)-catalyzed stereo- and chemoselective borylative radical cyclization of alkyl halides bearing an alkene moiety. Org Lett 19:2614–2617
18. Fu GC (2017) Transition-metal catalysis of nucleophilic substitution reactions: a radical alternative to SN1 and SN2 processes. ACS Cent Sci 3:692–700
19. Cherney AH, Kadunce NT, Reisman SE (2013) Catalytic asymmetric reductive acyl cross-coupling: synthesis of enantioenriched acyclic α, α-disubstituted ketones. J Am Chem Soc 135:7442–7445
20. Schmidt J, Choi J, Liu AT, Slusarczyk M, Fu GC (2016) A general, modular method for the catalytic asymmetric synthesis of alkylboronate esters. Science 354:1265–1269
21. Jin M, Adak L, Nakamura M (2015) Iron-catalyzed enantioselective cross-coupling reactions of α-chloroesters with aryl grignard reagents. J Am Chem Soc 137:7128–7134
22. Langlois JB, Alexakis A (2009) Dynamic kinetic asymmetric transformation in copper catalyzed allylic alkylation. Chem Commun 3868–3870
23. Kainz QM, Matier CD, Bartoszewicz A, Zultanski SL, Peters JC, Fu GC (2016) Asymmetric copper-catalyzed C-N cross-couplings induced by visible light. Science 351:681–684
24. During the course of the study, Fu and co-workers reported pioneering work on a nickel(II)-catalyzed reaction with moderate enantioselectivity (<88 % ee); see: Wang Z, Bachman S, Dudnik AS, Fu GC (2018) Nickel-catalyzed enantioconvergent borylation of racemic secondary benzylic electrophiles. Angew Chem Int Ed 57:14529–14532
25. In the stoichiometric reaction between a boryl copper(I) species and a benzyl halide, no extra amount of an alkoxide base furnished the borylation product in low yield with low enantioselectivity (39%, 50% ee), while addition of one equivalent extra amount of the base produced the product in high yield with high enantioselectivity (71%, 84% ee). Thus, the formation of cuprate would be necessary for high reactivity and selectivity in this radical borylation
26. Brock AJ, Whittaker JJ, Powell JA, Pfrunder MC, Grosjean A, Parsons S, McMurtrie JC, Clegg JK (2018) Elastically flexible crystals have disparate mechanisms of molecular movement induced by strain and heat. Angew Chem Int Ed 57:11325–11328
27. Elmagbari FM, Hammouda AN, Jackson GE, Bonomo RP (2019) Stability, solution structure and X-ray crystallography of a copper (II) diamide complex. Inorganica Chim Acta 498:119132
28. Orio M, Jarjayes O, Kanso H, Philouze C, Neese F, Thomas F (2010) X-ray structures of copper(II) and nickel(II) radical salen complexes: the preference of galactose oxidase for copper(II). Angew Chem Int Ed 49:4989–4992
29. Mori S, Nakamura E, Morokuma K (2000) Mechanism of S_N2 alkylation reactions of lithium organocuprate clusters with alkyl halides and epoxides. Solvent effects, BF_3 effects, and trans-diaxial epoxide opening. J Am Chem Soc 122:7294–7307
30. Ribelli TG, Matyjaszewski K, Poli R (2018) The interaction of carbon-centered radicals with copper(I) and copper(II) complexes. J Coord Chem 71:1641–1668
31. Zhang Q, Wang T, Zhang X, Tong S, Wu YD, Wang MX (2019) Radical reactivity, catalysis, and reaction mechanism of arylcopper(II) compounds: the missing link in organocopper chemistry. J Am Chem Soc 141:18341–18348
32. Johnson ER, Keinan S, Mori-Sánchez P, Contreras-García J, Cohen AJ, Yang W (2010) Revealing noncovalent interactions. J Am Chem Soc 132:6498–6506
33. Contreras-García J, Johnson ER, Keinan S, Chaudret R, Piquemal JP, Beratan DN, Yang W (2011) NCIPLOT: a program for plotting noncovalent interaction regions. J Chem Theory Comput 7:625–632
34. Schwarzer MC, Fujioka A, Ishii T, Ohmiya H, Mori S, Sawamura M (2018) Enantiocontrol by assembled attractive interactions in copper-catalyzed asymmetric direct alkynylation of α-ketoesters with terminal alkynes: OH$\cdots$O/Sp3-CH$\cdots$O two-point hydrogen bonding combined with dispersive attractions. Chem Sci 9:3484–3493

35. Wagner JP, Schreiner PR (2015) London dispersion in molecular chemistry—reconsidering steric effects. Angew Chem Int Ed 54:12274–12296
36. Liptrot DJ, Power PP (2017) London dispersion forces in sterically crowded inorganic and organometallic molecules. Nat Rev Chem 1:0004
37. Gaussian 09, Revision D.01, Frisch MJ, Trucks GW, Schlegel HB, Scuseria GE, Robb MA, Cheeseman JR, Scalmani G, Barone V, Petersson GA, Nakatsuji H, Li X, Caricato M, Marenich A, Bloino J, Janesko BG, Gomperts R, Mennucci B, Hratchian HP, Ortiz JV, Izmaylov AF, Sonnenberg JL, Williams-Young D, Ding F, Lipparini F, Egidi F, Goings J, Peng B, Petrone A, Henderson T, Ranasinghe D, Zakrzewski VG, Gao J, Rega N, Zheng G, Liang W, Hada M, Ehara M, Toyota K, Fukuda R, Hasegawa J, Ishida M, Nakajima T, Honda Y, Kitao Y, Nakai H, Vreven T, Throssell K, Montgomery JA Jr, Peralta JE, Ogliaro F, Bearpark M, Heyd JJ, Brothers E, Kudin KN, Staroverov VN, Keith T, Kobayashi R, Normand J, Raghavachari K, Rendell A, Burant JC, Iyengar SS, Tomasi J, Cossi M, Millam JM, Klene M, Adamo C, Cammi R, Ochterski JW, Martin RL, Morokuma K, Farkas O, Foresman JB, Fox DJ (2009) Gaussian, Inc., Wallingford CT
38. Gaussian 16, Revision B.01, Frisch MJ, Trucks GW, Schlegel HB, Scuseria GE, Robb MA, Cheeseman JR, Scalmani G, Barone V, Petersson GA, Nakatsuji H, Li X, Caricato M, Marenich AV, Bloino J, Janesko BG, Gomperts R, Mennucci B, Hratchian HP, Ortiz JV, Izmaylov AF, Sonnenberg JL, Williams-Young D, Ding F, Lipparini F, Egidi F, Goings J, Peng B, Petrone A, Henderson T, Ranasinghe D, Zakrzewski VG, Gao J, Rega N, Zheng G, Liang W, Hada M, Ehara M, Toyota K, Fukuda R, Hasegawa J, Ishida M, Nakajima T, Honda Y, Kitao O, Nakai H, Vreven T, Throssell K, Montgomery JA Jr, Peralta JE, Ogliaro F, Bearpark MJ, Heyd JJ, Brothers EN, Kudin KN, Staroverov VN, Keith TA, Kobayashi R, Normand J, Raghavachari K, Rendell AP, Burant JC, Iyengar SS, Tomasi J, Cossi M, Millam JM, Klene M, Adamo C, Cammi R, Ochterski JW, Martin RL, Morokuma K, Farkas O, Foresman JB, Fox DJ (2016) Gaussian, Inc., Wallingford CT
39. Becke AD (1993) J Chem Phys 98:5648
40. Weigend F, Ahlrichs R (2005) Phys Chem Chem Phys 7:3297
41. Weigend F (2006) Phys Chem Chem Phys 8:1057
42. Seeger R, Pople JA (1977) J Chem Phys 66:3045
43. Bauernschmitt R, Ahlrichs R (1996) J Chem Phys 104:9047
44. Fukui K (1981) Acc Chem Res 14:363
45. Funes-Ardoiz I, Paton RS (2018) GoodVibes, Version 2.0.1
46. Grimme S (2012) Chem Eur J 18:9955
47. Chai J-D, Head-Gordon M (2008) Phys Chem Chem Phys 10:6615
48. Scalmani G, Frisch MJ (2010) J Chem Phys 132:114110
49. Lu T, Chen F (2012) J Comput Chem 33:580
50. Johnson ER, Keinan S, Mori-Sánchez P, Contreras-García J, Cohen AJ, Yang W (2010) J Am Chem Soc 132:6498
51. Contreras-García J, Johnson ER, Keinan S, Chaudret R, Piquemal JP, Beratan DN, Yang W (2011) J Chem Theory Comput 7:625
52. Humphrey W, Dalke A, Schulten K (1996) J Mol Graph 14:33

Chapter 6
Summary

Organoboron compounds represent a versatile synthetic building block in organic synthesis. Recently, their applications to functional organic materials, such as fluorescent materials, electron transport materials, and pharmaceuticals, also gathered attention from researchers. In this thesis, I focused on the development of novel functionalized organoboron compounds and their stereo- and enantioselective synthetic methods using copper(I)/diboron systems. Furthermore, for a deeper understanding of the selectivity-determining mechanisms, DFT calculations were exhaustively performed.

In Chap. 2, I developed a new series of chiral QuinoxP*-type 1,2-bisphosphine ligands. The modification was implemented into the parent ligand QuinoxP* by introducing silyl groups onto the ligand backbone; the silyl groups interact with substituents on the phosphorous atoms through non-covalent interactions. The new ligands showed higher reactivity in the copper(I)-catalyzed direct enantioconvergent borylation of six-membered cyclic allyl substrates, including heterocycles, than the parent ligand without a drop of the enantioselectivity. Furthermore, I achieved the first development of borylative kinetic resolution of linear allyl substrates using the new ligand. This positive effect of the silyl groups was computationally explored and found to be attributable to the suppression of dimer formation of boryl copper(I) species which is assumed to be a dormant species in catalysis. Moreover, the detailed mechanism of the direct enantioconvergent borylation reaction was comprehensively investigated. During this computational study, I developed a new descriptor for describing the conformation of 1,2-bisphosphine ligands and analyzed the enantiodetermining transition state.

In Chap. 3, I developed the first intramolecular alkylboration reaction of allenes. The reaction proceeded in an exo-cyclization manner and produced alkenyl boronates having a four-membered ring structure. Furthermore, the products were obtained with high diastereoselectivity. Through a computational mechanistic study on the whole reaction pathways revealed that a facile allylic isomerization of allylcopper(I) species generated in situ is the key for the high stereoselectivity.

Y. Ozawa, *Copper(I)-Catalyzed Stereoselective Borylation Reactions*, Springer Theses,
https://doi.org/10.1007/978-981-99-1098-4_6

In Chap. 4, I synthesized multi-substituted allylic boronates, which contain a tetra-substituted alkene moiety having four different substituents, via a copper(I)-catalyzed regio- and stereoselective alkylboration of *gem*-disubstituted allenes. The borylation products were successfully applied to the allylboration reaction of aldehydes to furnish the densely substituted homoallylic alcohols with high diastereoselectivity. A computational study revealed an unprecedented regio- and stereoselectivity-determining mechanism, involving the boryl copper(I) coordination to allenes in addition to the borylcupration of the allene. Furthermore, the structural analysis of ligands indicated that the ligands having two pocket-like structures realized the regioselective borylcupration of the allene substrate.

In Chap. 5, I conducted the first computational mechanistic investigation of enantioselective borylation of radical species with a boryl copper(II) intermediate. The transition state analysis revealed that the reaction proceeds via the concerted mechanism instead of the stepwise mechanism via copper(III) intermediate. Furthermore, a non-covalent interaction analysis indicated that the enantioselectivity is controlled by a combination of small interactions: an attractive C–H/π interaction and a repulsive steric repulsion between the substrate and catalyst.

Through this thesis, I developed new densely substituted and functionalized allylic and alkenyl boronates under high stereo- and enantiocontrol using copper(I) catalyst. As I demonstrated, the borylation products could be used as versatile building blocks in organic synthesis. Also, I investigated and revealed the detailed mechanism of the copper(I)-catalyzed borylation reactions. Those highly reactive and selective catalyst systems I developed and the deep understanding of the mechanism would benefit the further development of new reactions and catalyst systems.